FUNDAMENTALS OF PHYSICAL VOLCANOLOGY

Fundamentals of Physical Volcanology

Elizabeth A. Parfitt
and Lionel Wilson

Blackwell
Publishing

© 2008 by Blackwell Science Ltd
a Blackwell Publishing company

BLACKWELL PUBLISHING
350 Main Street, Malden, MA 02148–5020, USA
9600 Garsington Road, Oxford OX4 2DQ, UK
550 Swanston Street, Carlton, Victoria 3053, Australia

First published 2008 by Blackwell Publishing Ltd

Library of Congress Cataloging-in-Publication Data

Parfitt, Elizabeth.
    Fundamentals of physical volcanology / Elizabeth Parfitt and Lionel Wilson.
        p. cm.
    Includes bibliographical references and index.
    ISBN 978-0-63205443-5 (pbk. : alk. paper)   1. Volcanism.   2. Volcanoes.   I. Wilson, Lionel, 1943–
II. Title.

    QE522.P378 2007
    551.21—dc22

                                                                                          2007038441

A catalogue record for this title is available from the British Library.

Set in 9.5/12pt Garamond by Graphicraft Limited, Hong Kong

For further information on
Blackwell Publishing, visit our website:
www.blackwellpublishing.com

Contents

Preface

Our knowledge of the physics of how volcanoes work has expanded enormously over the past 40 years, as have our methods of studying volcanic processes. In the late 1960s, George Walker conducted experiments into the fall-out of volcanic particles from eruption clouds by using stop-watches to time the fall of pieces of tephra dropped down a stairwell at Imperial College, London. Now, technology exists which uses RADAR interferometry from satellites to monitor tiny changes in the shape of volcanoes, and broadband seismometers can detect the "heartbeat" (and "indigestion") of volcanoes as magma moves around deep inside them. Sometimes, however, as we gain increasingly in-depth knowledge of a subject, it becomes all too easy to focus on the minute details and hard to see the fundamental principles underlying all the complex behaviours that we observe. In this book we have attempted to step back from the details, and to view volcanoes as systems governed by some basic physical principles. Our approach is to consider the physical processes that control the formation, movement and eruption of magma, starting in the source region and following the magma upwards. Our intention is to show that, for all the apparent complexity of volcanoes, a little basic physics can go a long way in explaining how they work, and that often eruptions that may at first sight look remarkably different from one another are, in fact, physically much the same.

Acknowledgments

EAP: This book developed out of an undergraduate course that I taught for a number of years at the University of Leeds. That course benefited from and evolved due to the feedback of the students who took it, and I thank them all for their enthusiasm. A number of colleagues at Leeds – Joe Cann, Sue Bowler, Jane Francis, Mike Leeder and Pete Baker – offered their time and thoughts to that course and I thank them very much for their input and support. My knowledge of Strombolian activity was greatly improved by a trip to Stromboli with staff and students at Leeds, and I thank Jurgen Neuberg, Graham Stuart and Roger Clark for taking me along with them. Much of the writing of the book took place while I was working at the State University of New York at Buffalo, and I thank Marcus Bursik, Tracy Gregg and Mike Sheridan for discussions and ideas shared during that time. Nigel Burrows offered invaluable help converting my many old slides to digital format. This book would never have been finished without the treatment I received from the Chronic Fatigue Syndrome Service at Ysbyty Eryri, Caernarfon: a huge thank you to Dr Helen Lyon Jones, Marian Townsend, Anne the nutritionist and Dr Paul Nickson for their dedication and encouragement. Many thanks to my mother for the constant nagging to get this book finished, and for the belief that it could be done. To my son, James, a big thank you for putting up with my hours at the computer. Most of all to David, who always reminded me of the importance of punctuation and who enthusiastically applied red ink to the many drafts of this book, my innumerable thanks for providing the support, emotional and financial, needed to make the completion of this book possible.

LW: When my co-author suggested that we collaborate on this book I was very happy to agree, as I have devoted most of my time since 1968 to understanding the physics of volcanic processes. During that year, while I was working on a pre-Apollo study of the mechanical structure of the surface of the Moon, we began to get spacecraft images showing very long lava flows in the lunar *mare* areas, and in an effort to learn more about lava eruptions I visited the eminent British volcanologist George Walker, then at Imperial College London, to ask what was known about the physics of eruptions. George patiently, and with some amusement, explained to me just how little was known about this subject at that time, and by the end of my visit my career path was decided. My interest in the Moon remained, and broadened as spacecraft visited other solar system objects. Indeed, the study of how planetary environments control the boundary conditions (e.g., acceleration due to gravity, atmospheric pressure) under which volcanoes operate has been a major source of ideas. An equally important source of inspiration for me has been my interaction with the more than 30 graduate students who have worked with me on volcanic topics over the last 35 years. I must also thank my immediate colleagues at Lancaster, Harry Pinkerton, Steve Lane, Jennie Gilbert and Ray MacDonald, for their unfailing willingness to enlighten a mere physicist on the finer points of geology and geochemistry, and I am indebted to the numerous other scientists with whom collaboration has been so stimulating over the years, especially Stephen Sparks, James Head, Peter Mouginis-Mark and, of course, my co-author Elisabeth Parfitt. Thanks for logistic help in locating and manipulating images go to Peter Neivert at Brown University and Ian Edmondson at Lancaster University. Last, but very much not least, thanks to my wife Dorothy for her unfailing tolerance and support.

Both: We wish to thank John Guest, Michael Branney and Tracy Gregg for their helpful comments on parts of this book; as ever, errors and omissions remain our responsibility.

Dedication

In memory of George Walker, who was an inspiration to us both.

Glossary

'a'a A type of lava having a very rough surface texture.

absolute temperature The thermodynamic temperature of a substance, measured on a scale where zero corresponds to the molecules forming the substance having no motion. The unit of absolute temperature is the Kelvin (K); the Kelvin has the same size as the degree Celsius (°C) and 0K corresponds to –273.15°C.

accidentally breached An accidentally breached lava flow is one in which an overflow or breakout from the main channel occurs when the channel is blocked by material breaking off its walls.

accretionary lapilli Small (between 4 and 32 mm in diameter) rounded particles formed by the accretion of large numbers of smaller particles in a volcanic eruption plume. The small particles may be held together by water, ice, or electrostatic forces.

acidity spikes Localized high concentrations of sulfuric acid found at certain depths in ice cores drilled in polar regions. These correspond to the deposition of snow soon after volcanic eruptions which released large amounts of sulfate aerosols into the atmosphere.

active continental margin The edge of a continental land mass at which one of the plate-tectonic processes of subduction or faulting is taking place.

aerosols Small droplets of water in the atmosphere in which volatile species, especially sulfur dioxide, are dissolved.

air drag The force exerted on a particle moving through the atmosphere as a result of the friction between the surface of the particle and the air.

andesite A rock type of intermediate silica content, commonly associated with subduction zones.

arachnoids Tectonic structures up to many hundreds of kilometers in diameter, with complex central parts and radiating fracture systems, seen in the crust of Venus. The name suggests that they look like spiders.

ash cluster A collection of ash particles loosely held together by moisture, ice, or electrostatic forces.

ash pellet A collection of ash particles strongly held together by moisture, ice, or electrostatic forces.

basalt A rock type with a low silica content, commonly associated with ocean floor spreading.

basaltic andesite A rock type of low to intermediate silica content.

base surge A cloud of hot gas and entrained particles flowing out close to the ground from the site of a volcanic (or nuclear) explosion.

billion Used in this book in its American context to mean the number 10^9, i.e., one thousand million.

Bingham plastic A liquid that has not only a viscous resistance to deformation and flow, but also a finite strength that must be overcome by any applied force before any flow takes place.

block lava A type of viscous lava in which the surface fractures into large blocks.

block-and-ash flow deposit A mixture of coarse and fine pyroclasts. deposited from a pyroclastic density current.

bulk modulus The property of a material expressing the way the density changes with pressure.

buoyancy The phenomenon whereby a low-density body surrounded by a higher-density fluid in a gravitational field experiences an upward force.

caldera A steep-walled depression, commonly found at the summit of a volcano, formed when a large volume of magma is removed quickly from an underlying magma reservoir and the overlying rocks slide down along faults to fill the vacated space.

canali Long (many hundreds of kilometers), narrow (a few kilometers wide) channels seen on the surface of Venus.

carbonatite A rare type of magma consisting mainly of liquid carbonates rather than liquid silicates, produced in the mantle beneath some continental areas.

channelized lava flow A lava flow in which liquid lava moves in a central region bordered on either side by a bank (called a levée) of stationary lava.

choked flow The flow of a fluid under conditions such that the speed of the fluid is equal to the speed of sound within the fluid. This is the maximum speed that can be reached by the fluid unless special conditions apply.

coalescence The joining together of two separate gas bubbles within a liquid (or of two droplets of liquid in a gas).

cock's tail plume The distinctive "feather-edged" jet of ash and steam formed in an explosion when a large amount of water gains access to a vent.

co-ignimbrite ash fall deposit A fine-grained deposit of pyroclasts settling out on the ground from a co-ignimbrite cloud.

co-ignimbrite cloud An eruption cloud of gas and small pyroclasts formed as gas rising through an ignimbrite carries small particles upward with it as it escapes.

column collapse The condition in which an eruption column fails to be positively buoyant in the atmosphere, so that a lower fountain of gas and entrained particles forms over the vent instead.

compound lava flow field A region containing many lava flow units, most of which have formed by new flows breaking out from the margins of earlier-emplaced flows.

conservation of momentum The physical law that asserts that the momentum of a system (the product of mass and velocity) cannot be destroyed, only redistributed among the components of the system.

continental arc An arcuate chain of volcanoes at the margin of a continent.

convective region Part of the interior of a fluid or plastic solid within which convection (relative movement of different parts of the fluid due to density differences) is taking place.

convergent margin A location where the edges of two tectonic plates are being driven into collision.

cooling unit One or more layers of pyroclastic particles (or lava) emplaced so soon after one another that they cool as though they had been emplaced at the same time.

cooling-limited Description of a lava flow that ceases to move because the front of the flow has cooled to the point of effectively being a solid.

corona (plural: **coronae**) One of a number of large (many hundreds of kilometers) roughly circular regions on Venus where tectonic forces have fractured and folded the surface rocks.

cryovolcanism A volcanic process in which liquid water rather than liquid rock is the moving fluid.

dacite A type of magma with intermediate to high silica content.

dark halo deposit A roughly circular region on the Moon where dark pyroclasts are deposited around an explosive vent.

de Lavalle nozzle Part of a volcanic conduit or dike where the shape changes from converging upward to diverging upward, thus allowing magma flowing through the conduit to accelerate from subsonic to supersonic speeds.

decompression The expansion of a material, especially a gas, when the pressure acting on it decreases.

decompression melting The process whereby solid rock begins to melt when the pressure acting on it decreases even if the temperature does not change.

diapir A body of plastic material that rises buoyantly as a coherent mass within a larger body of plastic material.

diffusion The process whereby the atoms or molecules of a volatile compound migrate by moving between the atoms or molecules of a host material.

dike A fracture filled with volcanic material cutting through earlier-emplaced host rocks.

dilatant Description of a type of nonNewtonian fluid in which the viscosity increases as the applied stress increases.

divergent margin A boundary between two tectonic plates, almost always on the ocean floor, at which new crustal material is being supplied by volcanic eruptions or intrusions as the plates move apart.

dome A deposit of (commonly viscous) lava where the width and maximum thickness are of the same order because the lava has not spread far from the vent.

dusty gas A gas containing solid particles so small that frictional drag forces effectively force the particles to travel at the same speed as the gas.

ejecta Any material thrown out from a vent (rather than flowing away from it) during volcanic activity.

elastic Description of any material that changes its shape when a stress is applied to it but recovers its original shape when the stress is removed.

energy equation An equation describing the law that says that the total energy of a system cannot be destroyed, only redistributed among its parts.

entrainment The process whereby the flow of one material through or past another surrounding material causes some of the surrounding material to be mixed into the flowing material.

equivalent diameter A geometric property of a flowing fluid in a channel equal to four times the cross-sectional area at right angles to the direction of flow divided by the length of the perimeter in contact with the floor and walls.

exit velocity The speed at which volcanic materials emerge through a surface vent.

expansion wave A moving zone within a fluid across which the pressure decreases significantly.

explosive Description applied to any process that takes place suddenly, or that involves a very large pressure change.

exsolution level The depth at which gas dissolved in a rising magma first starts to come out of solution and form gas bubbles in response to the decreasing pressure.

exsolve Release a gas from solution in a liquid. The opposite of dissolve.

fiamme Pumice clasts in the interior of an ignimbrite deposit that have become stretched sideways while still hot as the deposit is compressed by the weight of overlying material.

filter-pressing The process whereby magma is squeezed out of its partly molten source rocks when stress causes compaction of the unmelted material.

fissure A fracture in rock, more particularly an elongate surface vent from which magma is erupted.

flood basalt eruption A rare kind of basaltic eruption in which a very large volume (thousands of cubic kilometers) of basalt is erupted in a geologically short space of time.

flow unit A lava flow which is the product of a single eruptive event from a single vent.

focus The location beneath the surface where an earthquake takes place.

fractional crystallization The formation of crystals in a cooling liquid. One or more types of crystal may form at any one time, but each forms over its own characteristic temperature range.

fragment A piece of material broken from a larger piece of material.

fragmentation The process of breaking a material into smaller pieces. The material may be a solid or a liquid.

fragmentation level The depth below the surface at which shearing stresses tear a magma containing gas bubbles apart into clots of liquid carried along by the gas released from the bubbles broken by the tearing process.

friction A force that opposes the motion of any two materials in contact and sliding past one another.

frost ring A layer inside the trunk of a tree, forming parallel to the bark and marking a time when the growth of the tree was inhibited by unusually cold weather.

fumaroles Places where volatile compounds being released from the interior of a volcanic deposit reach the surface and settle to form deposits on the ground as they cool.

Gas Laws The laws describing the way the pressure, temperature, density, and internal energy of gases are related.

gas-thrust region The lowest part of an eruption column, where the inertia of the erupted materials, which has been determined by the expansion of volcanic gases beneath the surface, is the main control on their motion.

giant dike swarm A group of dikes radiating for great distances (at least many hundreds of kilometers) away from some region where a very large magma reservoir has existed at some time.

graben A trench-like depression formed in an area of extensional forces. The crust is forced apart and breaks along two parallel normal faults dipping toward one another, with the ground between the faults moving downward.

grading The variation of the average grain size of a deposit with vertical position within it. In normal grading the mean size increases downward, whereas in reverse (or inverse) grading it increases upward in the deposit.

granular flow The flow of a body of material consisting of discrete solid clasts in which only the interaction between clasts controls the motion – any gas or liquid between the clasts has no important effects.

Hawaiian Description applied to eruptions like those common in Hawai'I, where basaltic lava is erupted, commonly explosively.

heat pipes Regions where heat is transported upward through the crust mainly by the frequent passage of magma through the surrounding rocks.

heterogeneous nucleation The process of the formation of gas bubbles in a liquid supersaturated in a dissolved volatile compound when the bubbles nucleate on crystals in the liquid or irregularities in the boundary between the liquid and its solid surroundings.

hindered settling The settling of solid particles in a fluid where the particles are so close together that they either collide with one another or interfere with the smooth flow of the fluid around them.

homogeneous nucleation The process of the formation of gas bubbles in a liquid supersaturated in a dissolved volatile compound when the bubbles have no solid surfaces on which to nucleate and so appear at random within the liquid.

hot spot Place where there is an unusually large upward flow of heat from the mantle toward the surface. Generally a location of significant volcanic activity.

hyaloclastite A type of fragmental and chemically altered rock produced when erupting lava interacts strongly, generally explosively, with surface water.

hyaloclastite ridge A ridge composed of fragmental and chemically altered rock produced when a fissure eruption occurs in shallow water, most commonly beneath a glacier.

hydromagmatic Description of any eruption process in which magma or lava interacts with external water.

hydrothermal Description of any process involving the circulation of water at shallow depths in the crust as a result of heat supplied by intruded magma.

ignimbrite A large body of rock formed from the deposition of pyroclasts that have traveled from a vent as a pyroclastic density current.

ignimbrite-forming A type of explosive eruption that produces large volumes of pyroclasts emplaced as pyroclastic density currents.

inertial region An alternative description (see "gas-thrust region") of the lowest part of an eruption column where the inertia of the erupted material dominates the motion.

inflation The word has two uses in volcanology: (i) the enlargement of a magma chamber as new magma is added to it from the mantle; (ii) the process whereby a lava flow gets thicker after it has been emplaced as a result of additional magma being forced into its interior.

intraplate Any process that occurs within, i.e., well away from the boundaries of, a tectonic plate.

inversely graded Description of a deposit of pyroclasts in which the average grain size increases upward in the deposit.

island arc An arcuate group of volcanic islands formed above a subduction zone at the edge of a tectonic plate.

isopach A contour line on the map of a volcanic deposit joining places where the thickness of the deposit is the same.

isopleth A contour line on the map of a volcanic deposit joining places where the grain size of the deposit is the same.

jökulhlaup The Icelandic word for a "glacier-burst," the sudden release of a very large volume of water that has accumulated under a glacier as a result of melting caused by an eruption there.

juvenile In volcanology, the word implies material that has come directly from the deep interior of the planet.

kimberlite A rare type of mafic rock resulting from the eruption or intrusion of magma coming from unusually great depth in the mantle. Economically important because some kimberlites bring with them diamonds from the mantle.

kinetic energy The type of energy associated with the movement of material.

komatiite A type of ultramafic magma forming low-viscosity lava flows, common in early Earth history.

laccolith An intrusion of magma that has a relatively large vertical extent compared with its horizontal width.

Large Igneous Province A region where large volumes of basaltic lava have been erupted – essentially a more general term for a region in which a flood-basalt eruption has happened.

lava breakout A place where lava breaks out from the edge of an existing lava flow deposit.

lava dome A relatively thick and short lava flow deposit.

lava flow An individual deposit of a discrete phase of an effusive eruption.

lava flow field A group of lava flow deposits emplaced in successive phases of a prolonged eruption.

lava fountain A jet of hot pyroclasts ejected from an explosive volcanic vent, rising to a significant height, and then falling back to the surface. Also called a fire fountain.

lava tube The interior of a lava flow where the surface layers of the flow have ceased to move and thus form an insulating roof reducing heat loss from lava still flowing beneath.

levée The stationary edge of a lava flow.

level of neutral buoyancy See "neutral buoyancy level."

linear rille A type of graben found on the Moon.

liquidus The temperature at which a magma is completely molten.

lithic clast A fragment of rock broken from the rocks through which a volcanic event has taken place and incorporated into the erupted volcanic materials.

lithosphere The outer part of a planet where the rocks behave as brittle solids, consisting of the crust and the upper part of the mantle.

lithostatic load The pressure at a given depth below the surface due to the weight of the overlying layers of rock.

littoral cone A cone-shaped accumulation of pyroclasts on land close to the ocean, built up by explosions when lava enters the water.

maar A crater formed by an explosive interaction between magma approaching the surface and surface or near-surface water.

magma Molten or partly molten rock beneath the surface of a planet.

magma ocean A layer of molten rock on the surface of a planet, formed when the outer layers of the planet accumulate so fast that heat from the impact of each added asteroid cannot be radiated away completely before the next impact happens.

magma reservoir A long-lived body of magma beneath the surface that forms when new magma from the mantle is added faster than the existing magma body can cool.

magmon A localized concentration of magma in the pore space of host rocks. The rocks deform to allow the magma concentration to pass through, so that it moves like a wave through the host rocks.

mantle plume A part of the mantle where buoyancy causes the mantle rocks to rise toward the surface. Commonly the site of pressure-release melting.

maria (singular: **mare**) The latin name for the dark areas on the Moon, consisting of floods of basaltic lava filling very large impact craters.

mass extinction A biological event in which large numbers of species die out in a geologically relatively short space of time.

mass flux The mass of magma passing though a volcanic system every second.

meteoric Generally, description of any phenomenon associated with the atmosphere. In volcanology, applied to near-surface water that has collected as a result of rain or snow.

mid-ocean ridge The ridge formed along a constructive tectonic plate margin by the accumulation of lavas erupted onto the ocean floor.

mush column A zone within the crust where magma passes through so often that the host rocks are partly molten.

negatively buoyant Description of material that is denser than the material surrounding it, so that it will tend to sink through its surroundings.

neutral buoyancy level Any location above or below ground where volcanic materials have the same density as their surroundings.

Newtonian Description of a fluid with the property that any change in applied stress produces a directly proportional change in rate of deformation.

nonNewtonian Description of a fluid in which the rate of deformation is not directly proportional to a change in the applied stress.

normally graded Description of a pyroclastic deposit in which the grain size increases downward in the deposit.

nova (plural: **novae**) A type of tectonic structure on Venus in which fractures radiate out from a central zone.

nuclear winter A prolonged period of cooling when heat reaching the surface from the Sun is greatly reduced by dust in the atmosphere thrown up from the explosion of a large number of nuclear weapons.

nuée ardente French for "burning cloud," one of the possible names for a pyroclastic surge.

ophiolite A body of rock consisting of the subsurface part of an old spreading center now uplifted and exposed at the Earth's surface by tectonic forces.

p waves Primary waves, compressive waves spreading out from an earthquake focus into the surrounding rocks. These waves travel faster than any other type of seismic waves.

pahoehoe A type of lava where the surface is smoothly folded into a series of ripples called ropes.

pahoehoe toe A small lava flow unit consisting entirely of pahoehoe lava extending for a short distance from a larger flow unit.

partial melting The process in which part of a mass of rock melts, the liquid still containing the mineral grains that have not yet melted.

perched lava pond A lava flow that has spread out sideways so that it has a similar width to its length. The shape is sometimes controlled by pre-existing topography but can be self-generated by a suitable combination of eruption rate and very shallow ground slope.

peridotite A type of rock rich in olivine, found in the Earth's mantle.

petrology The general term for the study of all aspects of rocks.

phoenix cloud Alternative term for a co-ignimbrite cloud.

phreatomagmatic Description of an eruption involving interaction between magma and surface, near-surface or ground water.

phreato-Plinian Description of a sustained explosive eruption in which magma interacts with surface or ground water, generally resulting in more fragmentation of the magma than in a Plinian eruption.

pillow A lava flow lobe that is approximately as wide as it is thick, produced by a low lava extrusion rate under water.

pillow lava A lava flow consisting of a pile of pillows.

plastic Description of a fluid that is capable of deforming smoothly in response to an applied stress.

plastic viscosity The property of a non-Newtonian fluid expressing the ratio between a change in applied stress and a corresponding change in rate of deformation.

plate One of a series of sections of the Earth's lithosphere behaving as a rigid solid.

plate tectonics The term applied to our current understanding of the structure of the Earth's lithosphere, with rigid plates sliding as discrete structures on top of the plastic mantle and interacting at their edges.

Plinian Description of a sustained explosive discharge of volcanic gas and pyroclasts forming a large eruption cloud in the atmosphere.

Poisson's ratio An elastic property of a solid, specifically the ratio of transverse to longitudinal strain (i.e., fractional deformation) when a tensional force is applied.

potential energy Generally, a form of energy associated with the position of an object in a force field. In volcanology this is the planetary gravitational field.

pressure-balanced Description of a volcanic fluid emerging from a vent in such a way that the pressure within the fluid is equal to the atmospheric pressure at the level of the vent.

pressure-release melting An alternative term for decompression melting.

pumice A piece of volcanic rock containing vesicles.

pyroclastic density current A mixture of gas and suspended or entrained solids released in a sustained explosive eruption and forming a dense fluid that moves along the ground at high speed.

pyroclast General term for any fragment of volcanic material produced in an explosive eruption.

pyroclastic fountain A mixture of gas and pyroclasts erupted explosively though a vent, traveling upward, and then falling back to the surface.

pyroclastic surge A relatively short-lived form of pyroclastic density current.

regolith A fragmental layer on the surface of a planet. If a biological component is present, as on Earth, the regolith is called soil.

residence time The time that particles or aerosols spend in the atmosphere before settling to the ground. More generally the time taken for any particles suspended in a fluid to settle out.

rheology The study of the way fluids deform in response to applied stresses.

rhyolite A rock type with a high silica content.

Ring of Fire The regions around the rim of the Pacific Ocean dominated by volcanic activity.

rootless lava flow A lava flow formed by the accumulation and coalescence of hot pyroclasts falling from a lava fountain.

rootless vent The site of a volcanic explosion that is not directly underlain by a volcanic vent, for example a place where an explosion occurs in a lava flow advancing over waterlogged ground.

s waves Secondary waves, shear waves spreading out from an earthquake focus into the surrounding rocks. These waves can propagate in solids but not in liquids.

saltate To bounce over the ground, as when particles are almost suspended in a strong wind.

saturated Description of a fluid containing the maximum amount of volatiles allowed by the current pressure and temperature.

sedimentation The settling of particles from a fluid to form a layer at the base of the fluid.

seismic gap A subsurface region within which no sources of seismic waves occur because the region is occupied by magma.

seismic velocity The speed of a seismic wave, a sound wave in rock generated by an earthquake.

shear modulus The elastic property of a solid expressing the fractional amount by which it deforms in response to a shearing stress.

sheet flow A lava flow that is very wide compared with its thickness.

sill A sheet-like body of magma, often approximately horizontal, intruded at some depth below the surface along the interface between two pre-existing rock layers.

sinuous rille A meandering type of channel found on the Moon and Mars where hot turbulent lava has eroded the surface over which it has flowed.

slug A body of gas rising through a volcanic dike or conduit where the vertical extent of the gas is much greater than the width of the dike or conduit.

solidus The temperature below which a magma is completely solid.

soliton A solitary wave, i.e., a wave that travels without changing its size or shape.

solubility law The law specifying how much of a given volatile compound can be dissolved in a magma at a given pressure and temperature.

spatter rampart A ridge parallel to a fissure vent consisting of pyroclasts ejected from the fissure.

spreading center A boundary between two tectonic plates at which new crust is being created by volcanic eruptions and intrusions and the plates are moving apart.

stoping The process whereby blocks of country rock become detached from the roof or walls of a magma reservoir and fall into the magma.

strain rate The rate at which a solid or liquid changes its length, expressed as a fraction of its original length, as a result of an applied stress.

stratosphere The second layer of the Earth's atmosphere, lying above the troposphere.

Strombolian A style of explosive volcanic activity characterized by the intermittent arrival, at the surface of the magma in a vent, of giant gas bubbles that burst, throwing out the disrupted liquid skin of the bubble.

subaerial Description of any process taking place in an atmosphere.

subduction The process whereby some tectonic plates are forced down into the Earth's interior beneath other plates.

subPlinian Description of a class of sustained explosive volcanic activity producing relatively small eruption plumes in the atmosphere.

supersaturated Description of a fluid containing more of a dissolved volatile compound than it should be capable of dissolving under its current pressure and temperature conditions. This is an unstable state leading to bubble formation by exsolving gas.

surface tension A molecular attraction force acting parallel to any interface between two fluids and tending to reduce the area of contact.

tensile strength The strength of a material when subject to a tensional force.

tephra General term for relatively fine grained fragmented volcanic rock.

tephra jet Alternative name for a cock's tail plume.

terminal velocity The steady speed reached by an object falling through a fluid in a gravitational field when the upward drag force exerted on it by the fluid is just equal to the downward gravitational force.

tessera (plural: **tesserae**) A type of terrain on Venus characterized by extensive faulting in more than one direction.

thixotropic Description of a type of non-Newtonian fluid in which the viscosity decreases as the applied stress increases.

tholeiitic Description of a common type of basalt, rich in the minerals plagioclase and pyroxene, formed at mid-ocean ridges.

toothpaste lava A type of lava with a very rough surface texture consisting of many sharp spines roughly arranged in rows.

tractional Description of a process that involves a frictional force from an overlying fluid dragging a particle sideways.

triple junction A tectonically complex region where the boundaries between three tectonic plates meet.

tropopause The boundary between the troposphere and the overlying stratosphere, lying at a latitude- and season-dependent height of ~10–15 km.

troposphere The lowest part of the Earth's atmosphere.

ultraPlinian An unusually energetic form of Plinian explosive eruption.

umbrella region The uppermost part of a volcanic eruption cloud where vertical motion ceases and gas and pyroclasts spread sideways in the atmosphere.

undersaturated Description of a fluid containing less of a dissolved volatile than it is capable of dissolving under its current pressure and temperature conditions.

vein A relatively small and narrow fracture in a rock containing material that is different from the host rock. Commonly used in volcanology to describe a small off-shoot from a dike.

vesicles The holes in volcanic rocks showing where gas bubbles were present as magma rose to the surface.

viscosity The property of a fluid describing its resistance to deforming and flowing when a stress is applied to it.

volatile Generally, the description of any chemical compound with a low boiling point. Used in volcanology to refer to the gases commonly dissolved in magma at depth in the planet.

Volcanic Explosivity Index A single number between 0 and 8 giving a combined measure of the magnitude and intensity of a volcanic eruption.

volcanic tremor Relatively long-lived and steady seismic activity associated with the flow of magma through a volcanic dike or conduit.

volume-limited Description of a lava flow that ceases to advance because the supply of magma from the vent has ceased.

welding The development of mechanical strength between hot pyroclasts in close contact causing them to stick together.

xenolith A fragment of rock accidentally incorporated into rising magma and brought up from the interior of the planet.

yield strength The stress level that must be exceeded before a non-Newtonian fluid will begin to deform and flow.

1 Volcanic systems

1.1 Introduction

A volcanic eruption is an amazing event to watch: dangerous and frightening but also fascinating and awe-inspiring. While most people will never experience an eruption first-hand, accounts of volcanic eruptions in the media, television documentaries and Hollywood films all mean that even those living far from an active volcano have some idea of what volcanic eruptions are like.

Volcanic eruptions vary tremendously in style and in the deposits they produce, from lava fountaining eruptions in Hawai'I (Fig. 1.1), through moderately explosive eruptions, such as the 1980 eruption of Mount St Helens (Fig. 1.2) which devastated the area immediately around the volcano and

Fig. 1.1 An approximately 300 m high lava fountain eruption from the Pu'u 'O'o vent on the East Rift Zone of Kilauea volcano, Hawai'I. (Photograph by Pete Mouginis-Mark, University of Hawai'I.)

deposited ash as much as 1500 km downwind from the volcano, to huge explosive eruptions which have occurred in the geological past, such as the eruption 600,000 years ago at Yellowstone which covered half of the United States with ash. They vary in scale from tiny eruptions producing a few cubic meters of lava to eruptions which can produce up to ~2500 km^3 of ash or lava (enough to cover the whole of Great Britain with a layer more than 10 m thick, enough to bury all but the tallest buildings!). They vary in duration from a few seconds to years or decades. They vary tremendously in frequency – an observer at Stromboli volcano (in the Aeolian Islands, north of Sicily) usually has only to wait a matter of minutes to see an eruption, whereas an observer at Yellowstone National Park could wait 100,000 years or more to see a volcanic eruption!

Why does a volcanic eruption occur where it does and when it does, and what controls what the eruption is like? Physical volcanology is the branch of geology which seeks to answer these questions by applying basic physical principles to find out how volcanoes work. The study of volcanoes in this way over the last 30 years or so has shown that, despite the apparent complexity of individual volcanic eruptions, the basic physical processes which govern them are often surprisingly simple, and furthermore the processes can be very similar in eruptions which superficially appear very different from each other. In this book we seek to describe these physical processes and to underline their similarities in eruptions which are apparently so different in character.

Fig. 1.2 Eruption cloud from the sustained phase of the May 18, 1980 eruption of Mount St Helens volcano, Washington State, USA. The lower part of the cloud is almost vertical but near the top the cloud and the small particles falling from it are being carried away downwind. (Photograph taken by Robert Krimmel, courtesy of U.S. Geological Survey/Cascades Volcano Observatory.)

1.2 Styles of volcanic eruptions

Direct observations of volcanic eruptions, recorded on film, in photographs, and in written eye-witness accounts, combined with geological observations of older volcanic deposits, show that the styles, scales, and products of volcanic eruptions vary tremendously from volcano to volcano. In this section we will look at these different styles of volcanic activity, but we will start by drawing a basic distinction between effusive and explosive eruptions.

Volcanic eruptions can be divided into two main classes: effusive and explosive. An **effusive** eruption is one in which molten rock called **magma**, rising from the deep interior of the Earth, flows out of a vent as a coherent liquid called **lava**. Both while they are still molten and after they cool, the bodies of rock formed on the surface in this way are called **lava flows**. An **explosive** eruption is one in which the magmatic material is torn apart as it is erupted into pieces called **pyroclasts**. These can range from hot clots of still-molten liquid to much cooler, irre-

gularly shaped fragments of solid rock. Explosive activity happens when the magmatic liquid rising from great depth is torn apart by the expansion of gas bubbles formed by the release of volatile compounds dissolved in the liquid. The term effusive eruption is often used, incorrectly, to denote any eruption in which lava flows are produced. Hawaiian eruptions, in particular, are often referred to as effusive because they produce lava flows. However, in fact very many of these eruptions are explosive, and the lava flows they produce are formed indirectly, by coalescence of lava clots accumulating near the vent, rather than by the direct oozing of the magma from the vent.

We now present a broad overview of the range of styles of volcanic activity that occur on Earth. All but the first style – effusive – are explosive in character.

1.2.1 Effusive eruptions

An effusive eruption is an eruption in which lava flows away from a vent as a coherent liquid. For

Fig. 1.3 An approximately 160 m high, 400 m wide lava dome slowly growing in the vent of the May 18, 1980 eruption of Mount St Helens volcano, Washington State, USA. (Photograph taken on August 22, 1981 by Lyn Topinka, courtesy of U.S. Geological Survey/Cascades Volcano Observatory.)

Fig. 1.4 Part of a dense, sheet-like lava flow erupted on the ocean floor, where the high pressure suppresses gas release, minimizing explosive activity and the formation of gas bubbles in lavas. (Image courtesy of Monterey Bay Aquarium Research Institute, © 2001 MBARI.)

instance, after major explosive eruptions have finished their explosive phase it is common for viscous lava to ooze from the eruptive vent to form a lava dome (Fig. 1.3). In deep submarine eruptions, where the pressure of the overlying water is great enough to suppress the release of gas from the erupting magma, the dominant mode of eruption is effusion (Fig. 1.4). In other cases lava may effuse from a vent because the lava has previously lost the gas which was initially dissolved in it. This happens, for instance, at Stromboli, where repeated small explosions every few tens of minutes allow the

escape of most of the gas from a substantial volume of magma stored at shallow depth. That magma is then erupted every few years as lava flows.

1.2.2 Hawaiian-style eruptions

The **Hawaiian** eruption style is named after the predominant style of activity observed at the currently active volcanoes of the Hawaiian Island chain. The term **Hawaiian** can be applied, though, to any eruption exhibiting this same style regardless of where in the world it occurs. Hawaiian eruptions are characterized by their lava fountains (Fig. 1.1). These are composed of hot, incandescent clots of magma (often up to 1–2 m in diameter) which are ejected from the vent at speeds of ~100 m s^{-1} and typically rise to heights of only a few tens to hundreds of meters above the vent before falling back to the ground. The majority of the clots of magma fall close to the vent and are still very hot upon landing (~1135°C), hot enough that the clots coalesce on the ground forming fluid lava flows which may travel several kilometers or even tens of kilometers from the vent (Fig. 1.5). Some of the clots and smaller clasts landing close to the vent are cooled enough during flight and after landing that they are too cool to form lava flows but instead weld together forming a spatter cone or spatter rampart around the vent (Fig. 1.6). A small amount of the erupted material is sufficiently fine grained

Fig. 1.5 Lava fountain forming lava flows at the Pu'u 'O'o vent on the East Rift Zone of Kilauea volcano, Hawai'I. Hot clots of magma falling within the cone have formed a lava pond that is overflowing from the lowest point on the rim of the cinder cone to form a lava flow. On the right side of the cone, clots are coalescing as they land to form a rootless lava flow. (Photograph taken on June 30, 1984 by J.D. Griggs, courtesy of Hawaiian Volcano Observatory.)

Fig. 1.6 Spatter ramparts along either side of a fissure vent on the East Rift Zone of Kilauea volcano, Hawai'I. Figure is standing at the location of the fissure itself. (Photograph by Lionel Wilson.)

that it is carried downwind from the vent forming a **tephra** blanket (Fig. 1.7). Hawaiian eruptions are sustained eruptions which can last for hours or days – in some cases for years. The magmas involved in Hawaiian eruptions are usually hot magmas called **basalts**. The combination of their chemical composition (especially the relatively low silica content) and high temperature gives these magmas a relatively

Fig. 1.7 Tephra blanket from the Pu'u Puai vent near the summit caldera of Kilauea volcano, Hawai'I. The blanket extends into the forest from the edge of the cinder cone at the bottom right of the image. (Photograph by Pete Mouginis-Mark, University of Hawai'I.)

low **viscosity**. Lavas erupted at Kilauea volcano in Hawai'I, for example, commonly emerge with temperatures of ~1140–1150°C and viscosities of 50–100 Pa s. Basaltic magmas come directly from zones of melting in the mantle with very little interaction with other rock types on their way to the surface.

1.2.3 Flood basalt eruptions

Another type of basaltic lava-forming eruption is the **flood basalt eruption**. Humans have yet to witness a flood basalt eruption because the most recent one occurred ~20 million years ago, but their deposits have been mapped out in many parts of the world (Fig. 1.8). These are eruptions which generate enormous volumes of basaltic lava. They occur in sequences, so that the volume of an entire flood basalt province can be as great as 10^6 km^3. Individual lava flows in such a province can be more than 600 km long and 100 m thick with volumes as great as 2000 km^3. There is considerable debate about the exact character of these eruptions but they appear to be similar to Hawaiian eruptions, though with individual events producing far larger volumes of lava and with the lava being erupted far more rapidly.

The closest equivalent to a flood basalt eruption yet observed by humans is the Laki or "Skaftár Fires"

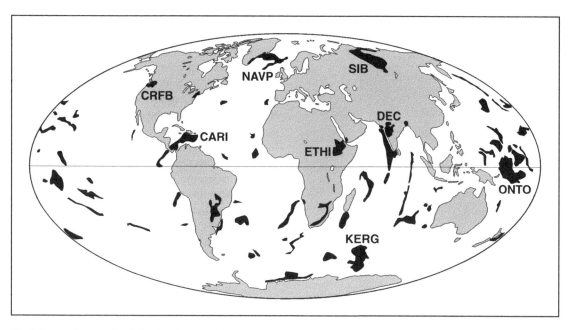

Fig. 1.8 Map showing the global distribution of flood basalt deposits, marked in black. The labeled deposits are: CRFB, Columbia River Flood Basalts; CARI, Caribbean Flood Basalts; NAVP, North Atlantic Volcanic Province; ETHI, Ethiopian Flood Basalts; SIB, Siberian Traps; DEC, Deccan Traps; KERG, Kerguelen Plateau; ONTO, Ontong Java Plateau. (Information derived from fig. 1 in Coffin, M.F. and Eldholm, O. (1994) Large igneous provinces: crustal structure, dimensions, and external consequences. *Reviews of Geophysics*, **32**, 1-36; and from fig. 5.4 in Courtillot, V. (1999) *Evolutionary Catastrophies. The Science of Mass Extinction*. Cambridge University Press, 173 pp.)

eruption which occurred in Iceland in 1783–85. This eruption produced ~15 km³ of lava, with individual lava flows exceeding 35 km in length. Eruption rates as great as 8700 m³ s⁻¹ have been estimated for the eruption. By contrast the 1983–86 eruption at Kilauea produced ~0.5 km³ of lava at eruption rates ranging from less than 20 to more than 350 m³ s⁻¹.

The Laki eruption is of special interest because of its effect on climate. The winter of 1783–84 was a particularly severe one in Europe and was associated with a "dry fog" which spread out across several countries reducing the amount of sunlight reaching the surface. The most plausible explanation for this fog is that it was caused by the release of mainly sulphurous gases during the Laki eruption. The effect of the fog or "haze" in Iceland itself was extremely severe: it stunted the growth of grass, and small amounts of fluorine released from the lava were taken up by grass and poisoned the ani-

mals in the pastures. This resulted in the death of half of all livestock in Iceland. The resulting "haze famine" caused the death of 22% of the human population through a combination of starvation, disease and severe cold. When the small scale of this eruption is considered relative to the sizes of flood basalt eruptions, it is clear that the environmental consequences of a flood basalt eruption are likely to be profound (see Chapter 12).

1.2.4 Plinian eruptions

Events such as the May 18, 1980 eruption of Mount St Helens in Washington State, USA (Fig. 1.2) are typical of **Plinian** eruptions. These eruptions were named after Pliny the Younger who wrote an account of the AD 79 eruption of Vesuvius which destroyed the cities of Pompeii and Herculaneum. In eruptions of this kind a jet of gas and magma

Fig. 1.9 A Plinian fall deposit from the ~3.3 ka Waimihia rhyolitic eruption of Taupo volcano, New Zealand. (Photograph courtesy of Stephen Self.)

clasts emerges from the vent at speeds of ~100–600 m s^{-1} and forms a convecting eruption plume in the atmosphere as the surrounding air is sucked into the jet (see Chapter 6). Clasts are carried upwards and progressively fall out from the eruption plume. Large clasts are carried upwards only a short distance and fall to the ground close to the vent, while smaller clasts are carried to greater heights and fall out further from the vent, forming part of an unconsolidated airfall deposit (Fig. 1.9, see Chapter 8). The eruptions are sustained, and can last for hours or days. Eruptions exhibiting this style of activity are subdivided into **sub-Plinian**, **Plinian**, and **ultra-Plinian** based on their mass flux and plume heights (factors which are linked). Mass fluxes across the three subtypes range from ~10^6 to 10^9 kg s^{-1}, rates which are generally greatly in excess of those for recent basaltic eruptions. Mass fluxes during recent eruptions at Kilauea are $\leq 10^5$ kg s^{-1}, whereas the highest mass flux during the (exceptional) Laki eruption was ~10^7 kg s^{-1}. Generally, Plinian eruptions involve magmas which are relatively rich in silica and in dissolved gases, are very viscous, and erupt at lower temperatures than basalts. They are the products of various kinds of interaction of more basaltic melts with other rocks in the crust. The commonest magma types involved in Plinian eruptions are called **dacites** and **rhyolites**, but rare basaltic Plinian eruptions are known.

1.2.5 Ignimbrite-forming eruptions

The **Hawaiian** and **Plinian** eruption styles were defined based on observations of modern and historic eruptions. By contrast the eruption style referred to by the (rather cumbersome) term **ignimbrite-forming** was defined based on eruption deposits: geologists observed and mapped **ignimbrite** deposits long before they witnessed or understood what type of volcanic eruption formed them. Decades went by after the term was defined before theoretical modeling of how eruptions work, together with observations of certain modern eruptions, gave volcanologists an understanding of the types of eruption in which ignimbrites form. Ignimbrites (originally known as ash-flow tuffs in the USA) are the deposits produced by very large-scale **pyroclastic density currents**. A pyroclastic density current is a hot cloud of volcanic ash, magmatic gas and air which flows along the ground at a very high speed. Possibly the fastest example yet documented in detail was produced during the AD 189 eruption of Taupo, in New Zealand. Deposits from this flow are found on top of a ~1600 m high mountain, which appears to imply a speed of at least ~180 m s^{-1}. More typical, much thicker, ignimbrites include the Bandelier Tuff in the USA and the Campanian Tuff in Italy.

Pyroclastic density currents form in various ways, one of which involves an eruption column, like that formed in a **Plinian** eruption, becoming unstable and collapsing (Fig. 1.10). No human being has ever experienced the formation of a really large pyroclastic density current. However, a tragic example of the devastating effect of even a small event of this kind occurred during the 1902 eruption of Mount Pelée, a volcano on the Caribbean island of Martinique. On May 8, 1902, Mount Pelee erupted producing a large black cloud that rolled down the flanks of the volcano and spread out, generating a kind of pyroclastic density current called a **pyroclastic surge** that engulfed the main town on the island, St Pierre, 6 km away from the volcano. The surge cloud moved rapidly through the town, setting anything combustible on fire. In the space of 2 to 3 minutes about 28,000 people were killed. Only a handful of people survived, possibly as few as two or three; one survivor was a prisoner locked in the

Fig. 1.10 A pyroclastic density current formed by partial collapse of an eruption column during the January 1974 eruption of Ngauruhoe volcano, New Zealand. The eruption column is approximately 4 km high. (Photograph credit: University of Colorado, courtesy of the National Oceanic and Atmospheric Administration, National Geophysical Data Center.)

Fig. 1.11 The type of pyroclastic density current called a **pyroclastic surge** or **nuée ardente**. This surge is one of a series from Mount Pelée volcano, Martinique, that destroyed the town of St Pierre in 1902. (Photograph taken on December 16, 1902 by A. Lacroix.)

town jail. The kind of dilute, fast-moving cloud which destroyed Mount Pelée is sometimes called a *nuée ardente* (Fig. 1.11).

1.2.6 Strombolian eruptions

The eruptions described thus far are all types of sustained eruption. Other eruptions, though, are transient in character and consist of discrete explosions of short duration. One example is the type of eruption called **Strombolian**, named after the volcanic island Stromboli in the Mediterranean Sea, which usually involves basaltic magma. Strombolian

Fig. 1.12 Approximately 120 m high ash-rich eruption cloud from a Strombolian explosion at Stromboli volcano, Italy. (Photograph by Matt Patrick, University of Hawai'I.)

eruptions consist of transient explosions typically lasting 1–2 seconds which occur in sequences. At Stromboli itself, for example, explosions generally happen at time intervals of a few minutes to a few hours. Each explosion generates a small ash plume which is usually less than 200 m high (Fig. 1.12) and throws out large incandescent blocks which, by night, can be seen following **ballistic** trajectories (Fig. 1.13). The volume of material produced in each explosion is small: it can be as little as a few cubic meters. Though this type of activity is characteristic of Stromboli itself, the term **Strombolian** is used more broadly to denote a range of styles of transient explosive activity. In active lava lakes, for example, gas rising to the lake surface causes updoming and bursting of large bubbles (which can exceed 1 m in diameter) resulting in very weak explosions (Fig. 1.14). These eruptions can also be described as Strombolian. In Strombolian activity at Heimaey in Iceland in 1973 explosions occurred at time intervals of 0.5 to 3 seconds. The close spacing in time of each explosion meant that the eruption was effectively sustained, not transient, and was able to generate an eruption plume up to 10 km high.

So the term **Strombolian** can be applied to eruptions as weak as the bubble bursting events in Hawai'I through to ones which generate high eruption plumes and deposit ash over a relatively large area. A further problem arises in understanding

Fig. 1.13 Pyroclasts following ballistic trajectories from a Strombolian explosion at Stromboli volcano, Italy. (Photograph by Pete Mouginis-Mark, University of Hawai'I.)

Fig. 1.14 Clots of fragmented magma ejected as gas bubbles burst on the surface of a lava lake on Kilauea volcano, Hawai'I. (Photograph by Pete Mouginis-Mark, University of Hawai'I.)

Fig. 1.15 Walker (1973) devised this classification scheme for different types of explosive eruption based on the degree of fragmentation, F, a function of the range of sizes of pyroclasts produced, and the dispersal, D, defined as the area within the boundary where thickness of the deposit decreased to 1% of its maximum thickness. The circled asterisk indicates where the deposits of the 1959 Kilauea Iki eruption plot using this classification system. After fig. 6.2 in Cas, R.A.F. and Wright, J.V. (1988) *Volcanic Successions. A Geological Approach to Processes, Products and Successions.* Chapman and Hall, 528 pp. With kind permission of Springer Science and Business Media.)

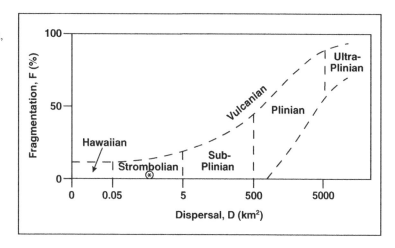

what is meant when an eruption is described as Strombolian, because the most widely used classification system defines an eruption as Strombolian if the dispersal of ash during the eruption lies within certain specific limits (Fig. 1.15). This leads to problems because it is common for the deposits of modern Hawaiian eruptions to lie within the Strombolian field (Fig. 1.15). Thus an eruption could be defined as Hawaiian based on observation of the eruption style but Strombolian based on the deposits generated. Extreme caution is needed, therefore, in classifying an eruption as Strombolian on the basis of ash dispersal alone. In this book we always use terms such as Strombolian to refer to the style of an eruption, not the extent of the ash dispersal involved. Characteristic features of Strombolian eruptions are that they consist of transient explosions, closely spaced in time, involving low viscosity, generally basaltic magma, and occurring within volcanic systems which are open to the surface.

1.2.7 Vulcanian eruptions

Vulcanian eruptions are like Strombolian eruptions in consisting of discrete or transient explosions, but are associated with more evolved magmas than those causing Strombolian eruptions, typically ranging from intermediate compositions (e.g., **basaltic andesite** to **andesite**) through to

more evolved compositions (e.g., **dacite**). Typically Vulcanian explosions last a number of seconds or minutes and often occur in sequences with repose times between explosions varying from tens of minutes to hours. It is common for the violence of the explosion to be related to the repose time, so that longer repose times result in more violent explosions. Vulcanian explosions can be extremely violent events: typical eruption velocities range between 200 and 400 m s^{-1} and the eruptions can eject blocks a few meters in size out to distances of as much as 5 km from the vent. One explosion at Ngauruhoe in New Zealand in 1975 ejected a dense block 27 m long and 15 m wide which is estimated to weigh around 3000 tons! In addition to ejecting large blocks and volcanic bombs ballistically, the eruptions commonly generate eruption plumes which carry finer material to heights of several kilometers. In eruption sequences in which explosions occur with relatively short repose times, higher plumes can be generated. These can be up to ~20 km in height. The eruption plumes generate a fall deposit and, as the eruption columns are commonly unstable, they can collapse to generate pyroclastic density currents. For example, during the 1975 Ngauruhoe eruption the eruption plume rose to heights of 4–5 km and deposited ash as much as 20 km from the vent, and the partial collapse of the plume generated pyroclastic currents which

Fig. 1.16 Coarse pyroclasts deposited by a pyroclastic density current formed during the January 1974 eruption of Ngauruhoe volcano, New Zealand. (Photograph by Lionel Wilson.)

traveled down the flanks of the volcano to distances of up to ~2 km (Fig. 1.16). Vulcanian eruptions vary widely in the proportion of **juvenile** (i.e., derived directly from the magma) and nonjuvenile (i.e., incorporated from the surrounding rocks) material that they eject. The volumes of material produced by individual Vulcanian explosions are far greater than those of Strombolian explosions. At Ngauruhoe the individual explosions produced up to 10^5 m^3 of **ejecta**; larger volumes have been inferred for other Vulcanian explosions, such as the 1968 eruption of Arenal in Costa Rica.

1.2.8 Hydromagmatic eruptions

Hydromagmatic eruptions involve interaction between magma or lava and external water. Such interactions can occur in a wide range of environ-

ments: in deep marine settings as a volcano grows on the ocean floor; in shallow water; between lava and water as lava enters a lake, river or the sea; between lava and ice where an eruption occurs beneath a glacier; or where magma comes into contact with groundwater before erupting (in which case the term **phreatomagmatic** is often used). The range of ways in which magma can interact with external water is reflected in the great diversity of the types of eruption which can occur. The following descriptions are designed to give some feel for this diversity.

DEEP MARINE ENVIRONMENTS

It is estimated that 70–80% of the annual volcanic output on Earth occurs at **mid-ocean ridges** (MORs). This means that the styles of volcanic activity which occur in this deep marine environment are, in fact, the dominant eruption styles on Earth. The inaccessibility of these regions, however, means that they are far harder to study than **subaerial** eruptions. Studies using a range of techniques, including sonar surveys and submersible dives, show that MOR volcanism is basaltic and effusive. There are two reasons why the eruptions are effusive rather than explosive. One is that when lava is erupted on the sea floor the pressure of the overlying water limits the formation of gas bubbles within the lava and thus suppresses the explosivity of the magma. In other words an eruption which would have generated Hawaiian activity on land will consist of passive effusion at a MOR. Interactions between magma and water can also generate explosions in some settings, but at a MOR the interaction between lava and water is generally nonexplosive. When lava oozes out onto the sea floor it is rapidly cooled and quenched by the surrounding water. **Pillow lavas** are a distinctive form of basaltic lava flow which form because of the rapid cooling of the outer skin of the lava flow (Fig. 1.17).

LAVA FLOWS ENTERING WATER

A different type of behavior occurs when lava enters water rather than being erupted beneath it. For example, Kilauea volcano in Hawai'I frequently generates lava flows which are long enough to reach

Fig. 1.17 Basaltic pillow lavas on the northeast rim of Lo'ihi volcano, a submarine volcano located south of the island of Hawai'I. Area shown is about 10 × 14 m. (Photograph taken in 1980 by A. Malahoff, University of Hawai'I, courtesy of U.S. Geological Survey.)

the coast and spill lava into the ocean. How the lava enters the sea affects the nature of the interaction which occurs. In some cases the lava oozes out into the sea and is rapidly cooled without an explosion occurring, but in other cases the interaction is explosive and the fragments produced as the lava is torn apart in the explosion are thrown into the air and deposited on the shoreline forming a **littoral cone** (Fig. 1.18).

SHALLOW MARINE AND CRATER LAKE ERUPTIONS

More dramatic hydromagmatic eruptions can occur during eruptions which occur through water, either through shallow seawater or through lakes which are found in the craters at the summits of some volcanoes.

A classic example of hydromagmatic interaction in a shallow marine environment occurred at Surtsey off the south coast of Iceland between 1963 and 1965. Eruptive activity was first noticed on November 14, 1963 when the top of the volcano was about 10 m below the water surface. A black eruption cloud was initially seen just rising above the sea surface. The cloud gradually grew to a height of ~65 m. By the next day Surtsey had grown above sea level and was erupting fairly constantly. Scientists observing the eruption noted two dominant styles of eruption. If seawater could flow into an erupting vent, intermittent explosions occurred at intervals of a few seconds to a few minutes. These explosions produced dark clouds of ash and steam which rose rapidly upwards and outwards from the vent producing what are known as **cock's tail plumes** or **tephra jets** (Fig. 1.19). The plumes rose to heights of ~500 m, with individual bombs reaching heights of ~900 m. During the larger explosions, **base surges** were generated. These are a kind of pyroclastic cloud consisting of ash and steam which

Fig. 1.18 Lava from Kīlauea volcano, Hawai'I, entering the sea through a lava tube and being torn apart explosively. Chilled pyroclasts are deposited on the shoreline forming a littoral cone. Photograph by Pete Mouginis-Mark, University of Hawai'I.

Fig. 1.19 Image of cock's tail plumes, transient explosive tephra jets generated when sea water and magma interact, produced during the eruption of Surtsey that began off the south coast of Iceland in 1963. The plume is approximately 500 m high. Photograph by Sigurdur Thorarinsson, used by permission of Sven Sigurdsson.

forms in a ring around the base of the eruption column and spreads out radially away from it at high speed (the term was originally coined for a similar phenomenon seen in some surface tests of nuclear weapons). Base surges are a common feature of many hydromagmatic eruptions.

As tephra was constantly being deposited around the vents at Surtsey, the easy access of seawater to the vents would be stopped at times and then a more continuous eruption would occur. These eruptions often lasted several hours at a time and produced a more stable eruption plume, which was dark in color and rose to heights of ~2 km. Incandescent clasts were common within the eruption plume and fell out close to the vent. By April 1964 the vent had grown to the point where seawater could no longer gain access to it, and the styles of activity changed to the more conventional forms of basaltic eruptions – Hawaiian and Strombolian.

Eruptions through crater lakes are often similar in style to those in shallow marine settings. A spectacular example of this type of eruption occurred at Taal in the Philippines in 1965. The eruption occurred at Volcano Island, which sits within Lake Taal, the lake itself occupying Taal **caldera** (a caldera is a crater or surface depression formed by the collapse of the ground above a magma chamber). The erup-

tion started at 2 a.m. on September 28, 1965 with a Strombolian phase but changed dramatically in style after ~90 minutes. The eruption began to produce large, ash-laden eruption clouds which rose to heights of 15–20 km and large base surges which spread out from the base of the eruption column to distances as great as 6 km. The speeds of these surges were estimated to be as great as 100 m s^{-1} (i.e., ~220 miles per hour). The surges appear to have been cold, as there is no evidence that they scorched or burned anything in their paths, but they knocked down all the trees within about 1 km of the vent and sand-blasted trees in the area beyond this. The explosions occurred at irregular intervals from a number of vents for a period of about 7 hours. The change in character of the eruption from the initial Strombolian phase to the violent explosions seems to have occurred when the lake water started to gain access to the active vents. After the initial 7 hours of violent activity the explosions gradually declined in vigor, and by 4 pm on September 30 the eruption was over. This eruption killed 190 people.

A further type of hydromagmatic activity is associated with eruptions through lakes or shallow seawater. These are **phreato-Plinian** eruptions, in which the fine-grained nature of the ash and other aspects of the associated deposits (e.g., the presence of **accretionary lapilli**, small clumps of particles stuck together because they were damp) suggest that the eruption was hydromagmatic. However, in other respects the deposits are like those of "normal" Plinian eruptions. For example, the 1875 eruption of Askja in Iceland lasted about 6.5 hours during which time it exhibited subPlinian, Plinian and phreato-Plinian phases. The phreato-Plinian phase appears to have lasted about an hour and deposited the Askja-C ash which has a volume of about 0.2 km^3. Phreato-Plinian eruptions differ from the other eruptions involving interaction of lake water with magma in that they appear to generate steady, continuous eruption phases in which the external water simply enhances the fragmentation of the magma without dramatically altering the eruption style. During the AD 181 eruption at Taupo (the eruption for which the term ultra-Plinian was coined), the eruption style also varied in this way between Plinian and phreato-Plinian depending on whether lake water could gain access to the magma.

SUBGLACIAL ERUPTIONS

These are eruptions in which the vent is situated beneath a glacier or ice sheet. The eruptions can be effusive or explosive depending mainly on the thickness of the overlying ice, and usually involve the production of large amounts of water as ice is melted. Often the melting produces so much water that the eruption is, in effect, submarine, and pillow lavas form, creating a pillow mound around a localized vent or a pillow ridge if the vent is an elongate **fissure**. In some fissure eruptions the **dike** (the fracture through which magma is rising to feed the vent) essentially overshoots the interface between the rock surface and the base of the glacier so that magma actually penetrates some distance into a crack in the ice. This leads to rapid chilling and fragmentation of the magma as it melts the surrounding ice, and the water produced undergoes chemical reactions with the magma fragments to form a rock called **hyaloclastite**. The fragments then fall through the water to accumulate alongside the vent forming a **hyaloclastite ridge**.

Iceland is one place where it is quite common for fissure eruptions to occur beneath glaciers, and in

September and October 1996 a spectacular subglacial eruption occurred along a fissure beneath the extensive Vatnajökull ice cap. The fissure was located between two well-known subglacial volcanic centers – Grímsvötn and Bárðarbunga. Initially the evidence for the eruption was a depression 2 km wide and 200–300 m deep which formed in the ice above the fissure. Eventually the eruption melted through the 400–600 m thick layer of ice above the vent and a steam column rose into the atmosphere. This changed into a dark eruption column which rose about 500 m and was produced by rhythmic explosions at the vent. The eruption column height increased to a maximum of 9 km late in the day on October 2 and a new depression formed in the ice 3 km away from the first one as a new subglacial fissure started erupting. The eruptions continued for a week with alternating periods of quiet and explosive activity. The length of the active fissure gradually increased to ~9 km with eruptions occurring through pockets of melt water as much as 50 m deep. The large amount of meltwater generated in subglacial eruptions gives rise to a phenomenon called a **jökulhlaup** (glacier-burst, an Icelandic word), a giant flood of glacial meltwater. The 1996 eruption produced a jökulhlaup which reached peak flow rates of ~45,000 m^3 s^{-1} and was able to transport ice blocks the size of houses. This did considerable damage to bridges along the southern coast of the island and affected an area of ~750 km^2. It is estimated that the eruption produced about 0.7–0.75 km^3 of volcanic products, including a new subglacial ridge ~7 km long, and melted ~4 km^3 of ice.

INTERACTIONS WITH GROUNDWATER

So far we have talked about interactions of magma with surface water or ice. Magma can also interact, however, with water beneath the Earth's surface, i.e., with groundwater. Some Vulcanian explosions seem to be caused in this way. Other eruptions involving interactions with groundwater form a type of wide, shallow, low-rimmed volcanic crater called a **maar** (Fig. 1.20). Two such craters formed during the 1977 eruption at Ukinrek in Alaska. The first (West) maar formed over a period of 3 days and was 170 m wide and 35 m deep, whereas the

Fig. 1.20 Maar crater in Death Valley, California, USA, formed during phreatomagmatic explosive activity. The crater is approximately 800 m in diameter. (Photograph by Elisabeth Parfitt.)

second (East) maar took 7 days to form and was 300 m wide and 70 m deep. Both centers exhibited intermittent explosions which produced eruption columns a few kilometers high, the maximum height being ~6.5 km recorded for an explosion from the East Maar. These eruption clouds deposited ash as much as 160 km downwind, though the bulk of the ash was deposited within 2–3 km of the vents. Base surges were observed during one violent explosion from the East Maar. Although the eruptions were hydromagmatic throughout, there was a gradual transition to less violent explosions and activity which was more nearly Strombolian in character as the eruption progressed. This was thought to be due to the gradual depletion of the aquifer supplying water to the eruption.

1.2.9 Diatreme-forming eruptions

Diatremes are conical to elongate zones of shattered rocks extending downward from the surface, often to depths of at least many hundreds of meters. They contain fragmental volcanic rocks called **kimberlites**. The mineralogy of most kimberlites implies that their parent magmas left source regions deep in the mantle containing large amounts of carbon dioxide. The violent release of this gas as the magma reached the surface caused the intense shattering of crustal rocks that characterizes the diatreme, and also rapidly fragmented and chilled the magma. The most recent diatreme-forming

eruption took place ~50 Ma ago, and so the surface deposits from such eruptions are not well preserved, though we assume that because the magmas forming the kimberlites contained large amounts of gas the eruptions were very explosive and possibly similar to Plinian eruptions. Diatreme formation is in many ways the least-well understood of all the types of volcanic activity, but these features are economically very important. The reason is that the magmas coming to the surface in these eruptions accidentally carry with them various minerals from the source regions deep in the mantle, and one of these minerals is the crystalline form of carbon called diamond.

1.3 Volcanic systems

The descriptions of volcanic eruptions given in the previous section are intended to give an indication of the diversity of volcanic activity which occurs on Earth and for the types of deposits that eruptions produce. Because we can observe eruptions as they occur and map and analyze the deposits that eruptions produce, it is inevitable that we know more about the mechanisms of volcanic eruptions than we do about the processes which occur beneath the ground before an eruption starts. From the point of view of a physical volcanologist, however, a volcanic eruption is an end-point, the culmination of a sequence of processes which have their origin deep within the Earth in the zone in which the magma initially formed. To fully understand the volcanic eruption itself, then, we need to understand the subsurface processes which led up to it, as well as the processes involved in the eruption itself.

In this book we consider the sequence of physical processes which lead up to and take place during a volcanic eruption as occurring within an overall volcanic system (Fig. 1.21). The physical processes operating within the system can be divided into four stages: Stage 1 is the formation of magma; Stage 2 involves the movement of magma away from the source zone; Stage 3 involves storage of magma at depth; and Stage 4 involves the movement of magma to the surface and the eruption process itself. The path taken by magma through any specific volcanic system is not always the same. For instance, in some cases magma may be formed (Stage 1) and be transported directly to the surface to be erupted (Stages 2 and 4) thus bypassing storage at depth (Stage 3). In other cases magma may form and move to a zone of storage (Stages 1–3) but never be erupted. This division of a volcanic system into just four stages is something of an oversimplification of what can happen, but it is very useful as a way of denoting the key processes that magma must go through in order to be erupted, and forms a basis for examining the successive physical processes experienced by magma as it travels to the surface.

1.4 The structure and aims of this book

In line with the stages shown in Fig. 1.21, we start in Chapter 2 by looking at Stage 1 processes: the generation of magma. In Chapter 3 we look at the ways in which magma moves within the mantle and crust (Stage 2). In Chapter 4 we look at magma storage (Stage 3). Successive chapters are then concerned with aspects of Stage 4 – the movement of magma at shallow levels within the crust and with the processes and consequences of eruptions. The bulk of the book is concerned with Stage 4 processes because these are the events that volcanologists know most about, and because eruption processes have the most direct impact on humans. Our intention throughout, however, is to emphasize how subsurface processes (Stages 1–3) affect

the nature of volcanic activity (Stage 4), and to show the importance of increasing our knowledge of these processes in order to improve our understanding of the eruptions themselves and the effects that they can have on our lives.

1.5 Further reading

GENERAL

Dobran, F. (2001) *Volcanic Processes: Mechanisms in Material Transport.* Plenum, 590 pp. ISBN: 9780306466250.
Schmincke, H.-U. (2004) *Volcanism.* Springer-Verlag, 324 pp. ISBN: 9783540436508.

HAWAIIAN ERUPTIONS

Head, J.W. & Wilson, L. (1989) Basaltic pyroclastic eruptions: influence of gas-release patterns and volume fluxes on fountain structure, and the formation of cinder cones, spatter cones, rootless flows, lava ponds and lava flows. *J. Volcanol. Geotherm. Res.* **37**, 261–71.
Wolfe, E.W., Neal, C.A., Banks, N.G. & Duggan, T.J. (1988) Geological observations and chronology of eruptive events. In *The Puu Oo Eruption of Kilauea Volcano, Hawaii: Episodes 1 through 20, January 3, 1983, through June 8, 1984. U.S. Geol. Soc. Prof. Pap.* **1463**, 1–97.

FLOOD BASALT ERUPTIONS

Self, S., Thordarson, T. & Keszthelyi, L. (1997) Emplacement of continental flood basalt lava flows. In *Large Igneous Provinces: Continental, Oceanic, and Planetary Flood Volcanism* (Eds J.J. Mahoney & M.F. Coffin), pp. 381–410. Geophysical Monograph 100, American Geophysical Union, Washington, DC.
Swanson, D.A., Wright, T.L. & Helz, R.T. (1975) Linear vent systems and estimated rates of magma production and eruption for the Yakima basalt on the Columbia Plateau. *Am. J. Sci.* **275**, 877–905.
Thordarson, T. & Self, S. (1993) The Laki (Skafta Fires) and Grimsvotn eruptions of 1783–1785. *Bull. Volcanol.* **55**, 223–63.

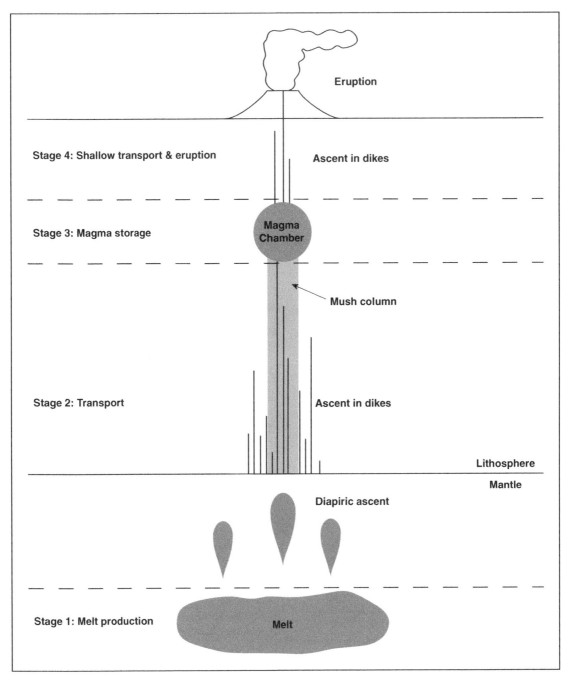

Fig. 1.21 Four stages involved in the formation of a volcanic system. Stage 1: generation of magma by partial melting of rock. Stage 2: movement of magma away from the source zone. Stage 3: storage of magma at depth. Stage 4: movement of magma from storage to either erupt at the surface or form intrusions.

PLINIAN ERUPTIONS

Carey, S.N. & Sigurdsson, H. (1982) Influence of particle aggregation on deposition of distal tephra from the May 18, 1980 eruption of Mount St. Helens volcano. *J. Geophys. Res.* **87**, 7061-72.

Sparks, R.S.J., Wilson, L. & Sigurdsson, H. (1981) The pyroclastic deposits of the 1875 eruption of Askja, Iceland. *Phil. Trans. Roy. Soc. Ser. A* **299**, 241-73.

Walker, G.P.L. (1980) The Taupo pumice: product of the most powerful known (ultraplinian) eruption? *J. Volcanol. Geotherm. Res.* **8**, 69-94.

Walker, G.P.L. & Croasdale, R. (1970) Two Plinian-type eruptions in the Azores. *J. Geol. Soc. London* **127**, 17-55.

IGNIMBRITE-FORMING ERUPTIONS

Branney, M.J. & Kokelaar, P. (2002) *Pyroclastic Density Currents and the Sedimentation of Ignimbrites.* Memoir 27, Geological Society Publishing House, Bath, 143 pp.

Fisher, R.V., Smith, A.L. & Roobol, M.J. (1980) Destruction of St. Pierre, Martinique, by ashcloud surges, May 8 and 20, 1902. *Geology* **8**, 472-6.

Sparks, R.S.J. & Wilson, L. (1976) A model for the formation of ignimbrite by gravitational column collapse. *J. Geol. Soc. London* **132**, 441-51.

STROMBOLIAN ERUPTIONS

Blackburn, E.A., Wilson, L. & Sparks, R.S.J. (1976) Mechanisms and dynamics of Strombolian activity. *J. Geol. Soc. London* **132**, 429-40.

Chouet, B., Hamisevicz, N. & McGetchin, T.R. (1974) Photoballistics of volcanic jet activity at Stromboli, Italy. *J. Geophys. Res.* **79**, 4961-76.

McGetchin, T.R., Settle, M. & Chouet, B. (1974) Cinder cone growth modeled after Northeast Crater, Mount Etna, Sicily. *J. Geophys. Res.* **79**, 3257-72.

VULCANIAN ERUPTIONS

Morrissey, M. & Mastin, L.G. (2000) Vulcanian eruptions. In *Encyclopedia of Volcanoes* (Ed. H. Sigurdsson), pp. 463-75. Academic Press, San Diego, CA.

Nairn, I.A. & Self, S. (1978) Explosive eruptions and pyroclastic avalanches from Ngauruhoe in February 1975. *J. Volcanol. Geotherm. Res.* **3**, 39-60.

HYDROMAGMATIC ERUPTIONS

Batiza, R. & White, J.D.L. (2000) Submarine lavas and hyaloclastite. In *Encyclopedia of Volcanoes* (Ed. H. Sigurdsson), pp. 361-81. Academic Press, San Diego, CA.

Gudmundsson, M.T., Sigmudsson, F. & Björnsson, K. (1997) Ice-volcano interaction of the 1996 Gjálp subglacial eruption, Vatnajökull, Iceland. *Nature* **389**, 954-7.

Kienle, J., Kyle, P.R., Self, S., Motyka, R.J. & Lorenz, V. (1980) Ukinrek Maars, Alaska, I. April 1977 eruption sequence, petrology and tectonic setting. *J. Volcanol. Geotherm. Res.* **7**, 11-37.

Moore, J.G., Nakamura, K. & Alcaraz, A. (1966) The 1965 eruption of Taal volcano. *Science* **155**, 955-60.

White, J.D.L. & Houghton, B. (2000) Surtseyan and related phreatomagmatic eruptions. In *Encyclopedia of Volcanoes* (Ed. H. Sigurdsson), pp. 495-511. Academic Press, San Diego, CA.

SUBMARINE ERUPTIONS

Head, J.W. & Wilson, L. (2003) Deep submarine pyroclastic eruptions: theory and predicted landforms and deposits. *J. Volcanol. Geotherm. Res.* **121**, 155-93.

SUBGLACIAL ERUPTIONS

Wilson, L. & Head, J.W. (2002) Heat transfer and melting in subglacial basaltic volcanic eruptions: implications for volcanic deposit morphology and meltwater volumes. In *Volcano-Ice Interaction on Earth and Mars* (Eds J.L. Smellie & M.G. Chapman), pp. 5-26. Special Publication 202, Geological Society Publishing House, Bath.

1.6 Questions to think about

1 What is the difference between an effusive and an explosive eruption, and what factors can make an eruption truly effusive rather than explosive?

2 What types of eruption are common and what types are rare on Earth, using a typical human life span as a scale? Is there any correlation between eruption style and rarity?

2 Magma generation and segregation

2.1 Introduction

We saw in Chapter 1 that a volcanic eruption can be viewed as the culmination of a series of physical and chemical processes (Fig. 1.21). The initial stage in the sequence of events which ultimately cause a volcanic eruption is always the generation of magma deep within the planet (Stage 1 in Fig. 1.21) – without magma there can be no eruption! In this chapter we will look at what we know about magma generation and then go on to consider the very earliest stages of magma movement in which the magma first starts to segregate and move away from the region in which it formed.

2.2 Rock-melting mechanisms

All rocks contain a mixture of different minerals and, as a result, they melt over a range of temperatures rather than at one specific temperature. The temperature at which melting first starts in a rock is called the **solidus** temperature; the temperature at which the last bit of solid vanishes and all of the rock is liquid is called the **liquidus** temperature. There are three main processes by which melting of rock can occur within the Earth.

1 By heating the rock and raising its temperature above the solidus temperature.
2 By reducing the confining pressure on the rock while keeping its temperature nearly constant. In most cases reducing the pressure acting on a rock reduces the solidus and liquidus temperatures (Fig. 2.1). So although the actual temperature of the

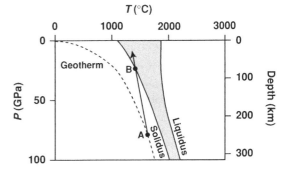

Fig. 2.1 The temperatures and pressures (and hence depths below the surface) at which mantle rocks begin to melt (the curve labeled **solidus**) and at which they are completely melted (the curve labeled **liquidus**). At any point in the shaded zone a rock is a mixture of unmelted minerals and liquid. The typical variation of temperature with depth in the Earth is the **geotherm**. If convection in the mantle brings a body of rock up from a depth corresponding to A to a depth corresponding to B, it cools slightly, but even so reaches a state in which it starts to melt. If the same body of rock continues to rise to shallower depths, the melt fraction increases.

rock does not change much, it is nevertheless at a much higher temperature relative to its solidus. This kind of melting is known as **decompression melting** or **pressure-release melting**.
3 By changing the composition of the rock, usually by the addition of water. The "dry" melting temperature of a rock is considerably higher than its "wet" melting temperature, i.e., the temperature at which it will melt when abundant water is present (Fig. 2.2). As a result, the addition of water to an initially dry rock can induce melting if the initial temperature of the rock is sufficiently high (Fig. 2.2).

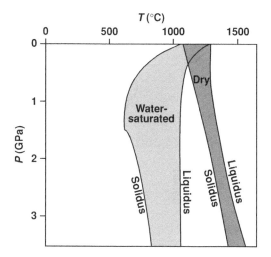

Fig. 2.2 The solidus and liquidus curves, and the zones of partial melting (shaded) are compared for a mantle rock containing no water (dry) and containing abundant water (wet). The addition of water to a rock moves the entire temperature range over which it melts to lower temperatures and also increases the melting temperature range. Thus addition of water can allow a rock to melt even if its actual temperature and pressure do not change. (After fig. 2 in Lambert, I.B. and Wyllie, P.J. (1972) Melting of gabbro (quartz eclogite) with excess water to 35 kilobars, with geological applications. *Journal of Geology*, **80**, 692–708. Copyright University of Chicago Press.)

Next we look at the evidence for where melting occurs on Earth, how this relates to the structure of the planet, and what kinds of melts are produced in different settings.

2.3 Volcanism and plate tectonics

Some fundamental information comes from looking at a simple map of the locations of volcanic activity on the Earth (Fig. 2.3). Volcanoes are not distributed randomly around the planet but instead occur in well-defined zones. The most famous of these is the Pacific **Ring of Fire** – this is a narrow band or ring of volcanic centers which circles the Pacific, running from New Zealand, up through the Tonga and Solomon islands and the New Hebrides to the Philippines, through Japan and the Kamchatka peninsula in Russia, through the Aleutian islands, then through the Cascades in western Canada

and the USA, through Central America, and finally down the west coast of South America to Deception Island at the tip of the Antarctic peninsula (Fig. 2.3). The Ring of Fire is associated with many of the largest volume and most energetic volcanic eruptions to have occurred in human history, including the largest eruptions of the 20th century – the 1912 eruption of Katmai in Alaska and the 1991 eruption of Mount Pinatubo in the Philippines, and also the 1883 eruption of Krakatau and the 1815 eruption of Tambora in Indonesia (the latter being the largest eruption to have occurred in modern history).

Less obvious but just as important are the long narrow chains of volcanoes lying beneath the Earth's oceans. Observations using sonar imaging systems and manned and remote-controlled submersibles show the existence of **mid-ocean ridges** (MORs), actually volcanic mountain chains which mark the sites of repeated eruptive activity (Fig. 2.4). As eruptions along the MORs occur at great depths beneath the ocean (typically 1–4 km) we are rarely aware of activity there and even less often able to observe it. In June 1993, however, a newly emplaced network of **hydrophones** detected seismic activity along part of the Juan de Fuca ridge (a spreading center located ~400 km off the west coast of Oregon in the USA). Subsequent investigations using various types of equipment showed that the seismic activity had been the precursor to an eruption along the ridge which produced a basaltic lava flow 3.8 km long and up to 500 m wide.

Volcanism in these long narrow zones is intimately associated with the large-scale structure of the Earth, specifically the fact that the outermost layer of the planet consists of a series of separate slabs called **plates**. The study of the relationships between these plates and the deeper interior of the Earth is called **plate tectonics**, and the narrow volcanic zones mark some of the boundaries where the plates meet. In some cases these are places where two plates are moving apart. These are known as **divergent margins** or **spreading centers** and are the sites of MOR volcanism (Fig. 2.4). The fact that plates are moving apart at these locations and new material is reaching the surface makes it clear that some slow, upward movement of deeper material in the mantle must be taking place here.

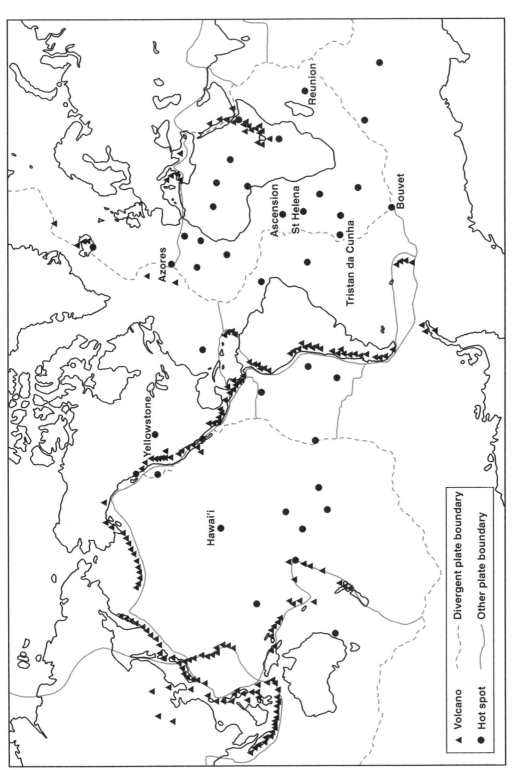

Fig. 2.3 The distribution on Earth of subaerial volcanoes and of hot spots (sites of high heat flow from the mantle). The boundaries of the tectonic plates are also shown. Note the concentration of subaerial volcanoes at plate margins where subduction of oceanic plates beneath continental plates is taking place. (Diagram drafted using data from fig. 1 in Duncan, R.A. and Richards, M.A. (1991) Hotspots, mantle plumes, flood basalts, and true polar wander. *Reviews of Geophysics*, **29**, 31–50, and fig. 2 in Perfit and Davidson (2002) *Plate Tectonics and Volcanism in Encyclopedia of Volcanoes*. Sigurdsson, H., Houghton, B.F., McNutt, S.R., Rymer, H. and Stix, J. (Eds), Academic Press, pp. 89–113, copyright Elsevier (2002).)

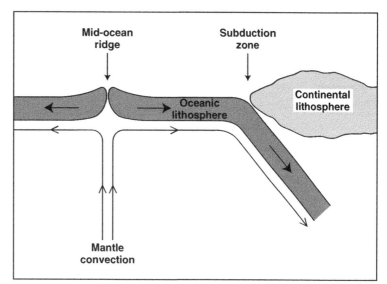

Fig. 2.4 Cross-sectional representation (vertically exaggerated) of the topography at divergent (mid-ocean ridges) and convergent (subduction zones) plate margins. New oceanic lithosphere is created in the volcanoes of the mid-ocean ridges and this material is recycled into the mantle beneath the continental margins via subduction zones.

In other places two plates are moving together and are in collision; such places are known as **convergent margins**. Convergent margins associated with volcanism are those where **subduction** is occurring, i.e., where one plate is descending beneath another (Fig. 2.4); it is this process that dominates the formation of the Ring of Fire around the Pacific. The fact that the descending plate is moving into the mantle, which must be deforming to allow its entry, again shows that on some long time scale the mantle is capable of behaving as though it were a fluid. In fact we now understand that the temperature difference between the hot interior of the mantle and the cool surface layer above it drives the mantle into a slow but continuous state of convection.

The best estimates of magma production on Earth suggest that 88% of magma is generated at plate margins – 62% at MORs and 26% at subduction zones. Figure 2.3 shows that the remaining 12% of magma production occurs in much more widespread **intraplate** settings, well away from plate boundaries. For instance, there are centers such as Hawai'I in the Pacific Ocean and Réunion in the Indian Ocean which are located near the middle of oceanic plates. There are other centers, such as Yellowstone in the western USA, which are located within continental landmasses, also well away from

any plate margin (Fig. 2.3). These centers are associated with **hot spots**, features which appear to represent zones in which temperature variations in the deep mantle have led to upwellings forming **mantle plumes**. Although the current locations of hot spots seem unrelated to the main plate tectonic system, hot spots appear to play an important role in initiating new spreading centers. When the head of a mantle plume first impinges on the lithosphere it causes updoming and rifting (Fig. 2.5) and the eruption of huge volumes of basaltic lava to form **Large Igneous Provinces** (LIPs). If rifting continues then sea-floor spreading starts (Fig. 2.5) and the gradual movement of the Earth's plates causes the new spreading center to move away from the original hot spot. The hot spot continues to generate magma, although not in the enormous quantities associated with the initial rifting stage, and produces the isolated intraplate magmatic centers. A number of such hot spots occur close to the mid-Atlantic ridge, reflecting their relatively recent role in initiating spreading there (Fig. 2.6). One obvious example of where rifting related to a hot spot is occurring at the present time is in the East African Rift Valley (Fig. 2.7). Here the Afar hot spot has caused flood basalt volcanism, up-doming and faulting of the continental crust, forming a **triple junction**, of which the Red Sea and Gulf of Aden

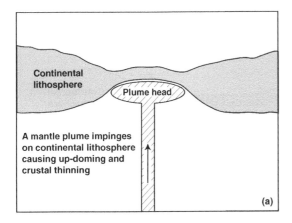

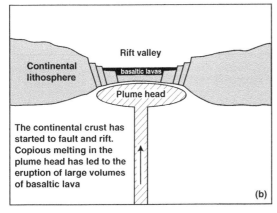

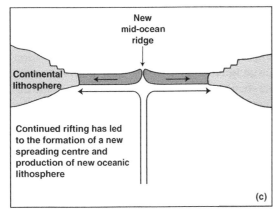

Fig. 2.5 The consequences of the head of a mantle plume impinging on continental lithosphere. Initially (a) the lithosphere is up-domed and the crust is thinned and stretched. Soon (b) the stretching exceeds the strength of the rocks, leading to formation of faults that assist the process, forming a rift valley. Basaltic lavas erupt into the valley as part of the plume head melts due to the decompression (see Fig. 2.1). In (c) the continental lithosphere has been rifted to the point where a new mid-ocean ridge has formed and the eruption of basaltic lavas is generating new oceanic crust.

mark two arms and the East African Rift Valley the third (Fig. 2.7). Along the Red Sea and Gulf of Aden rifting has advanced to the point where sea-floor spreading has started. Along the East African Rift Valley extension and normal faulting have occurred in association with continuing volcanism but the process has not advanced to the point of sea-floor spreading.

2.3.1 Tectonic settings, melting processes and magma composition

Not only can we link volcanic activity with very specific tectonic settings, but the types of magma generated in these different settings are distinct from each other in terms of composition and physical properties.

MID-OCEAN RIDGES AND OCEANIC INTRAPLATE SETTINGS

The dominant magma type at mid-ocean ridges is basalt, and mid-ocean ridge basalts (MORBs) are derived directly from the partial melting of the upwelling mantle beneath the spreading ridge. When a mantle hot spot interacts with oceanic crust well away from a spreading ridge, melting in the plume head first gives rise to very large volumes of basaltic magma which erupt to form a Large Igneous Province (LIP). After this initial phase of flood volcanism, magmatism continues at a lower rate and is dominated by ocean island basalts (OIBs).

We know that both of these settings are zones in which mantle upwelling is occurring and it is thought, therefore, that the dominant melting mechanism here is decompression melting. Figure 2.1 illustrates this process. The first thing to note is that the geothermal gradient – the variation of temperature with depth – is such that the temperature of the mantle is lower than the solidus temperature of the mantle material, and so we would expect the mantle to be solid. However, as we have seen, over very long time scales the mantle convects and so mantle material from deeper, hotter levels is gradually brought up to shallower, cooler zones. What happens to this upwelling mantle depends critically on the rate at which it ascends. When the

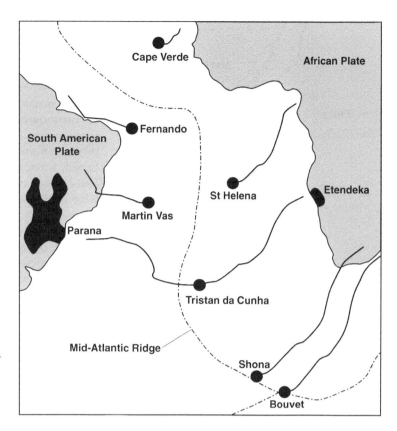

Fig. 2.6 Hot spots near the mid-Atlantic ridge that helped to initiate spreading there. The Parana basalts in South America and the Etendeka basalts in southwest Africa were both erupted from the same hot spot, now marked by Tristan da Cunha, before the Atlantic ocean opened. (After fig. 3 in Duncan, R.A. and Richards, M.A. (1991) Hotspots, mantle plumes, flood basalts, and true polar wander. *Reviews of Geophysics*, **29**, 31–50.)

mantle material rises it experiences lower confining pressures and so will expand slightly in volume as it ascends. This expansion is adiabatic (i.e., occurs without the addition of any external heat) and so causes a decrease in temperature of the rising mantle of ~0.5–1.0°C km^{-1}. In addition, the material is moving into a zone of lower temperature and so there is the opportunity for heat loss to the surrounding mantle. If the rise of the mantle material is slow then heat loss to the surroundings will dominate and the rising mantle material will cool sufficiently to ensure that, even at the progressively lower pressures it experiences, no melting will occur. However, if ascent is sufficiently rapid, conduction of heat to the surrounding mantle will be minimal and the cooling which occurs is limited to that caused by the adiabatic expansion of the plume material. In practice it seems that most commonly the rise rate of mantle material, while not strictly adiabatic, is sufficiently rapid to minimize loss by conduction and so the rising mantle material

follows a path like that between points A and B in Fig. 2.1. Consider what happens to mantle material rising from point A to point B. The rise speed is sufficiently great that the temperature declines only slightly between points A and B, but during this ascent the reduction in confining pressure experienced by the rising mantle material means that its melting temperature is considerably lower at depth B than it was at depth A. At depth B the temperature of the rising mantle material matches its solidus temperature and melting commences. Further ascent and reduction in confining pressure causes further melting.

Not all the material in the upwelling mantle melts. Experimental studies in which mantle material is melted at high pressures suggest that 20–25% melting of "typical" mantle material produces **tholeiitic** basalts like those produced at MORs and leaves behind a depleted mantle residuum from which it would then be hard to produce further melts. Such studies thus support the idea that the

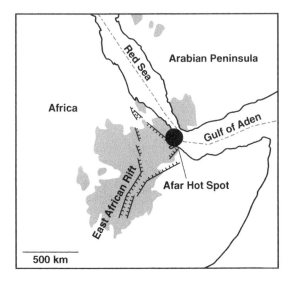

Fig. 2.7 Up-doming and faulting of the African continental crust by the Afar hot spot. Flood basalt volcanism dominates the area (shaded area), and the Red Sea, the Gulf of Aden and the East African Rift Valley form the three arms of a **triple junction**. Sea-floor spreading has already started along the Red Sea and the Gulf of Aden, but not in the East African Rift Valley. (After fig. 3 in Ernst, R.E. and Buchan, K.L. (1997) Giant radiating dyke swarms: their use in identifying pre-Mesozoic large igneous provinces and mantle plumes. *Geophysical Monograph*, **100**, 297–333.)

basaltic magmas which dominate in these tectonic settings are generated by **partial melting** of the mantle.

CONTINENTAL INTRAPLATE SETTINGS

These are settings which, like the MORs and oceanic hot spots, are zones of upwelling mantle and so zones in which melting is likely to occur as the result of decompression. As in the oceanic setting, the initial interaction of a mantle plume with continental crust results in the production of flood basalts to form LIPs. After this initial phase, magmatism continues generating basaltic magma but the interaction of these basalts with the continental crust creates a range of magma compositions. In some instances, for example at Yellowstone, the magmatism is bi-modal, producing both basaltic and rhyolitic magmas. The large volumes of rhyo-

lite are generated by melting of the crustal rocks induced by heat transfer from the basaltic magma ponded at the base of the crust. In other settings, notably in the East African Rift Valley, a tremendous diversity of compositions is found. The most unusual magmas found in continental rift settings are **carbonatites**. Carbonatites are magmas which contain greater than 50% carbonate minerals and they have been observed being erupted at Oldoinyo Lengai, a volcano in the East African Rift Valley. Initial observers thought the carbonatite flows were mudflows as they are jet black on eruption and barely incandescent even when observed at night. These lavas have the lowest eruption temperatures (typically 500–590°C at Oldoinyo Lengai) and the lowest viscosities of any known terrestrial lava. They behave like extremely fluid basalts (see Chapter 9).

SUBDUCTION ZONES — ISLAND ARCS AND CONTINENTAL ARCS

Two types of subduction zone occur on Earth. In the first type, one oceanic plate subducts beneath another oceanic plate generating an **island arc**; in the second type an oceanic plate subducts beneath a continental plate generating a **continental arc** or **active continental margin** (Fig. 2.8). Island arcs show a wide range of magma types, ranging from basalts through basaltic andesites and andesites to dacites and rhyolites. Although there is considerable variation between different island arcs, the dominant magma type is andesite, with basalts and basaltic andesites being fairly common and the more evolved dacites and rhyolites being rarer. Continental arcs show an even greater diversity of magmas than island arcs and typically produce a greater proportion of evolved magmas than island arcs. Although andesite is still a common composition, there are more dacites and rhyolites and less basalts and basaltic andesites than are found in island arc settings.

On the face of it, subduction zones present an unlikely setting for magma production because they are zones in which the descent of cold oceanic lithosphere causes cooling of the surrounding mantle, and in which mantle material is descending as part of the large-scale convection system. So this is

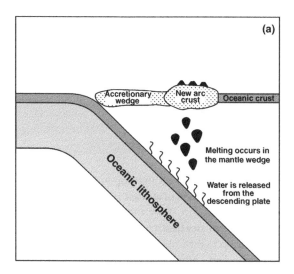

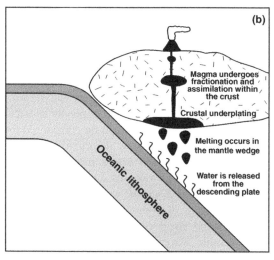

Fig. 2.8 The two types of subduction zone that can occur on Earth. In (a), one oceanic plate subducts beneath another oceanic plate producing an **island arc**. In (b) an oceanic plate subducts beneath a continental plate generating a **continental arc** or **active continental margin**. In both cases melting occurs in the mantle as water released from the descending plate infiltrates the mantle and changes the solidus and liquidus temperatures (see Fig. 2.2). However, differences in the surface volcanism occur because the mantle melts interact with rocks of differing composition as they rise, oceanic crust in case (a) and continental crust in case (b).

not a setting in which melting by decompression or by heating is likely to occur. Instead it is thought that melting occurs due to the release of volatiles, especially water, from the descending lithospheric slab. The slab is topped by water-saturated sediments, some of which will descend with the slab. More importantly, the rocks of the slab itself contain many hydrous mineral phases as the result of chemical reactions between the rocks and the **hydrothermal** water that was circulating through the rocks while they formed the ocean floor.

As the slab descends, reactions occurring within the rocks lead to the dehydration of the rocks and the release of this bound water. It appears that the water rises into the mantle wedge which overlies the descending slab (Fig. 2.8), reducing the solidus of the mantle material sufficiently to induce melting (Fig. 2.2). As in the MOR setting, melting of the mantle is likely to produce basaltic magmas but, although present, they are not the dominant magma type seen in subduction settings. In island-arc settings it is thought that ponding of the basaltic melts at the base of the **lithosphere** and within the crust itself allows fractional crystallization to occur, with the residual liquids giving rise to the more common basaltic andesites and andesites found there (Fig. 2.8).

In continental arcs, where any magma reaching the surface must travel through a considerable thickness of continental crust, there is the potential for a wide range of processes to operate. Here, then, although the primary magma produced is basalt from the mantle wedge, interaction of the magma with the continental crust creates the diversity of magma erupted at the surface (Fig. 2.8). There is great potential here for melting of crustal rocks, for assimilation of crustal material during magma ascent, for **fractional crystallization** (where crystals form and are left behind by the remaining liquid magma), and for mixing of magmas at different depths beneath the surface. The relative importance of each of these processes is still a source of considerable debate amongst igneous petrologists and geochemists, and is well beyond the scope of our discussion here. We are more concerned with the physical properties of the resulting magmas, and these, together with the properties of less evolved magmas such as basalts,

Magma type	Eruption temperature (°C)	Viscosity (Pa s)	Water content (%)
Basalt	1050–1200	30–300	0.25–2
Basaltic andesite	950–1200	100–1000	0.5–2
Andesite	950–1100	300–1000	1–4
Dacite	800–1100	10^4–10^6	2–4
Rhyolite	700–900	10^5–10^{10}	3–6

Table 2.1 Summary of the main types of magma and the trends of their physical properties. The viscosity tends to increase as the water content increases.

can be characterized quite well in terms of the magma silica contents (Table 2.1), although the amounts of other chemical compounds in the minerals forming the rock are important too. The key issue is that the differing chemical compositions of the magmas lead to them having very different abilities to contain dissolved volatile compounds such as water and carbon dioxide, and very different abilities to flow under a given set of stress conditions, i.e., they have very different viscosities.

2.4 Melting and melt segregation in the mantle

We have seen that basaltic magmas generated by melting of upwelling parts of the mantle are by far the most common magma type on Earth, and it is their production that ultimately leads, however indirectly, to the diverse range of magmas found in the various tectonic settings on Earth. To understand a volcanic system, then, we have to start by understanding how basaltic melts form in the mantle and how they segregate from the region in which they form.

2.4.1 Nature of the mantle

The typical composition of the mantle is known from various lines of evidence. Most significantly, certain geological processes bring samples of mantle to the surface. **Ophiolite** complexes are parts of the oceanic lithosphere which, as a result of tectonic processes, have been uplifted, rotated, and incorporated into continental crust, and where subsequent erosion sometimes exposes them at the surface. Such complexes contain sections of rock from the upper mantle. Other samples of the mantle are brought up as **xenoliths** during volcanic

eruptions. Xenoliths brought up in kimberlites are derived from depths of at least 100–200 km. Direct samples of this sort show that the mantle consists dominantly of **peridotite**, a crystalline rock composed of up to 50% olivine, ~40% ortho- and clinopyroxene and ~10% garnet, spinel or plagioclase, the exact mixture depending on the pressure.

Next we need to consider how the mantle deforms when it is stressed, i.e., when a force is applied to it: this introduces the concept of **rheology**, the way materials change shape when stressed. On very long time scales (tens of millions of years) the rocks forming the mantle can deform very slowly in a plastic manner, typically at rates of centimeters per year, and can be considered to behave as a liquid. The slow deformation rate in response to the applied forces means that this liquid has a very high viscosity. On the very short time scales it takes a seismic wave to pass through the mantle (tens of minutes), both the compressional (p) and shear (s) waves are transmitted. When seismic waves encounter a normal liquid such as water, however, only the p waves are transmitted, because the shearing force of the s waves causes the liquid to flow; the viscous resistance to the shearing transforms kinetic energy to heat and the s waves are rapidly damped out. So, on the short time scale of passage of a seismic wave, the mantle behaves as a solid. This is called an **elastic** solid because, although it deforms under stress, the shape returns to normal after the stress is removed (interestingly, what we call "elastic bands" are very inelastic – they do not go back to exactly the original length after a stretching force is removed). It is the interaction between these properties of the mantle, apparently liquid on long time scales and apparently solid on short ones, that controls the melting process within it.

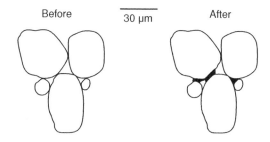

Fig. 2.9 The first stages in the formation of melt as the temperature of a rock exceeds the solidus. The first liquids form along the contacts between pairs of mineral grains of different composition, for which the energy needed for melting is a minimum. The shape of the growing melt body depends on the geometry of the nearby grains and on the surface tension of the liquid–solid contacts.

2.4.2 Onset of mantle melting

As pressure release melting begins in some slowly ascending part of the mantle, the first liquid to appear will form along the contact between a pair of mineral grains (Fig. 2.9). As the amount of melt increases, the shape of the melt body will change; it may spread out along a grain–grain contact as a thin film or it may concentrate into the cusp where three grains meet (Fig. 2.9). The main control on this process is the **surface tension** of the liquid–solid contact, and the factors controlling this are complicated, so that our knowledge of what happens in particular mixtures of minerals comes mainly from high-pressure melting experiments in the laboratory. It seems clear that there will be no connections between the individual pockets of melt until some minimum amount of melting has taken place. Estimates of this critical level of melting vary from only one or two percent of the whole volume to at least several percent, with the control again being the exact mixture of minerals present.

There is a process which acts to help nearby pockets of melt to make connections with one another. The liquid that forms when most minerals melt is less dense than the solid crystalline material from which it forms. Thus the liquid will try to occupy a greater volume than it had when it was solid. The only way in which space can be made for it is for the surrounding unmelted solid crystals to occupy a smaller volume. They could in theory do this by moving away from the site of the melting, but the viscosity of the solid crystal network is so high that it cannot deform fast enough. The only alternate is for the crystals surrounding a new melt pocket to be compressed into a smaller volume. To compress a solid, however, pressure must be applied to it, and so as soon as any melt forms it compresses all of the mineral grains in contact with it. The total pressure in the melt is then equal to the **lithostatic load** pressure (i.e., the weight of all of the overlying rocks) that was present in the region before melting started plus the excess pressure needed to make enough space for the melt. As more melt forms, the excess pressure increases.

This excess pressure causes stresses to form at cusps where grains meet and anywhere that there is an irregularity on the surface of a grain. If the stress becomes large enough the chemical bonds between the atoms will be broken and a crack will form; thus on the short time scale of the build-up of these stresses the bulk of the mantle rocks are behaving as elastic solids, even though there are small pockets of true liquid in the spaces where mineral grains are not in contact. The shape of the crack will be controlled by the elastic properties of the solid crystals in ways discussed in Chapter 3. For the moment we note that all elastic cracks tend to have dimensions at least a few hundred times longer and wider than their thickness in the third dimension (Fig. 2.10). In Fig. 2.10a, a cross-section through four different grains in contact is shown. The roughly triangular region between grains A, B and C, like the one between grains B, C and D, is an empty pore space. In Fig. 2.10b the pressure and temperature conditions have reached the point where melting starts at the contact between grains B and C, and a film of melt has formed between these two grains. The film extends for about 10 μm (microns) both in the plane of the diagram and at right angles to it. In Fig. 2.10c the film has grown in length and partly invaded the previously empty pore spaces. Surface tension has stopped the melt from getting into the sharp corners of the pore spaces. Grains B and C have moved very slightly closer together, and stress has been exerted on the faces of grains A and D. In Fig. 2.10d, a fracture 100 μm long has formed, which extends into the

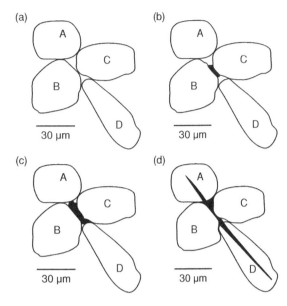

Fig. 2.10 Increasing amounts of melt formation in the spaces between mineral grains lead to stresses in the grains. These stresses eventually cause cracks to form in the grains, and liquid under pressure can migrate into the cracks, causing them to grow further.

interiors of grains A and D. Magma from the continuing melting of grains B and C has migrated into this fracture, which is now a magma-filled **vein**, and grains B and C are by now significantly smaller, causing deformation of the surrounding grains. The vein extends a similar distance at right angles to the plane of Fig. 2.10d as the 100 μm length shown. Thus there will be a strong tendency for the pockets of melt from which veins form to get thinner in one direction but much longer in the other two directions, and this gives them a much greater chance of intersecting other melt pockets or veins nearby. The excess pressure in the melt forming a vein will generally get smaller as the vein grows, but will never be reduced to zero, and so on average we expect there to be excess pressures of some sort in all accumulating bodies of partial melt.

2.4.3 Melt migration

Once connections between individual melt pockets occur, it becomes possible for the melt to start to move. There are two reasons why the melt should do this. The first relates again to the density of the liquid: not only is the density of the liquid less than that of the solid that has melted, it is also less than that of the solid crystals that have not yet started to melt, and so the *buoyancy* of the liquid exerts a force on it which tries to drive it upward. Second, the rocks which are melting are always acted on by a stress due to the weight of the overlying rocks, and in general are also subjected to nonuniform stresses due to the convection currents slowly moving them around in the mantle. The combination of these stresses will tend to cause compaction of the solid minerals and so will try to force the liquid upward out of whatever space it occupies. This process is called **filter-pressing**, and it leads to the volume fraction of melt increasing in the upper part of a zone of partial melting and decreasing in the lower part (Fig. 2.11).

The rate at which melt segregates upward in such a system depends on the total force driving it, its viscosity, and the typical width of the pathways between the unmelted mineral grains. As more melting occurs and more melt pockets join up, the average pathway width increases and the melt moves faster. Some studies have shown that the compaction and segregation are probably not a steady process but may occur in the form of waves of changing liquid fraction traveling up through the region where melting is occurring. These waves travel in such a way that their shapes are preserved, and are called **solitons**; by analogy with this the waves of magma liquid concentration are called **magmons**. Whatever the details of the melt separation process, the net effect is the upward con-

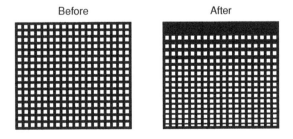

Before   After

Fig. 2.11 Settling of mineral crystals in a partially molten region concentrates liquid upward, so that eventually a layer containing no crystals may form at the top of the region.

centration of melt. The point may come at which the total buoyancy of the region in which the melt is now concentrated is large enough that this melt-rich region begins to accelerate ahead of the rest of the rising plume that gave rise to it, and can be thought of as a separate entity – a **diapir**. This term can be applied to any very buoyant region containing melt, from the upper part of a large (150–200 km wide) plume in the upper mantle to a much smaller body of melt only a few kilometers in size generated in a subduction zone, or where a mantle plume rises under the base of a continent and causes melting of overlying crustal rocks.

By definition, any diapirically rising body of rock is moving from hotter into cooler surroundings, and the viscosity of almost all liquids increases as they get cooler. Eventually, therefore, the increasing viscosity of the rocks surrounding the diapir must drastically slow and eventually stall its ascent. Around the time that happens, or earlier in some cases, a new process that moves magma to shallower depths sets in. That new method is the extensive joining together of many of the veins of melt within the diapir to create larger fractures, through which melt moves much more efficiently. These large fractures are called **dikes**, and the relationship between diapiric rise and dike formation is the subject of the next chapter.

There is one tectonic setting where these categorizations into diapiric rise and dike formation become blurred. Under some mid-ocean ridges, the rate at which the overlying plates spread apart may be fast enough to allow the upwelling of mantle material to proceed more or less continuously. As a result, instead of thinking of a mantle plume head stalling at some level we should think of the plume splitting in the middle as it rises, with half of it heading off horizontally under each of the two plates being formed at the ridge. It is, of course, the melt separating from the plume that is forming the new crust. Although it is clear from ophiolite complexes that large numbers of dikes do form in these settings, the chemical compositions of magmas arriving at shallow depths imply that some (probably small) fraction of the magma has traveled most of the way from its melt source by percolating slowly through the network of narrow veins between the dikes.

2.5 Summary

- Most volcanoes on Earth are located along the boundaries of tectonic plates and thus are related to the large-scale convection of the mantle which drives plate tectonics. Other volcanoes are found far from plate boundaries and mark the locations of **hot spots** – mantle plumes which are thought to form due to thermal anomalies in the deep mantle.

- The composition of the lava erupted from volcanoes is related to the tectonic setting. At mid-ocean ridges, and where hot-spot plumes occur under ocean-floor crust, melting of the mantle by decompression produces hot, fluid basalts. Hot spots beneath continents also produce basalt but interaction of these basalts with the continental crust generates a wide diversity of magmas. At subduction zones the presence of water carried down with the descending plate lowers the melting temperature, causing melting of the mantle wedge to generate basalts. Interaction of the basalt with the overlying crust, and the fractional crystallization of the resulting magma as it ascends, generates a great diversity of magmas, particularly where oceanic crust descends beneath continental crust. Although magma compositions are diverse in subduction-zone settings, the dominant magma type is andesite.

- Although a great diversity of magmas are found on Earth these magmas all ultimately owe their generation to the production of basalts by mantle melting. This melting can occur by decompression of upwelling mantle material or by melting induced by the release of water from subducting oceanic lithosphere, decompression melting being the dominant mode.

- Melting starts in rock at favorable contacts between mineral grains when the combination of slowly decreasing temperature and more rapidly decreasing pressure reaches the solidus. When enough melting has taken place, individual melt pockets between grains start to connect together. The natural buoyancy of the melt then causes it to start to move by percolation through the existing melt pathways. Because the melt is less dense than the rocks from which it is generated, it tries to occupy a larger volume than its parent materials. This causes a pressure increase in the melt as it forms, and extra space for melt movement may be created by this fluid pressure fracturing some of the grain contacts.

- As melt moves upward, the mineral grains in the lower part of the region where melting is taking place are compacted, and the melt fraction in the upper part of the region increases. Eventually the whole upper part of the melt-rich region may begin to rise as a separate body called a diapir. The rise of diapirs is limited by the increasing viscosity of the surrounding rocks, and subsequently the melt is likely to rise through magma-filled fractures called dikes, initiated when the strain rate induced in the host rocks becomes so large that they respond in a brittle, rather than plastic, fashion.

2.6 Further reading

Kelemen, P.B., Hirth, G., Shimizu, N., Spiegelman, M. & Dick, H.J.B. (1997) A review of melt migration processes in the adiabatically upwelling mantle beneath oceanic spreading ridges. *Phil. Trans. Roy. Soc. Lond. Ser. A* **355**, 283–318.

Maaloe, S. (2003) Melt dynamics of a partially molten mantle with randomly oriented veins. *J. Petrol.* **44**(7), 1193–210.

McKenzie, D.P. (1985) The extraction of magma from the crust and mantle. *Earth Planet. Sci. Lett.* **74**, 81–91.

Niu, Y.L. (1997) Mantle melting and melt extraction processes beneath ocean ridges: Evidence from abyssal peridotites. *J. Petrol.* **38**, 1047–74.

Perfit, M.R. & Davidson, J.P. (2000) Plate tectonics and volcanism. In *Encyclopedia of Volcanoes* (Ed. H. Sigurdsson), pp. 89–113. Academic Press, San Diego, CA.

Rubin, A.M. (1998) Dike ascent in partially molten rock. *J. Geophys. Res.* **103**, 20 901–19.

Scott, D.R. & Stephenson, D.J. (1986) Magma ascent by porous flow. *J. Geophys. Res.* **91**, 9283–96.

Sleep, N.H. (1988) Tapping of melt by veins and dikes. *J. Geophys. Res.* **93**, 10 255–72.

Spiegelman, M. (1993) Flow in deformable porous media. *J. Fluid. Mech.* **247**, 17–63.

Winter, J.D. (2001) *An Introduction to Igneous and Metamorphic Petrology.* Prentice-Hall, Upper Saddle River, NJ.

2.7 Questions to think about

1 What are the main differences, and how do these arise, between magmas generated at mid-ocean ridges and hot spots on the one hand and subduction zones on the other?

2 What factors cause magmas to separate from their parent rocks when partial melting occurs in the mantle?

3 Magma migration

3.1 Introduction

This chapter is concerned with how magma separates from source regions at depth and moves upward toward the surface. We first consider the slow, wholesale movement of large, melt-rich regions by convection and then the formation of large fractures, in which magma can rise very much faster to form dikes. The density and temperature structures of the crust and mantle control the locations where diapirs and dikes are most likely to dominate magma movement. The density structure also determines the ranges of depths at which magma may be stored in long-lived reservoirs before eventually erupting to the surface or forming shallow intrusions. We focus on these shallow reservoirs in the next chapter, but here we are concerned mainly with the general upward movement of magma at depth.

3.2 Diapiric rise of melt

In Chapter 2 we saw that if part of a body of rock is molten, the buoyant melt can be concentrated upward into a smaller region that becomes even more buoyant relative to its surroundings. This region may then rise a considerable distance through the surrounding rocks as a **diapir**. The melt content of a diapir may range from only 1 or 2% to as much as 25% by volume, the remainder being unmelted mineral grains. Just how much melt is present in any part of the diapir at any stage depends on how efficiently the melt is moving through the veins between the mineral grains. The

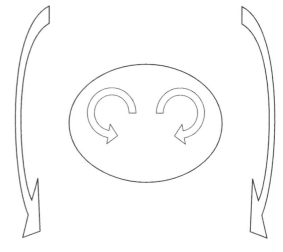

Fig. 3.1 The arrows indicate the directions of convective movement occurring in rocks of the interior of a rising diapir and in the surrounding rocks through which it is rising. The host rocks are still solid and have a very high viscosity as they deform in a plastic fashion, whereas the interior of the diapir is partly molten, and so its viscosity is very much smaller. Transfer of heat from the diapir to its surroundings helps to reduce the viscosity of the immediately surrounding host material, but increases the viscosity of the diapir fluid.

buoyancy of the melt in the diapir causes the surrounding rocks to make way for it by flowing round it. At the very small deformation rates involved, the host rocks can act as a very viscous liquid. The word **plastic** is used to describe this kind of viscous behavior.

Quite a complicated flow pattern is set up, not only in the surrounding rock but also within the diapir itself (Fig. 3.1). Heat is conducted from the body of the diapir into the surrounding rock, and this process becomes ever more significant as the diapir rises because the surrounding rock is cooler at shallower depths. The heat added to the host rocks reduces their viscosity and makes the ascent of the diapir somewhat easier than it otherwise would be. However, the diapir itself pays a price for this in terms of the heat it loses: the cooler it becomes, the less buoyant it will be.

To put some of these relationships in perspective we can consider two cases of diapiric rise: the upward movement of the head of a large mantle plume that would produce a major hot spot such as that forming the Hawaiian volcano chain, and the rise of a much smaller diapiric body above a subduction zone at a continental margin or in an island arc. We first need to specify the buoyancy force acting on the diapir, which we approximate as a sphere of radius R, and the drag force that it experiences as a result of deforming the plastic mantle rocks surrounding it. The buoyancy force is $F = (4/3)\pi R^3 g\Delta\rho$, where $(4/3)\pi R^3$ is the volume of the sphere, g is the acceleration due to gravity, and $\Delta\rho$ is the amount by which the sphere is less dense than its surroundings. The drag force is $D = 4\pi\eta R U$, where η is the viscosity of the host rocks and U is the rise speed. This second formula is similar to the more familiar version $D = 6\pi\eta R U$ giving the drag force on a rigid sphere moving through a fluid, but the differing constant is due to the fact that there is significant circulation of the fluid inside the sphere (see Fig. 3.1). If we equate D to F we find that the diapir rise speed is

$$U = (R^2 g \, \Delta\rho)/(3 \, \eta) \qquad (3.1)$$

For the diapiric head of a large mantle hot spot plume, R might be 400 km. With a mantle density, ρ, equal to 3300 kg m^{-3}, a typical volume expansion coefficient of rock, α, equal to 3×10^{-5} K^{-1} and a temperature difference, ΔT, between the inside of the plume and its surroundings of 200 K, the density difference would be $\Delta\rho = (\rho\alpha\Delta T) = {\sim}20$ kg m^{-3}. The plastic viscosity of the mantle, η, is of order 10^{21} Pa s, giving a diapir rise speed, U, of ${\sim}0.3$ meters per year. In marked contrast to this, a diapir in a subduction zone setting might have a radius of order 5 km. It could well be driven upward by

a larger density contrast than in the previous example if its composition had been influenced by incorporation of hydrated ocean floor sediment, in which case $\Delta\rho$ might be about 100 kg m^{-3}. In this case the diapir rise speed U would be about 0.25 millimeters per year, more than 1000 times slower than the mantle plume head.

3.3 The change from diapir rise to dike formation

There is good evidence that in continental environments diapirically rising mantle plumes stop rising when they get close to the base of the continental crust. This is because they are no longer buoyant: the higher silica content of the crustal rock means that it is less dense than the mantle rocks in the plume even though the plume is hotter than the crust. In a case like this, where a body of rock reaches a level at which it is less dense than the rocks below it but more dense than the rocks above it, we say that a **neutral buoyancy level** has been reached. When the rise of material is halted in this way, a significant amount of heat can be transferred by conduction into the crustal rocks above the plume and, because their solidus temperature is significantly lower than that of the plume material, the crustal rocks may melt to form rhyolites, and the rhyolitic melt produced may itself ascend diapirically some way into the shallower crust above.

In other cases, however, it may not be the change in rock density at the base of the crust that limits the rise of a plume; instead it is the **strain rate** that the plume imposes on the surrounding rocks through which it rises. The strain rate is a measure of the rate of deformation of the host rocks, and is found by dividing the speed of the rising plume head by its diameter. Equation 3.1 shows that large diapiric bodies rise faster than small ones in proportion to the square of their diameter. Thus, other things being equal, the strain rate is just proportional to the diameter. Using the values from the previous section, the strain rate imposed on the surrounding rocks by a mantle plume head would be about 1.25×10^{-14} s^{-1}, whereas that due to a small subduction zone diapir would be about 1.7×10^{-15} s^{-1}. These values are roughly ten times and two times larger, respectively, than the average strain rates elsewhere in the mantle where convection is taking place without any melting.

The way in which a rock responds to being deformed, i.e., its rheology, depends on its composition, its temperature, and the strain rate imposed on it. If the temperature of a given hot rock is kept constant and the strain rate applied to it is increased, its response eventually changes from that of a very viscous liquid, i.e., a plastic solid, to that of a brittle elastic solid. Alternately, keeping the strain rate constant and decreasing the temperature will also trigger the change from plastic to elastic behavior and initiate fracturing. Diapirically rising masses of rock are always moving from hotter into cooler surroundings, and so the onset of fracturing will occur whenever the rheological boundary between the plastic and elastic responses of the host rocks is crossed, and this is likely to happen at higher temperatures, and therefore at greater depths, for larger, more rapidly rising diapiric bodies than smaller, slower ones.

It is very difficult to predict exactly the conditions under which elastic to plastic transitions in rock behavior will occur in the Earth because it is hard to simulate mantle conditions in the laboratory. The problem is not so much in producing the required temperatures and pressures but in dealing with the time scales. If we are prepared to wait one year, that is $\sim 3 \times 10^7$ s, to conduct an experiment in which a rock sample is deformed by 100%, i.e., compressed to about half its thickness and double its cross-sectional area, in a high-pressure, high-temperature apparatus in the laboratory, the average strain rate will be $\sim 3 \times 10^{-8}$ s^{-1}, which is 30 million times faster than in the mantle. Extrapolations made across this enormous difference in time scale are not very reliable.

However, it is clear that the heads of mantle plumes reaching the base of the overlying crust are particularly likely to cause their host rocks to fracture. Two factors combine to cause this. First, a change in bulk composition occurs, with the crust being less dense than the mantle in general, including the hot material in the plume. Thus the plume buoyancy is lost and the top of the plume ceases to rise. The material below it continues to rise, however, and the plume head flattens out and grows sideways along the boundary, which locally increases the average strain rate. Also, the crustal rocks are cooler than those in the mantle, and therefore closer to their plastic–elastic transition temperature for any given strain rate. Thus both the lower temperature of the overlying rocks and the increase in strain rate acting on them will act together to make the creation of fractures particularly likely as a plume head reaches the base of the crust. Note that the increased strain rate applies to the rocks inside the top of the plume as well as to the crustal rocks above; thus the process of coalescence of existing large melt veins into dikes can occur within the plume head itself.

3.4 Dike propagation

Wherever the change from plastic deformation to brittle fracture of rocks occurs, fractures start to grow from any feature that can concentrate the stress – this could be a particularly sharp-cornered interface between three crystals as described in section 2.4.2 or some accidental defect in the internal structure of a single mineral grain. As soon as a fracture forms in the crust above a region containing magma, the liquid will start to flow into the fracture to fill the space created. A major property of brittle fractures is that the rate at which the sharp tip of the fracture can propagate into the unfractured rock ahead of it is limited only by the speed of sound in the rock (a sound wave is just a wave of deformation of the material in which it is traveling). In practice the crack-tip propagation speed is somewhat less than the actual sound speed, but it is still on the order of a few kilometers per second, and at first sight it is tempting to predict that dikes should grow at this rate. However, there is a limit to the speed at which magma can flow into a narrow, opening crack. The speed is controlled by the crack width and the magma viscosity, and also by the widths of the network of smaller veins that are feeding melt toward the crack. This magma flow speed will be very much less than the speed of sound in rock – typically more than a thousand times less. Thus in practice it is the flow speed of the liquid magma in the narrow veins that determines the speed at which the dike as a whole can grow upward.

The above discussion tacitly assumed that because the magma in a dike is buoyant the fracture will grow upward. In fact the orientation of the fracture will be strongly controlled by the state of stress in the rocks that are fracturing. As discussed earlier, the hot rocks in the mantle behave as fluids on long time scales. A fluid cannot sustain differential stresses – the fluid flows to remove the differences in stress, and so the force acting within a body of fluid is the same in all orientations. However, the cooler rocks in the crust can easily resist deformation on short time scales, and so can support nonzero stress differences. The basic stress acting on any body of rock is the vertical compressive force representing the weight of the overlying rocks. The horizontal forces acting may be significantly different from this, since they include the consequences of horizontal movements within the crust due to, for example, plate tectonics. The result is that horizontal stresses in the deep crust are generally found to be about 30% less than the vertical stress. When a fracture opens, the amount of work that it has to do in deforming the surrounding rocks will be minimized if it opens in a plane that is oriented at right angles to the least compressive stress, and since this is very likely to be horizontal, the plane of the resulting dike will almost certainly be vertical. The situation can be quite different at very shallow depths in the crust, where the presence of local layers of rock with low densities can cause the least compressive stress to be vertical. In that case, a vertically rising fracture may cease to extend upward and may instead spread sideways, especially if it finds a weak boundary between two rock layers. The approximately horizontal layer of magma that eventually fills such a fracture is called a **sill**.

Much of the study of the propagation of dikes is concerned with two critical issues: what conditions allow brittle fractures to start forming, and what conditions cause the resulting dikes to cease propagating. The rock property controlling these processes is called the fracture toughness, and it is a measure of the maximum intensity of the stress that the mineral grains at the dike tip can withstand before failing. For many years scientists treated fracture toughness as a basic rock property such as

density. However, it is now understood to be an expression of the combination of a number of rock properties, including the tensile strength of the rock (i.e., its ability to resist its mineral grains being pulled apart from one another), the typical size of the mineral grains, and the difference between the pressure in the molten rock waiting to enter the fracture when it opens and the compressive stress trying to force the fracture shut. The compressive stress is partly due to the weight of the overlying rock in the place where the fracture is about to form and partly due to any other tectonic stresses that are present, such as those linked to plate tectonics. The surface tension between the molten rock and the mineral grains is also important. We saw in section 2.4.2 that surface tension prevents liquid rock spreading in an infinitely thin film over the surfaces of mineral grains so that, when melting first begins, a finite amount of melt is needed before an interconnected melt network can exist. In the same way, melt cannot enter an extremely narrow crack at all, and so the extreme tip of an opening fracture will not contain liquid rock.

One might imagine this implying that there must be a vacuum in the fracture tip, but in practice all molten rocks contain dissolved volatiles. At low pressures, some of these volatiles will be released from the liquid rock as gases and can enter the fracture tip, providing a pressure that is greater than zero, but less than the compressive effect of the overlying rocks. The pressure difference, strictly speaking a stress difference, between the inside and outside acts to try to push the tip of the crack shut while at the same time the pressure of the underlying molten rock trying to get into the fracture is exerting a leverage acting to force it open. It is the competition between these two effects that determines the value of the apparent fracture toughness of the host rocks. However, because the apparent fracture toughness depends on the amount of volatiles exsolved, which in turn depends on both the amount of volatiles originally dissolved in the magma in its source zone and the exact conditions within the crack tip, it is very hard to anticipate the exact value of apparent fracture toughness to be expected in a given situation. The units in which fracture toughness is measured are stress multiplied by the square root of length, and in practice values can range from a few million Pa m$^{1/2}$ for rocks being fractured in the laboratory with no volatiles present inside the fracture to at least many tens of millions of Pa m$^{1/2}$ for rocks at the tips of dikes within the Earth.

Tens of millions of Pa m$^{1/2}$ sounds like an impressive number, but the size of the fracture toughness can be put in context by considering the stress intensity at the tip of a dike due to a given internal excess pressure. Consider the head of a mantle plume in which melting is occurring over a vertical distance of $H = 3$ km. This would cause a few percent of melt to be produced, enough to form a network of interconnected veins which would allow the melt to be extracted into a fracture if one formed. The effective excess pressure, ΔP, defined as the pressure in excess of the local lithostatic rock load, in the melt in the middle of the 3 km high zone of liquid is proportional to the buoyancy of the melt and is given by

$$\Delta P = \Delta \rho_m \, g \, (H/2) \tag{3.2}$$

where g is the acceleration due to gravity, ~ 9.8 m s^{-2}, and $\Delta \rho_m$ is the amount by which the melt is less dense than the surrounding mantle rocks from which it is forming, typically ~ 300 kg m^{-3}. Thus ΔP would be about 4.4 MPa. If a fracture starts to form at the top of the melt zone, the stress intensity at this point, available to overcome the apparent fracture toughness of the solid rock in front of the fracture tip, would be K, which is given by

$$K = \Delta P \, (H/2)^{1/2} + 0.5 \, g \, \Delta \rho_m \, (H/2)^{3/2} \tag{3.3}$$

Using the above values, $K = \sim 255$ MPa m$^{1/2}$, more than enough to guarantee that fracture growth, and hence the formation of a dike, does in fact get started, even for the largest likely value of the effective fracture toughness. Furthermore, as the fracture extends and H gets larger, then as long as ΔP and $\Delta \rho_m$ do not change significantly the stress intensity at the fracture tip gets larger, making continued growth of the dike ever easier. Of course in practice, if the dike grows very far, both ΔP and $\Delta \rho_m$ will begin to change. Movement of magma into the dike removes that magma from the source region, and so ΔP must eventually decrease slowly. More important is the fact that $\Delta \rho_m$ will change a great deal if the dike tip rises into the relatively low density crust, and so we need to consider the issue of the density difference between the magma and the surrounding rocks in more detail.

If a fracture starts some way below the top of a mantle plume head, then as it grows upward the surrounding rocks get slightly denser for a while due to their decreasing temperature (although this is offset to some extent by the opposite effect of the decreasing pressure). The result is a slight increase in the buoyancy of the magma. Eventually, when the base of the crust is reached (or at once if the fracture starts at the base of the crust), the surrounding rocks become much less dense than the rocks of the mantle. However, this does not automatically mean that they are less dense than the magma in the dike. Thus there are two possibilities: if the crustal rocks are denser than the magma then the magma simply becomes less buoyant than it was in the mantle, but if the crustal rocks are less dense than the magma then it ceases to be buoyant at all, in other words it becomes **negatively buoyant**. There are two ways, which we now discuss, in which this loss of magma buoyancy can lead to the cessation of the upward growth of the dike, in other words, to its trapping in the crust.

3.5 Trapping of dikes

1 *Stress traps.* As soon as the magma in a growing dike passes through the level at which the magma is neutrally buoyant, and thus becomes negatively buoyant, the stress at the upper tip of the dike will decrease as it continues to grow upward, because the second term in eqn 3.3 changes sign and becomes negative. The details are quite complicated because, now that $\Delta \rho_m$ is not constant along the dike, eqn 3.3 must be replaced by an integral equation which adds up the contributions to the stress at the tip from each small vertical segment of the dike using the local value of $\Delta \rho_m$. Eventually the stress intensity at the tip falls to a value that is smaller than the fracture toughness and upward growth of the dike must cease.

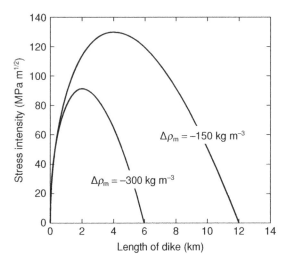

Fig. 3.2 The variation of the stress intensity at the growing tip of a dike as a function of the dike length, shown for two values of the amount $\Delta\rho_m$ by which the melt in the dike is less dense than the surrounding mantle rocks from which it has formed.

An order of magnitude for the distance to which a dike tip can rise after it becomes negatively buoyant can be illustrated by setting $\Delta\rho_m$ in eqn 3.3 equal to a negative value, i.e., the magma is now denser than the host rocks, and retaining $\Delta P = 4.4$ MPa. Initially, as H increases, the first term in eqn 3.3 greatly exceeds the second term and the stress intensity at the dike tip increases. However, as dike growth continues, the second term eventually increases faster than the first term and the stress intensity goes through a maximum and then decreases: Figure 3.2 shows two examples. With $\Delta\rho_m = -300$ kg m^{-3} the dike-tip stress intensity reaches a maximum of 92.7 MPa m$^{1/2}$ when H has grown to about 1995 m, and decreases to small values as H approaches about 6 km. With $\Delta\rho_m = -150$ kg m^{-3} the dike-tip stress intensity reaches a maximum of 131 MPa m$^{1/2}$ when H has grown to about 4000 m, and decreases to small values as H approaches about 12 km. Thus if the magma has a relatively low volatile content, so that as the dike grows there is not a great accumulation of vapor in the dike tip, and the effective fracture toughness of the surrounding rocks increases to, say, 30 MPa m$^{1/2}$, then for both of these magma density differences the dike will be able to grow to most of its maximum

potential vertical length, the values being about 5.3 and 10.8 km, respectively. However, if the magma is richer in volatiles, and the effective fracture toughness increases as the dike grows to, say, 70 MPa m$^{1/2}$, then for $\Delta\rho_m$ equal to -300 kg m^{-3} the dike would grow to only about 3.8 km and for $\Delta\rho_m$ equal to -150 kg m^{-3} the maximum vertical length would be about 9 km.

2 *Density traps.* Even if a magma does not become negatively buoyant at the base of the crust, it may well do so somewhere within the crust itself. This is because the density of the crust decreases as the surface is approached, as a result of the way in which the crust itself grows. In some places, especially in continental areas, the crust will consist mainly of sedimentary rocks, whereas in others, especially on the ocean floor, it will be formed of volcanic rocks. Sedimentary deposits are formed by the accumulation of clastic material, and subsequently these are compacted into denser rock by the accumulated weight of overlying deposits. Volcanic rocks are erupted either as fragmental pyroclasts or as lavas which contain gas bubbles, and in both of these cases compaction will occur as the volcanic rock layers pile up. The effect of the compaction process is to produce density profiles such as shown in Fig. 3.3. Figure 3.3a deals with a volcano built on typical oceanic crust and Fig. 3.3b shows an average continental interior.

Also shown in these figures are the ranges of depths at which one might expect to find two magmas, the denser one labeled D with a density of 3000 kg m^{-3} and a less dense one labeled L with a density of 2700 kg m^{-3}. Both magmas are produced when partial melting starts somewhere in the mantle. Magma D has its neutral buoyancy level at the base of the crust in both the oceanic (Fig. 3.3a) and continental (Fig. 3.3b) environments, because it is at this level that the density of the surrounding rocks decreases from a value greater than the magma density to a value smaller than the magma density. Thus if local buoyancy alone were the controlling factor, this magma should always be trapped at the base of the crust. In the continental case, magma L should apparently be able to penetrate all the way to the surface, being buoyant at all depths in both the mantle and the crust. In the oceanic case for magma L, however, a neutral buoy-

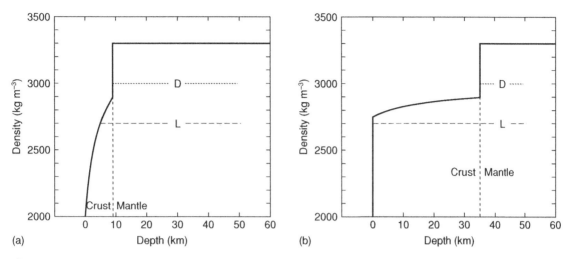

Fig. 3.3 The variation of density with depth through the crust and mantle in areas of (a) oceanic, and (b) continental lithosphere. In each case the range of depths is indicated at which one would expect to find dikes containing magmas of density 3000 kg m^{-3} (labeled D) and 2700 kg m^{-3} (labeled L) if density were the only factor controlling the movement of the magma.

ancy level exists at a depth of 5 km, and so at first sight we would not expect this magma to be able to rise to depths shallower than this level.

However, the situation is not quite as simple as this. Magmas produced by melting in the mantle are always buoyant relative to their surroundings, and their positive buoyancy in the mantle can compensate for the negative buoyancy conditions that they may encounter on their way to the surface. This can be quantified by realizing that the magmas will rise from a partially molten source region into the overlying rocks until the pressure exerted by the weight of the column of liquid magma is equal to the stress exerted on the source region by the surrounding unmelted rock. For sources in the mantle, the stress at a given depth z is essentially the same as the pressure P due to the weight of all the overlying rock, and this can be found by multiplying the depth z by the average density, ρ, of the overlying rocks exerting the pressure and the acceleration due to gravity, g. Figure 3.4 shows the general situation for a simple two-layer crust and mantle. The crust has thickness z_{crust} and average density ρ_{crust}; the mantle has average density ρ_{mantle} and the magma source is at a depth z_{source} below the base of the crust. The pressure at the depth of the magma source is thus equal

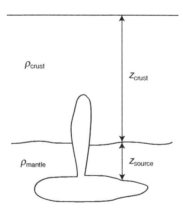

Fig. 3.4 Diagram showing magma in a dike penetrating crust of thickness z_{crust} and density ρ_{crust} from a source in an upper mantle of density ρ_{mantle}. The top of the source region is located a distance z_{source} below the crust–mantle boundary.

to $[(\rho_{crust}\, g\, z_{crust}) + (\rho_{mantle}\, g\, z_{source})]$. If the magma has density ρ_{magma}, the pressure at the base of a column of magma of height z_{magma} is $(\rho_{magma}\, g\, z_{magma})$. By equating these two pressures and solving for z_{magma} we find

$$z_{magma} = [(\rho_{crust}\, z_{crust}) + (\rho_{mantle}\, z_{source})]/\rho_{magma}$$

$$(3.4)$$

Table 3.1 The influences of the densities of magmas and the tectonic setting, which controls the density structure of the crust and mantle, on how close to the surface magma can rise as a function of the depth from which it starts. (a) Oceanic island volcano.

Depth of base of continuous magma pathway (km)	Depth below surface of shallowest dike top (km)	
	Magma density 3000 kg m^{-3}	Magma density 2700 kg m^{-3}
9.0	1.24	0.38
10.0	1.14	0.16
10.7	1.07	0.00
15.0	0.64	Magma
20.0	0.14	reaches
21.4	0.00	surface

(b) Typical continental crust.

Depth of base of continuous magma pathway (km)	Depth below surface of shallowest dike top (km)	
	Magma density 3000 kg m^{-3}	Magma density 2700 kg m^{-3}
35	2.3	Magma
40	1.8	always
45	1.3	reaches
50	0.8	surface

Then subtracting z_{magma} from the total source depth $(z_{crust} + z_{source})$ gives the depth below the surface of the top of the dike containing the magma.

Table 3.1 shows the result of carrying out this calculation for the density profiles shown in Fig. 3.3. For the oceanic island case, dikes containing the dense magma D can rise above the neutral buoyancy level at the base of the crust but are still trapped below the surface for all melt source depths down to about 21.4 km. Deeper sources allow this magma to reach the surface in an eruption. Table 3.1 shows that the lighter magma L, not surprisingly, finds it even easier to reach the surface in an eruption as a result of the compensating effect of its positive buoyancy in the mantle, and it is only if melting is confined to the very top of the mantle that dikes are trapped below the surface. For the continental case, the magma density has a more clear-cut influence on what happens. Dikes containing the lighter magma cannot only reach the surface, as expected because they are buoyant at all depths, but will always erupt vigorously because of the large excess pressure provided by the buoyancy. The denser magma is no longer trapped at the base of the crust but can rise into it, although this magma can never reach the surface.

There is a final point to be made here. In addition to all of the above issues, we should also apply the considerations of the previous section and check that for every scenario in Table 3.1 the stress intensity at the tip of the dike exceeds the likely fracture toughness of the crustal rocks. This involves carrying out a numerical integration of the effects of all of the local density differences, and when this is done we find that it is only in the case of the dense magma D rising into the continental crust that the stress traps are more important than the density traps. But in that case the effect is very important. Thus for a melt source at a depth of 35 km, even though on density grounds alone the magma should be able to reach within 2.3 km of the surface, it is found that the dike tip ceases to be able to fracture the crustal rocks after penetrating only a short distance into the crust. For a melt source at 40 km the magma apparently should be able to rise to a depth of 1.8 km as shown in Table 3.1, but in practice the dike tip ceases to propagate at a depth

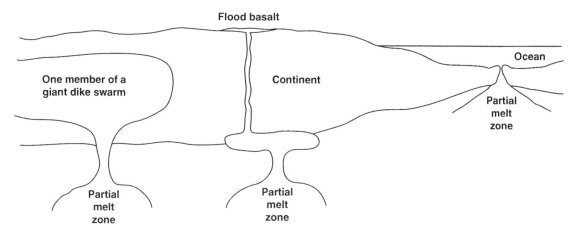

Fig. 3.5 Three likely scenarios for the fate of magma rising in a dike from a partial melt zone. In the first two, magma is produced in the mantle beneath continental lithosphere. A dike may penetrate the crust-mantle boundary but fail to reach the surface, at least immediately, in which case the dike propagates laterally, possibly for many hundreds of kilometers, to form one of the dikes of a giant dike swarm. In the second continental case, magma may stall at the crust-mantle boundary, lose some heat and undergo some chemical changes; subsequently a dike grows from this melt body and reaches the surface to form an extensive flood basalt province. The third case involves melting beneath oceanic lithosphere, in which case dike intrusions and lava eruptions on the ocean floor build new oceanic crust at a mid-ocean spreading centre.

of 19 km. For a melt source at 45 km, the magma should be able to rise to a depth of 1.3 km, but the dike tip would cease to fracture the crustal rocks at a depth of 5 km. Finally, for a melt source at a depth of 50 km, the situation appears to be particularly strange because the pattern described so far is reversed: the stress conditions will just allow the dike tip to reach the surface, even though Table 3.1 implies that the magma inside the dike should not be able to rise above a depth of ~800 m. In practice, when account is taken of the release of magmatic volatiles in the low-pressure region near the tip of the dike, it is found to be likely that some sort of short-lived eruption would take place.

3.6 Consequences of dike trapping

If the conditions controlling a growing dike can no longer allow it to grow upward, it may still be able to expand sideways to accommodate more magma. It is easy to show by considering the stresses acting along all the edges of the dike that it is most likely to grow sideways at the depth where it is neutrally buoyant, and this is the basic reason why **magma**

reservoirs form at neutral buoyancy levels within the crust. We discuss shallow magma reservoirs in the crust in detail in Chapter 4, but for the moment Fig. 3.5 shows some general examples of possible scenarios for dike growth and trapping.

By far the most impressive type of geological system that appears to be a consequence of the vertical trapping and lateral spreading of a large volume of magma at a deep neutral buoyancy level is called a **giant dike swarm**. Many of these structures are found in parts of the continental crust on Earth greater than 2 Gyr old, and even the youngest formed more than 10 Myr ago. Analogous features also exist on Venus and Mars (Chapter 13). The key characteristic of a giant dike swarm is that a series of dikes is seen to radiate outward from a relatively small region for distances of at least several hundred and up to 2000 km. The dikes do not necessarily propagate to all points of the compass – they may just occupy a sector as shown in Fig. 3.6. Exposures of these dikes in regions that have undergone great uplift and erosion suggest that the dikes commonly extend vertically for at least 10 km, and in some cases 20 km. Dike thicknesses range from several tens of meters to as much as 200 m, and the

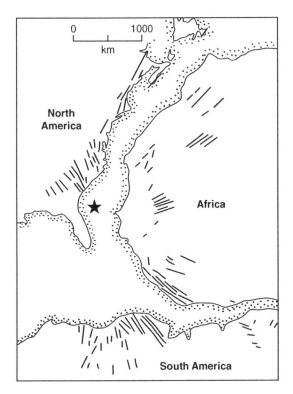

Fig. 3.6 A reconstruction of the North American, South American and African plates as the Atlantic was opening. Giant dike swarms radiated out from a hot spot, the position of which is indicated by the star. Although they are present on now widely separated continents, these dikes have essentially identical compositions and emplacement ages, confirming their common origin before the Atlantic formed. (Adapted from fig. 6 published in *Earth Science Reviews*, Vol. 39, Ernst, R.E., Head, J.W., Parfitt, E.A., Grosfils, E. & Wilson, L., Giant radiating dike swarms on Earth and Venus, 1–58, copyright Elsevier (1995).)

only simple explanation of the sizes and geometries of these dikes is that they grow from very large reservoirs of magma that have been extracted from the mantle but then trapped at or near the base of the crust. Detailed measurements of the fabric and magnetization of the dike rocks show that growth is mainly upward immediately above and near to the reservoir, and mainly lateral in the more distant parts of the swarm.

Are the widths of these giant dikes consistent with what we would expect? The average width W of a dike extending from the base of the continental

crust to a height z_{magma} and held open by a pressure in its source region in excess of the local lithostatic pressure by an amount ΔP_{d} is given by

$$W = [(1 - v)/\mu]z_{\text{magma}} \{\Delta P_{\text{d}} \\ + [z_{\text{magma}} \, g \, (\rho_{\text{magma}} - \rho_{\text{crust}})/\pi]\} \quad (3.5)$$

where v and μ are **Poisson's ratio** and **shear modulus**, respectively, two of the elastic properties of the host rocks. ΔP_{d} is the pressure at the inlet to the dike at the base of the crust, and so the buoyancy of all of the vertical extent H of the plume head providing magma contributes to this pressure, i.e., in this case

$$\Delta P_{\text{d}} = \Delta \rho_{\text{m}} \, g \, H \quad (3.6)$$

Table 3.2 shows the average widths of the dikes that would contain the magma columns listed in Table 3.1; the range, of order 100–200 m, seems to be quite consistent with the field observations. The great widths of these dikes means that magma may flow through them very easily at relatively high speed. This is probably directly linked to the formation of very large volume flood basalt flow fields on the continents and large igneous provinces on the ocean floors.

3.7 Summary

- When a new convection cell forms in the mantle, its rising upper part, a mantle plume, undergoes partial melting due to the pressure reduction, and the resulting body of low-density melt mixed with the residual solids constitutes a buoyant diapir rising through the surrounding mantle rocks. These rocks, although solid, act as a very viscous fluid on the long time scales and slow deformation rates involved.

- In a similar way, melting due to the temperature rise as wet sediments and ocean floor rocks are carried down in a subduction zone can also create a region of lower-density partial melt that then forms a slowly rising diapir. Subduction zone diapirs are expected to be much smaller than mantle plume diapirs.

Table 3.2 The average widths of the dikes containing the columns of magma described in Table 3.1.
(a) Oceanic island volcano.

Depth of base of continuous magma pathway (km)	Average width of dike (m)	
	Magma density 3000 kg m^{-3}	Magma density 2700 kg m^{-3}
9.0	5.8	1.6
10.0	7.7	5.1
10.7	9.1	7.6
15.0	17.8	24.0
20.0	29.1	45.1
21.4	32.4	51.4

(b) Typical continental crust.

Depth of base of continuous magma pathway (km)	Average width of dike (m)	
	Magma density 3000 kg m^{-3}	Magma density 2700 kg m^{-3}
35	103	28
40	143	102
45	184	179
50	226	257

- Diapirs slow down as they rise to shallower depths because decreasing temperature makes the host rocks more viscous. Diapirs cease to rise when they cease to be buoyant, which commonly happens when they reach the base of the crust, where the rock density is much less than in the mantle.

- The rise speed and size of a diapir control the rate at which it strains the host rocks as they move aside to make way for its passage. If the strain rate becomes too high, the response of the host rocks changes from viscous to elastic and they fracture, allowing a dike to start to grow. This will tend to happen at a greater depth for a large mantle plume diapir than for a smaller, slower-rising subduction zone diapir.

- The rate at which melt can drain from a partially molten region into a rapidly growing dike is limited by the viscosity of the magma rather than the potentially very rapid growth rate of the dike. The ability of a dike to start to grow, and the final limit on its growth, are controlled by the apparent fracture toughness of the host rock, which depends on the tensile strength of the rock, the compressive load due to the weight of overlying rocks, and the presence of any gas, released from the magma, that has accumulated in the tip of the growing fracture.

- Dikes can become trapped, i.e., cease to grow, when the stress at the dike tip becomes less than the local fracture toughness. If a dike ceases to grow in this way and there is still magma available in the source region, a new dike may propagate from the source region.

- The sizes and shapes of dikes propagating from deep magma sources are controlled by the distribution of stress and strength in the rocks through which they pass, and many configurations of the end products are possible. In oceanic environments these include the very localized feeding of mid-ocean ridge volcanoes and the formation of chains of shield volcanoes above mantle plumes well away from ridges. In continental environments dikes may be trapped to form giant dike swarms or may erupt to form flood basalt provinces.

3.8 Further reading

DIAPIRS

Griffiths, R.W. & Turner, J.S. (1998) Understanding mantle dynamics through mathematical models and laboratory experiments. In *The Earth's Mantle* (Ed. I. Jackson), pp. 191–227. Cambridge University Press, Cambridge.

Marsh, B.D. (1984) Mechanics and energetics of magma formation and ascension. In *Explosive Volcanism: Inception, Evolution, and Hazards* (Ed. F.R. Boyd, Jr.), pp. 67–83. National Academy Press, Washington, DC.

Rubin, A.M. (1993) Dikes vs. diapirs in viscoelastic rock. *Earth Planet. Sci. Lett.* **119**, 641–59.

DIKES

Lister, J.R. & Kerr, R.C. (1991) Fluid-mechanical models of crack propagation and their application to magma transport in dikes. *J. Geophys. Res.* **96**, 10049–77.

Rubin, A.M. (1993) Tensile fracture of rock at high confining pressure: implications for dike propagation. *J. Geophys. Res.* **98**, 15919–35.

Rubin, A.M. (1995) Propagation of magma-filled cracks. *Ann. Rev. Earth Planet. Sci.* **23**, 287–336.

GIANT DIKE SWARMS

Ernst, R.E., Grosfils, E.B. & Mège, D. (2001) Giant dike swarms: Earth, Venus, and Mars. *Ann. Rev. Earth Planet. Sci.* **29**, 489–534.

3.9 Questions to think about

1 What factors control the deformation properties, i.e., the rheology, of rocks? Why do we not understand the rheology of the rocks in the mantle as well as we would like?

2 What are the two main ways of moving large volumes of molten rock within the Earth?

3 Why are the rise speeds of diapirs and dikes so different?

4 Why does magma rising from the mantle not necessarily reach the surface to erupt?

4 Magma storage

4.1 Introduction

The previous chapter dealt with how magma moves within the mantle and crust. Although some magma may ascend directly to the surface, most magma experiences a period of storage during its ascent. This storage may be permanent, i.e., the magma cools and solidifies forming an intrusive body, or may represent only a temporary halt *en route* to the surface. Magma storage during ascent has a profound impact on the nature of magmatic and volcanic activity. Amongst other things it influences the composition of the erupting magma, the physical properties of the magma (such as its viscosity), and the size and frequency of volcanic eruptions. This chapter examines what is known about the geometry and size of crustal storage zones, how

they form, processes operating within them, and how they regulate volcanic activity.

4.2 Evidence for magma storage within the crust

We start by looking at the various lines of evidence which show that magma is commonly stored within the crust prior to eruption.

4.2.1 Calderas and magma chambers

Probably the most obvious evidence for magma storage within the crust comes from the morphology of volcanoes. Many volcanoes have a summit crater called a **caldera**. Figure 4.1 shows the sum-

Fig. 4.1 The summit caldera complex of Kilauea volcano, Hawai'I, seen from the southwest. The large pit crater, Halema'uma'u, on the floor of the main caldera is ~1000 m in diameter. (Photograph by Pete Mouginis-Mark, University of Hawai'I.)

Table 4.1 The size of calderas formed during a selection of volcanic eruptions.

Eruption date	Caldera	Caldera diameter (km)	Erupted volume* (km³)
1991	Pinatubo, Philippines	2.5	4–5
1968	Fernandina, Galapagos	5 × 6	0.1
1912	Katmai, Alaska	2.5 × 4	12
1883	Krakatau, Indonesia	8	10
1.4 ka	Rabaul, Papua New Guinea	10 × 15	11
1.8 ka	Taupo, New Zealand	35	35
3.6 ka	Santorini, Greece	7 × 10	25
75 ka	Toba, Indonesia	30 × 80	1500
600 ka	Yellowstone, USA	40 × 70	1000–2000
27.8 Ma	La Garita, USA	35 × 75	5000

Data taken from Lipman (2000) Calderas. In *Encyclopedia of Volcanoes*, pp. 643–662. Academic Press.
*Dense rock equivalent.

mit caldera of Kilauea volcano which is ~3 × 5 km across and up to ~120 m deep. This is a relatively small caldera. Observations of modern eruptions, and geological studies of older volcanoes, show that calderas often form during large-volume eruptions. In general, the larger the erupted volume the larger the size of the caldera which is formed (Table 4.1). Calderas are thought to form by the collapse of surface layers into an underlying magma reservoir as magma is erupted from it.

4.2.2 Petrological evidence for magma storage

Study of the **petrology** of igneous rocks allows us to distinguish between rocks formed from magma erupted directly from the mantle and those formed from magma which has been stored within the crust prior to eruption. A melt formed within the mantle has a composition and temperature which reflects the depth (and, therefore, pressure) at which it formed. Typically a mantle melt will form at temperatures of 1200–1400°C. The liquidus temperatures of melts depend strongly on the pressure conditions, and decrease with decreasing pressure (Fig. 4.2). For this reason, if a magma generated in the mantle is erupted without a period of shallow storage occurring, the magma will have a temperature and composition which reflect its origin in the mantle. If, however, the magma is stored prior to eruption then it has the opportunity to cool and begin to crystallize. Cooling and crystallization at

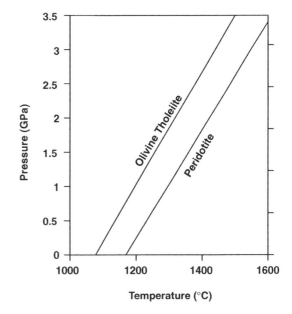

Fig. 4.2 Variation of liquidus temperatures of two melts with pressure. In each case the liquidus temperature increases with pressure and hence with depth beneath the surface. (After fig. 3 in Lambert, I.B. and Wyllie, P.J. (1972) Melting of gabbro (quartz eclogite) with excess water to 35 kilobars, with geological applications. *Journal of Geology*, **80**, 692–708. Copyright University of Chicago Press.)

lower pressures change the composition and temperature of the erupting magma. Study of igneous rocks shows that mantle-derived magmas are usually close to their low-pressure liquidus temperature upon eruption and thus that they have equilibrated to low-pressure conditions prior to eruption. (Typically eruption temperatures at Kilauea volcano are ~1150°C, for example.) In other words, igneous petrology suggests that it is very common for mantle-derived magma to experience crustal storage prior to eruption. Furthermore, detailed petrological study of the properties of the various minerals in volcanic rocks can indicate the actual depth at which storage occurred.

4.2.3 Geophysical observations

During the past 100 years scientists have developed increasingly sophisticated methods to monitor the activity of volcanic systems (see Chapter 11). Various geophysical techniques have been developed which can be used to "look inside" active volcanoes and these indicate the presence of stored magma beneath the summits of many active volcanoes. The two most widely used techniques employ seismic and deformation methods.

SEISMIC TECHNIQUES

A range of seismic techniques can be used to detect the presence and estimate the size of magma chambers. One method is to look for a **seismic gap** beneath the summit region of the volcano. Earthquakes occur continuously within active volcanic systems. The majority of these earthquakes are too small to be "felt" but they are readily detectable using seismometers and their source or **focus** can be located. When a magma chamber is being sup-

plied with fresh magma, earthquakes will tend to occur around the edges of the chamber because the influx of magma causes stress at the chamber walls. Similarly when magma is withdrawn from the chamber during an eruption or intrusion, changes in stress and the emplacement of feeder dikes generate more earthquakes. However, no earthquakes are generated within the magma itself, so when the locations of earthquakes are plotted in a diagram, a region (the seismic gap) may appear in which there is a scarcity of earthquakes and which coincides with the zone in which magma is stored (Fig. 4.3).

The time taken for a seismic wave to travel through the ground to a detector depends on the **seismic velocity** of the material through which the wave passes. Compressional seismic waves (**p waves**) will pass through magma but their speed in magma is much slower than in solid rock. So seismologists can use delays in the passage of seismic waves through the ground to detect zones of low seismic velocity which correspond to areas of magma storage. In some cases they may do this by using the natural seismic waves generated by earthquakes within the volcanic system, but they may also carry out seismic surveys in which the seismic waves are generated artificially using controlled explosions.

A further technique involves locating the source of **volcanic tremor** within a volcanic system. Volcanic earthquakes are recorded as discrete events by seismometers (trace A in Fig. 4.4). In contrast, volcanic tremor is a continuous type of seismic signal which can last minutes, days or even longer (trace B in Fig. 4.4). Tremor is a very com-

Fig. 4.3 Distribution with depth, and with position in a vertical plane oriented NW–SE, of the sources of some earthquakes around the summit magma chamber of Kilauea volcano. The earthquakes are generated by brittle fractures in response to stresses in rocks, and cannot be generated in the molten or partially-molten magma in the chamber. (Modified from fig. 14 in Decker et al. (1983) Seismicity and surface deformation of Mauna Loa volcano, Hawai'I. *EOS, Trans. Am. Geophys. Union*, **37**, 545–547.)

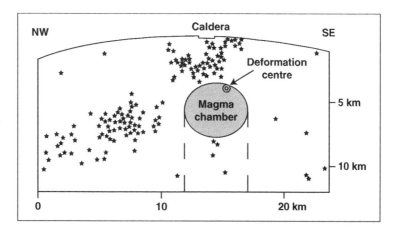

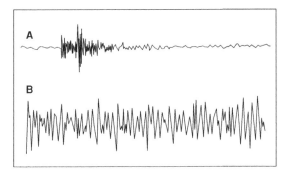

Fig. 4.4 Comparison of the seismic signals from a volcanic earthquake (trace A) and a period of volcanic tremor (trace B). The earthquake is a discrete event, finished in seconds to tens of seconds, whereas the tremor continues for as long as magma is moving beneath the surface, which can be from minutes to many tens of hours. (Modified from fig. 3 in McNutt, S.R. (2000) Seismic monitoring. *Encyclopedia of Volcanoes*. Academic Press, pp. 1095–1119, copyright Elsevier (2002).)

mon precursor to eruptions and some forms of it are generated by the movement of magma within the volcanic plumbing system. Thus monitoring of volcanic tremors provides a method of detecting magma movement beneath the ground before an eruption starts at the surface. If the place where the tremor starts can be located accurately it should indicate a boundary of the magma storage region from which a new dike is propagating.

A seismic study carried out during the eruption of Usu volcano in Japan in 2000 illustrates the use of all three of these techniques. Seismic tremor was detected originating at depths of 5–6 km beneath the volcano prior to eruption. This depth coincides with the location of a seismic gap and of a low-velocity region. The amplitude of the tremor was strongly correlated with measurements of the rate at which the ground surface was being uplifted prior to the eruption (Fig. 4.5a). This combination of seismic and ground deformation evidence suggests that a magma chamber was located at a depth of 5–6 km beneath the surface and that the tremor, ground deformation and subsequent eruption were caused by the upward movement of magma from this depth (Fig. 4.5b).

The power of seismic techniques in giving insight into the location and size of magma chambers can be illustrated by another recent study. In 1998 an eruption occurred from Axial Volcano, a basaltic shield volcano located on the Juan de Fuca ridge on the floor of the Pacific ocean. This is a volcano formed above a **hot spot** which happens to be located directly beneath a mid-ocean ridge. During the eruption, detection of earthquakes showed that a dike propagated laterally 50 km away from the summit caldera and fed a lava flow (the presence of which was subsequently detected on the sea floor by a small scientific submarine). The eruption was associated with 3 m of subsidence of the caldera floor, suggesting that magma withdrawal associated with dike propagation and eruption caused partial collapse of the roof of a magma chamber. A subsequent seismic study was carried out using an artificial seismic source. This study looked at the velocity structure beneath the volcano and indicated the presence of a low-velocity zone below the caldera which was most pronounced at a depth of 2.25–3.5 km but which extended to at least 6 km. The resolution of the survey was sufficient to show that the low-velocity zone covered an area 8×12 km when viewed from above. This is considerably larger than the size of the caldera itself (3×8 km). The volume of the magma chamber was estimated as being ~250 km^3 of which only ~5–21 km^3 was actual melt (Fig. 4.6). This illustrates an important point about magma chambers. It is common to think of a magma chamber as containing only molten magma. In reality cooling of the magma within the chamber means that crystallization is occurring all the time so the low-velocity area inferred to be a magma chamber will actually contain melt surrounded by a "mush" of liquid containing crystals. The amount of actual melt might be small compared with the total volume of the magma chamber, as appears to be the case at Axial volcano.

DEFORMATION TECHNIQUES

Volcanic activity is often associated with deformation of the volcanic edifice. There is a range of geophysical methods which can be used to monitor this deformation including the leveling, tilt measurements, GPS (Global Positioning System) and EDM (Electronic Distance Measurement) techniques which are described in detail in Chapter 11.

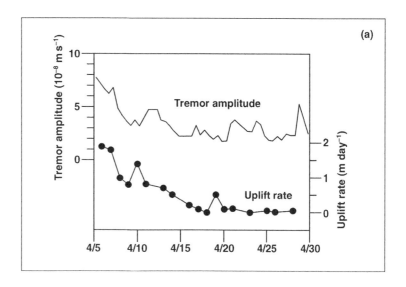

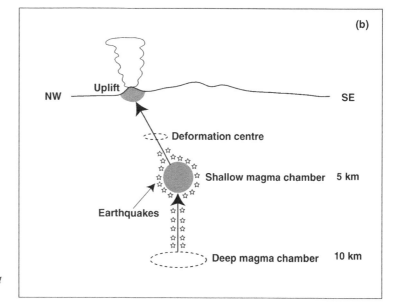

Fig. 4.5 (a) Measurements of seismic tremor and ground uplift rate over a 25 day period in April 2000, prior to an eruption at Usu volcano in Japan; (b) the calculated size and location of the magma chamber consistent with these measurements. (Modified from figs 6 and 7 in Yamamoto et al. (2002) Long-period (12 sec) volcanic tremor observed at Usu 2000 eruption: seismological detection of a deep magma plumbing system. *Geophysical Research Letters*, **29**, 1329.)

Figure 4.7a shows, for example, tilt measurements made at Krafla volcano in Iceland during 1976–77. Tiltmeters measure the angle through which the ground surface has tilted at a particular point over a period of time. The Krafla tilt pattern shown in Fig. 4.7a is typical of many active volcanoes and shows that eruptions and intrusions are often accompanied by rapid inward tilting of the ground (or "deflation") within the summit caldera whereas the period between intrusions or eruptions is asso-ciated with more gradual outward ground tilting (referred to as "inflation"). The inflation and deflation of the summit area coincide exactly with periods of vertical uplift and subsidence respectively (Fig. 4.7b). Such patterns of deformation within the summit region of a volcano are usually interpreted as being due to the influx of magma into a shallow magma chamber from deeper levels (causing inflation and uplift) and the subsequent removal of magma from the chamber during an intrusive

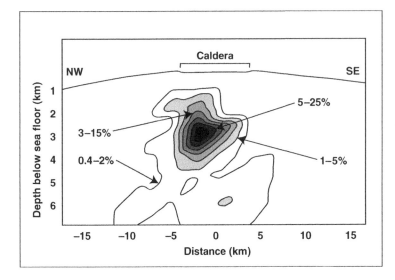

Fig. 4.6 Contours of the percentage of melt present in the region beneath the summit caldera of Axial Volcano, a basaltic shield volcano on the Juan de Fuca ridge on the floor of the Pacific ocean. The hottest, almost completely molten central region of the magma chamber is surrounded by progressively cooler "mushy" zones in which an ever larger proportion of solid crystals is present. (Adapted by permission from Macmillan Publishers Ltd: *Nature*, West, M., Menke, W., Tolstoy, M., Webb, S. and Sohn, R., Magma storage beneath Axial Volcano on the Juan de Fuca mid-ocean ridge. **413**, 833–836, copyright (2001).)

or eruptive event (causing deflation and subsidence). The magma chamber can thus be pictured simplistically as being like a balloon which is repeatedly blown up and let down again. In 1958 a theoretical model was developed by a scientist named Mogi which relates the changes in volume of a sphere buried at a depth, h, beneath the surface to the changes in the elevation of the ground surface, Δh,

Table 4.2 Magma chamber depths and sizes inferred from geophysical observations.

Volcano	Depth of magma chamber (km)	Chamber volume (km³)	Caldera diameter (km)	Geophysical data used
Axial, Juan de Fuca Ridge	2.25–6	250	3 × 8	Seismic
Hekla, Iceland	~9? (center)	145	–	Deformation
Kilauea, Hawai'I	2–7	14–65	3.5 × 5	Deformation and seismic
Krafla, Iceland	2.5–7	28–56	8 × 9	Deformation and seismic
Long Valley, USA	5–25		17 × 32	Seismic
Mauna Loa	3–8	65	2.5 × 5	Deformation and seismic
Mono Craters, USA	8–10 (top)	200–600	–	Seismic
Mount St Helens, USA	7–11, 9–14		–	Deformation and seismic
Yellowstone, USA	6–12		40 × 70	Magnetic

Data from Koyanagi, R.Y., Unger, J.D., Endo, E.T. and Okamura, A.T. (1976) Shallow earthquakes associated with inflation episodes at the summit of Kilauea Volcano, Hawai'I. *Bull. Volcanol.*, **39**, 621–631; Einarsson, P. (1978) S-wave shadows in the Krafla caldera in NE-Iceland, evidence for a magma chamber in the crust. *Bull. Volcanol.*, **41**, 1–9; Iyer, H.M. (1984) Geophysical evidence for the locations, shapes and sizes, and internal structures of magma chambers beneath regions of Quaternary volcanism. *Philos. Trans. R. Soc. Lond., Ser. A*, **310**, 473–510; Achauer, U., Greene, L., Evans, J.R. and Iyer, H.M. (1986) Nature of the magma chamber underlying the Mono Craters area, Eastern California, as determined from teleseismic travel time residuals. *J. Geophys. Res.*, **91**, 13,873–13,891; Ryan, M.P. (1987) Neutral bouyancy and the mechanical evolution of magmatic systems. In *Magmatic Processes: Physiochemical Principles*, pp. 259–287. Special Publication No. 1, The Geochemical Society; Sigurdsson, H. (1987) Dyke injection in Iceland: a review. In *Mafic Dyke Swarms*, pp. 55–64. Special Paper 34, Geological Association of Canada; Barker, S.E. and Malone, S.D. (1991) Magmatic system geometry at Mount St Helens modelled from the stress-field associated with posteruptive earthquakes. *J. Geophys. Res.*, **96**, 11,883–11,894; Lees, J.M. (1992) The magma system of Mount St Helens: non-linear high-resolution P-wave tomography. *J. Volcanol. Geotherm. Res.*, **53**, 103–116; Rutherford, M.J. and Gardner, J.E. (2000) Rates of magma ascent. In *Encyclopedia of Volcanoes*, pp. 207–217. Academic Press; and West et al. (2001).

at a distance, *d*, from the center of uplift (Fig. 4.7c). If the uplift Δh is divided by its value Δh_0 immediately above the buried sphere, the relationship is

$$(\Delta h/\Delta h_0) = 1/[1 + (d^2/h_0^2)]^{3/2} \tag{4.1}$$

So, when $d = h_0$, $(\Delta h/\Delta h_0) = 0.353$, and when $d = 2\,h_0$, $(\Delta h/\Delta h_0) = 0.089$, and so on. Figure 4.7c shows the results of using a Mogi-type model to fit a set of uplift patterns observed at Krafla. This indicates that the deformation recorded during activity at Krafla during the 1970s is consistent with the inflation and deflation of a magma chamber centered at a depth of ~3 km beneath the surface.

Similar patterns of deformation have been observed at other volcanoes and successfully modeled using Mogi-type models. For instance, Fig. 4.8a shows the uplift at Kilauea volcano in Hawai'I between January 1966 and October 1967. Figure 4.8b shows the uplift of a single benchmark within the caldera between August and October 1967 which has been fitted using a Mogi model. The best fit suggests that the center of deformation is located at a depth of ~3 km. Figure 4.9 shows similar modeling for Mauna Loa volcano, also in Hawai'I. The center of deformation at Mauna Loa is at a depth of ~3.1 km. Seismic activity there defines a storage zone at depths of ~3 to ~8 km. The center of deformation is thus within the seismically defined storage zone but is located towards its top. This is also the case at Kilauea where the deformation center is located at a depth of ~3 km while the seismically defined chamber extends from ~2 to ~6 km (Fig. 4.3).

Table 4.2 summarizes geophysical observations of magma chamber sizes and locations made on a number of currently active volcanoes.

4.2.4 Geological evidence for magma storage

While geophysical and petrological techniques allow us to infer the presence of magma chambers beneath active volcanoes, geological studies allow us to examine the intrusive bodies left behind when a volcanic system cools and is eroded. Numerous intrusive bodies have been identified and studied in detail by geologists. Of interest to us here are those which are large enough to have played a significant role in controlling the dynamics of a magmatic sys-tem. The simplest such bodies are **sills** and **laccoliths** (Fig. 4.10). Figure 4.11 shows an outcrop of the Whin Sill, which crops out across much of northern England and which varies in thickness from 2–3 m to greater than 60 m. Larger sills occur: the Basement Sill in Antarctica, for example, is typically 300–400 m thick with a maximum thickness of ~700 m. Occasionally sills can exceed 1 km in thickness.

Many large intrusive complexes may have started as sills but gradually evolved in shape and size as more magma was added to them. For example, the Skaergaard layered intrusion in Greenland appears to have started as a sill which was intruded along an unconformity but later developed a more laccolithic shape (Fig. 4.12). A **laccolith**, like a sill, intrudes between layers in the country rocks, but a laccolith has a less tabular shape because the intrusion causes updoming of the overlying rocks (Fig. 4.10b). Many large intrusions have a generally sheet-like shape (e.g., Bushveld in South Africa, Dufek in Antarctica, Newark Island in Canada) suggesting that the formation of sills often plays an important part in the initial formation of magma chambers. The largest known basaltic intrusions on Earth (Dufek and Bushveld) have volumes of ~10^5 km^3, which is far in excess of the ~10 km^3 volumes of the magma reservoirs beneath current basaltic centers such as Krafla, Kilauea and Mauna Loa.

Not all intrusions have this sheet-like form. Figure 4.13 shows, for example, a cross-section through the Cadillac Mountain Intrusive Complex in Maine which has a more basin-like shape. The top of this intrusion has been lost to erosion so it is not known what the geometry of the upper boundary was like, but the basin-like shape appears to be more like the relatively equant shape usually inferred for modern magma chambers from geophysical studies. The various Tertiary intrusive complexes of the west coast of Scotland also exhibit this more equant shape (Fig. 4.14). These centers were some of the earliest intrusive complexes to be studied in detail and are wonderful examples of what a currently active volcanic center may look like after final solidification and erosion. Figure 4.14 shows the association of the central complex of Mull with contemporaneous lavas

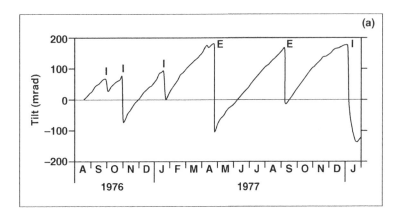

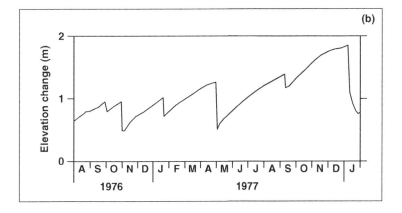

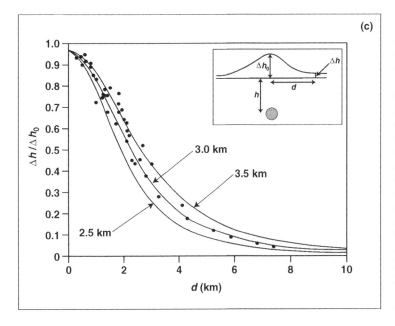

Fig. 4.7 Measurements of (a) ground tilt and (b) surface elevation made at Krafla volcano in Iceland during 1976-77. The timing and sense of the changes are perfectly correlated showing that they have a common source. Sudden deflation of the summit and an inward tilt accompany eruptive episodes (indicated by E) or intrusion (I) as magma leaves the summit chamber. Part (c) shows the results of fitting a Mogi model to the uplift of the summit as the chamber is recharged with magma between eruptions and intrusions. (Modified from figs 3, 9 and 10 in Bjornsson et al. (1979) Rifting of the plate boundary in North Iceland 1975-1978. *J. Geophys. Res.* **84**, 3029-3038.)

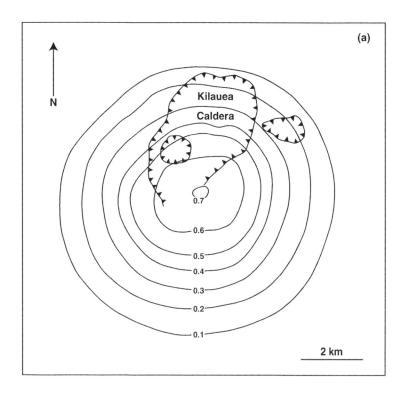

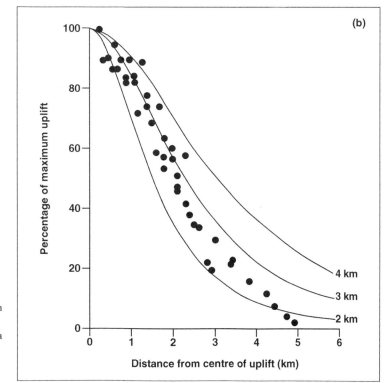

Fig. 4.8 Modified from SCIENCE: Figs 4 and 6. (a) Contours of surface uplift in meters in the summit region of Kilauea volcano in Hawai'I between January 1966 and October 1967. (b) A Mogi model fitted to the uplift of a single benchmark within the caldera between August and October 1967 suggests that the centre of deformation is located at ~3 km depth. (Modified from figs 4 and 6 in Fiske and Kinoshita (1969) Inflation of Kilauea volcano prior to its 1967–1968 eruption. *Science*, **165**, 341–349.)

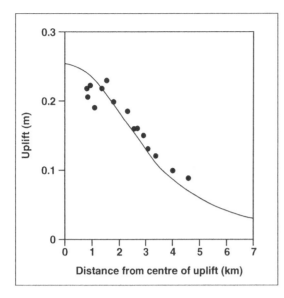

Fig. 4.9 The result of fitting a Mogi model to surface uplift at the summit of Mauna Loa volcano, Hawai'I, suggests that the centre of deformation is at a depth of ~3.1 km. (Modified from fig. 10 in Decker et al. (1983) Seismicity and surface deformation of Mauna Loa volcano, Hawai'I. *EOS, Trans. Am. Geophys. Union*, **37**, 545–547.)

while Fig. 4.15 shows how dikes were emplaced laterally from the various Scottish Tertiary centers. Centers such as those of western Scotland appear to be equivalent to small basaltic magma chambers such as those currently present beneath Kilauea, Mauna Loa and Krafla volcanoes.

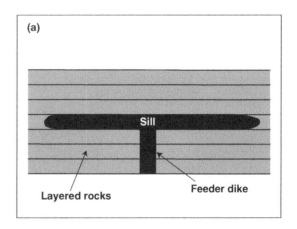

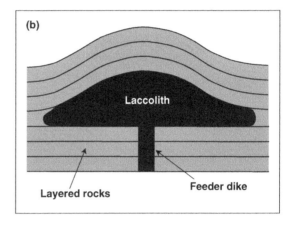

Fig. 4.10 Illustrations of the relationships between a feeder dike, the host rocks, and an intrusion when the intrusion is described as (a) a sill and (b) a laccolith.

Fig. 4.11 An outcrop of the Whin Sill beneath Bamburgh Castle, Northumbria. This sill crops out across much of northern England, varying in thickness from 2–3 m to > 60 m. Here the base of the sill is picked out by the white line. (Photograph by Elisabeth Parfitt.)

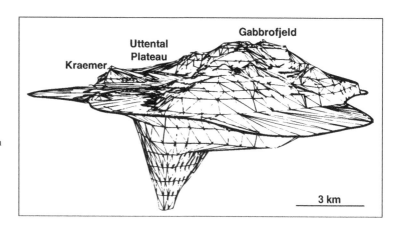

Fig. 4.12 Schematic of the shape of the Skaergaard layered intrusion in Greenland, which appears to have been intruded as a sill along an unconformity and later inflated into a more laccolithic shape. (Adapted from fig. 5 in Norton, D., Taylor, H.P. and Bird, D.K. (1984) The geometry and high-temperature brittle deformation of the Skaergaard intrusion. *J. Geophys. Res.*, **89**, 10178–10192.)

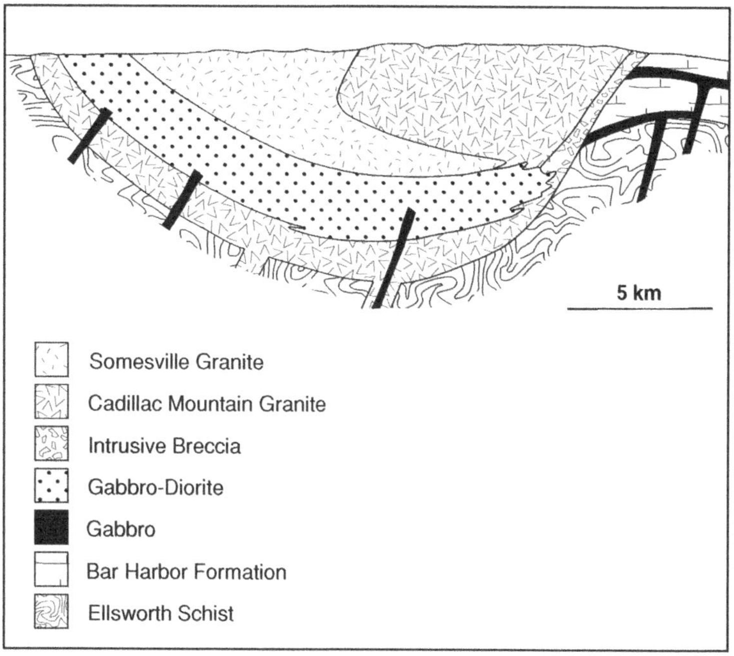

Fig. 4.13 Cross-section through the Cadillac Mountain Intrusive Complex in Maine. The top of the intrusion has been eroded away, but the basin-like shape of the lower part is similar to the shapes of modern magma chambers implied by geophysical studies. (Modified from fig. 2 in Wiebe, R.A. (1994) Silicic magma chambers as traps for basaltic magmas: the Cadillac Mountain intrusive complex, Mount Desert Island, Maine. *Journal of Geology*, **102**, 423–437. Copyright University of Chicago Press.)

Somesville Granite

Cadillac Mountain Granite

Intrusive Breccia

Gabbro-Diorite

Gabbro

Bar Harbor Formation

Ellsworth Schist

4.3 Formation and growth of magma chambers

A magma chamber is a zone of storage that receives multiple pulses of magma and stores magma within the crust for an extended period of time and, in doing so, significantly influences the character of the particular volcanic system. Study of intrusive bodies and active magma chambers shows that magma storage zones vary widely in shape and size. The bodies with the simplest geometries (i.e., sills

and dikes) probably represent the emplacement of one pulse of magma and thus are not magma chambers in the sense that we use the term here. A fundamental issue in the development of a magma chamber is that it must be resupplied intermittently or continuously with fresh magma. Any intrusive body within the crust is subject to cooling: a narrow dike intruded into the shallow crust is likely to cool and solidify within a matter of hours or days (Fig. 4.16). A thicker dike or a sill will take longer to cool but is still likely to do so within years to

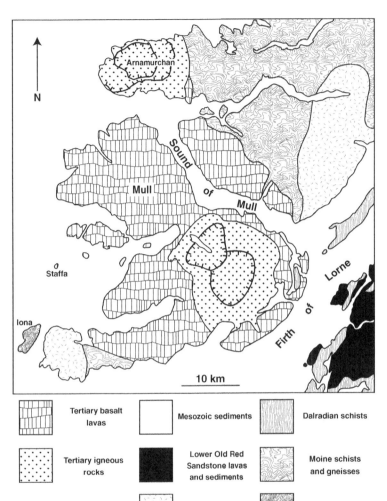

N

Arnamurchan

Sound of Mull

Mull

Staffa

Iona

Firth of Lorne

10 km

| | Tertiary basalt lavas | | Mesozoic sediments | | Dalradian schists |

| | Tertiary igneous rocks | | Lower Old Red Sandstone lavas and sediments | | Moine schists and gneisses |

| | Caledonian granite | | Lewisian gneiss |

Fig. 4.14 The Mull and Ardnamurchan Tertiary intrusive complexes on the west coast of Scotland. The rather equant shapes of these complexes are similar to the shapes implied by geophysical studies for modern magma chambers. (Modified from fig. 6 in Richey, J.E. (1961) *Scotland: the Tertiary Volcanic Districts*. HMSO, Edinburgh. Reproduced courtesy of the British Geological Survey. IPR/92-27C.)

decades to centuries depending mainly on the thickness of the intrusion and the temperature contrast between the magma and country rocks (Fig. 4.16). An initial intrusion can develop into a long-lived magma chamber only if the magma within it is prevented from solidifying. This can be achieved only by the repeated input of heat to the proto-chamber in the form of fresh magma. The most likely way for this to happen is if the initial feeder dike is reused by fresh magma traveling upwards from deeper levels. Reuse of dikes by fresh batches of magma is a well-documented phenomenon. This reuse sometimes occurs when the dike is almost totally solidified. For example,

Fig. 4.17 shows an example of a dike in which fresh magma has been emplaced through the middle of an earlier dike at a stage when the magma in the dike is cool enough to have almost solidified. In other cases fresh pulses of magma are injected into dikes which have experienced only small amounts of cooling and solidification.

The injection of fresh magma from deeper levels heats the magma already in the sill and the surrounding rocks and in doing so reduces the cooling rate and lengthens the time that the enlarged sill will take to solidify. At this stage in the evolution of the sill it is still highly susceptible to cooling and solidification. With each fresh pulse of magma which

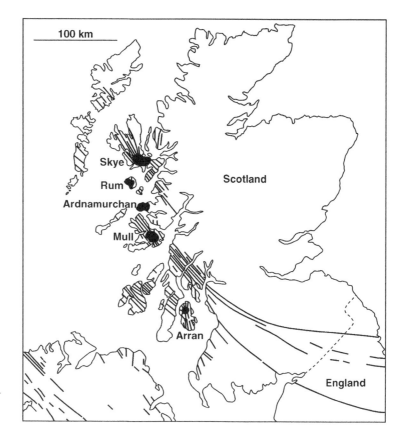

Fig. 4.15 The patterns of lateral dike emplacement from Tertiary intrusive complexes in western Scotland. Individual intrusive complexes are named and indicated by solid black areas. (Modified from fig. 2 in Richey, J.E. (1961) *Scotland: the Tertiary Volcanic Districts*. HMSO, Edinburgh. Reproduced courtesy of the Geological Survey. IPR/92-27C.)

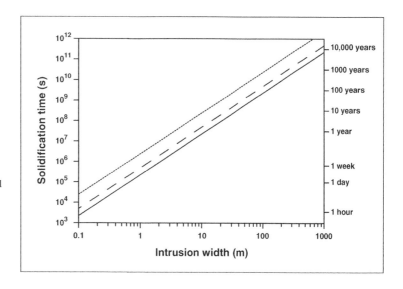

Fig. 4.16 The times needed for heat loss to cause the solidification of intrusions of various widths. The solid line represents an initial temperature contrast between the magma and country rock of 1000°C, the dashed line represents a 500°C contrast and the dotted line represents a 200°C contrast.

Fig. 4.17 An example of a dike intruding into the interior of a previously emplaced dike. Erosion has exposed these dikes in what was the interior of the Ko'olau volcano on O'ahu, Hawai'I. The longer dashed lines define the edges of the first dike and the shorter dashed lines mark the sides of the later dike. (Photograph by Lionel Wilson.)

arrives, however, the chances of the sill evolving into a long-lived chamber improve. This is because each new pulse of magma heats the walls of both the feeder dike and the sill (now a proto-chamber) and so reduces the temperature contrast between the magma and country rocks, thus slowing the cooling rate (Fig. 4.16). Once these critical early stages have been overcome then an established magmatic system is likely to evolve in which magma formed at depth has a well-defined pathway linking the zone of magma formation with the shallow magma chamber. Such a magmatic system is thought to underlie Kilauea volcano in Hawai'I (Fig. 4.18).

Even given the above arguments and the issues discussed in Chapter 3, the nature of the deep plumbing systems feeding magma chambers is not well understood. While reuse of dikes provides a mechanism for developing such a system, the repeated transfer of magma through the system and the higher temperatures prevailing in the lower crust mean that the deep plumbing system may evolve through time to something less dike-like. For instance, the volcanologist Bruce Marsh suggested the idea of **mush columns**. He pictures the plumbing system as a heated zone through which magma repeatedly passes and in which magma never completely solidifies so that the zone always contains a crystal mush through which fresh batches of magma may ascend. Other scientists have described these features as **heat pipes**. Whether magma ascends continuously or in discrete batches through this deep plumbing system is not clear, and both situations may prevail at different volcanoes or at one volcano at different times. For example, study of Kilauea volcano over the past 50 years has suggested that supply to its shallow magma chamber is fairly continuous and occurs at a rate of ~0.05 km^3 yr^{-1} or 1.6 m^3 s^{-1}. Prior to 1950 the rate of activity was lower and it has been calculated that supply rates were only 0.009 km^3 yr^{-1} or 0.03 m^3 s^{-1}, suggesting that even if supply is continuous the rate of supply may vary considerably through time. Measurements made at Krafla volcano in Iceland between 1975 and 1984 suggest continuous supply of magma at a rate of ~5 m^3 s^{-1}. Patterns of inflation and deflation during this time were very similar to those commonly seen at Kilauea. However, phases of activity at Krafla such as those of the 1970s and 1980s are punctuated by long periods of inactivity, suggesting that magma supply may be continuous for some years during periods of activity but is then interrupted or much slower during the periods of inactivity. Activity at Askja (also in Iceland) in 1875 also suggests that supply to chambers may be intermittent. At Askja, both a major rhyolitic Plinian eruption and a substantial basaltic fissure eruption seem to have been triggered by a sudden influx of basaltic magma from depth into a shallow magma chamber. Whether the supply is continuous or intermittent may affect the frequency and character of volcanic eruptions from a system (see section 4.4.2) but, in terms of survival of the magma chamber, the important factor is simply that supply is regular enough to prevent solidification of the chamber.

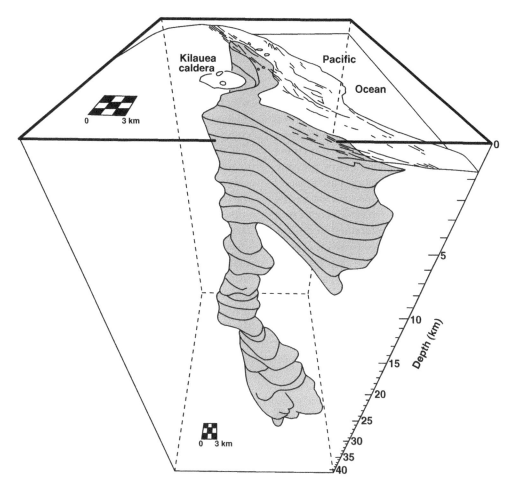

Fig. 4.18 The magmatic system thought to underlie Kilauea volcano in Hawai'I as derived mainly from seismic data. Not all of the shaded region is occupied by completely molten rock at any one time. Note the frequently-used pathway leading up from the mantle and the extension of the summit chamber into the volcano's two lateral rift zones. (Adapted from fig. 12 in Ryan, M.P. (1988) The mechanics and three-dimensional internal structure of active magmatic systems: Kilauea Volcano, Hawai'I. *J. Geophys. Res.*, **93**, 4213–4248.)

Once a deep plumbing system has become established, the magma chamber it feeds may progressively evolve in shape and size by a range of processes. At Skaergaard (Fig. 4.12), for example, the intrusion developed from a sill into a more laccolithic shape by the updoming of the overlying layers. Deformation of the surrounding rocks is therefore one method by which the chamber shape may change and its volume expand. As chambers inflate and the pressure inside them increases it becomes possible for fractures to develop in the chamber walls and to grow outwards from them. These offshoots from the main chamber may increase the complexity of the chamber shape. They also facilitate the **stoping** process in which blocks of wall rock or roof rock break away and sink into the magma to be melted, thus increasing the chamber volume at the expense of reducing its temperature.

A final issue here concerns the lifetimes of magma chambers. The application of deformation and seismic techniques shows that the volcano

Mount Etna, in Sicily, currently has no permanent shallow magma storage system of any significant size. This volcano is particularly interesting in that there have been major structural events during its history where parts of the eastern flank of the mountain have collapsed into the sea forming giant landslides. These processes have exposed ancient lava flows whose compositions can only be explained if the magmas that erupted to form them were stored for a time at the kind of low pressure found in a shallow magma chamber. Furthermore, there are signs of caldera collapse events having occurred, another indicator of shallow magma storage. Why should the internal plumbing of the volcano have changed? Any major collapse event must cause a very great alteration to the distribution of stress within the body of the volcano, and if this interrupts the internal pathways supplying magma to shallow levels for long enough, any existing magma chamber may solidify. Furthermore, if the same stress changes happen to make it easier for magma to rise directly to the surface, there is no reason for a new shallow chamber to form.

4.4 Magma chambers and their impact on volcanic systems

We have seen that magma chambers represent a staging post for magma on its journey from the mantle to the surface and are fed by semi-permanent, deep plumbing systems which represent the evolution of magma supply from simple transport in dikes to the ability of pulses of magma to travel upward in a mush column (Fig. 1.21). This section examines some of the profound effects that the presence of shallow magma chambers has on the nature and behavior of volcanic systems.

4.4.1 Fractionation in magma chambers

Magmas generated in the mantle are typically of basaltic compositions. Erupted magmas have a far wider range of compositions ranging from these same basaltic compositions up to highly evolved rhyolitic lavas. While there are many reasons for this compositional diversity, storage and evolution of magma in shallow magma chambers play an important role in generating it. Storage of magma within a magma chamber will inevitably lead to cooling and **fractional crystallization** of the magma. Fractional crystallization is the process in which a cooling magma crystallizes out successive minerals and in doing so progressively evolves in composition. The study of fractional crystallization falls within the realm of igneous petrology rather than physical volcanology and we are concerned here only with the ways in which crystallization affects the physical processes of the magmatic system.

Basaltic magmas will only undergo fractional crystallization if their residence time within a magma chamber is long enough for cooling to become significant. Two critical (and linked) factors controlling the cooling and crystallization history of a particular magma chamber are the nature of magma resupply to the chamber from depth, and the frequency with which magma is erupted from the magma chamber. If magma is supplied to the magma chamber continuously then cooling will be minimized and the magma chemistry is unlikely to alter greatly. In this situation eruptions are also likely to be frequent and so the residence time of magma in the chamber is minimized, also limiting the cooling experienced by a particular batch of magma. If, however, magma supply is intermittent, then the magma in the chamber is likely to cool between resupply events and so to evolve in composition.

These differences can be illustrated by comparing activity in Hawai'I with that in Iceland. In both Hawai'I and Iceland, basaltic magma is supplied to shallow magma chambers from the mantle. In Hawai'I the vast majority of eruptions produce only basaltic magmas. There is little evidence for the formation of significantly more evolved magmas by fractional crystallization (except during the very final stages of activity of a given volcano when the rate of supply of mantle magma to it becomes very small). In Iceland many eruptions are also purely basaltic in character but it is not uncommon for there to be eruptions of more evolved magmas or mixed-magma eruptions (eruptions in which both basaltic and more evolved magmas are released together). The difference between the Hawaiian and Icelandic systems seems primarily to be the way magma is supplied to the shallow magma chamber. In Hawai'I the supply is either continuous or sufficiently frequent that the magma in the shallow chamber never cools enough to allow significant evolution of the magma composition and so all eruptions are basaltic. In Iceland, which is located astride the Mid-Atlantic spreading ridge, the supply of magma to shallow chambers is influenced by the spreading process, which is intermittent rather than smooth. As a result, resupply can be infrequent enough to allow cooling and crystallization of the magma between resupply events, so that eruption of evolved magmas can occur.

The consequences of these different patterns of cooling and fractionation are profound because they fundamentally influence the style of volcanic activity which will occur. Considering the contrast between Hawai'I and Iceland again, the lack of significant fractionation in Hawaiian magma chambers means that eruptions from them are associated with the typical range of mild basaltic eruption styles (see Chapters 9 and 10). By contrast, an Icelandic chamber which has had an opportunity to cool may erupt highly evolved magmas in major explosive eruptions. The 1875 eruption of Askja in Iceland, for example, involved the eruption of both basaltic and rhyolitic magmas (a fresh influx of basaltic magma apparently triggering the eruption). The main eruption occurred from the Askja caldera and involved the predominantly rhyolitic magma. This eruption was Plinian in style (see Chapters 6 and 8) and generated a plume 26 km high, making it one of the largest explosive eruptions to have occurred in Iceland in historical times.

4.4.2 Regulation of eruption frequency and magnitude

Volcanic eruptions vary tremendously in magnitude and frequency. Some volcanoes erupt almost continuously, with individual eruptions producing very small volumes of material. For example, mildly explosive eruptions at Stromboli volcano typically occur every few minutes to tens of minutes, with eruptions producing 1–2 m³ of ejected material. By contrast, Yellowstone has major eruptions at

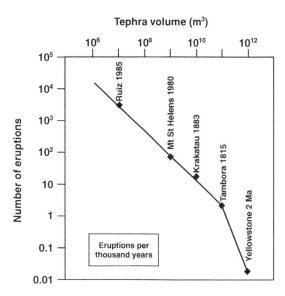

Fig. 4.19 The relationship between the magnitude of eruptions, expressed as volume of tephra, and their frequency, expressed as number of eruptions per thousand years, illustrated by some well-studied volcanoes. The pattern that small eruptions occur much more frequently than large eruptions is clear. (Modified from fig. 6 in Simkin, T. and Siebert, L. (2000) Earth's volcanoes and eruptions: an overview. *Encyclopedia of Volcanoes*. Academic Press, 249–261. Copyright Elsevier (2002).)

intervals of ~600,000 years but these can have volumes in excess of 2000 km³, making them 10^{12} (a trillion!) times larger than a typical Stromboli eruption! Detailed study of volcanic systems shows that:

• there is a link between the magnitude and frequency of eruptions such that small eruptions occur frequently and larger eruptions are less frequent (Fig. 4.19);
• there is a broad link between the magnitude of an eruption and the size of the magma chamber feeding it (Table 4.1).

These two points taken together suggest that the presence of a magma chamber within a magmatic system plays a crucial role in controlling eruption frequency and magnitude. To understand why this should be the case we can examine the findings of a simple mathematical model of magma chamber behavior.

Figures 4.7 and 4.8 illustrated geophysical studies of volcanoes showing that magma chambers inflate prior to eruption as fresh magma is added to them. When some critical point is reached the chamber ruptures and a dike is emplaced. As the dike propagates, magma is withdrawn from the chamber causing deflation. If the dike reaches the surface then an eruption will result. Various mathematical models have been developed to look at the nature and consequences of the failure or rupture of magma chambers. The general features of such models can be illustrated by the following one, due to Stephen Blake. The criterion for failure of the walls of a magma chamber (approximated as a sphere) is that the stress difference across the walls must exceed twice the tensile strength of the country rocks. The stress difference is $(P - P_L)$, where P is the pressure in the magma and P_L is the stress in the surrounding rocks due to the weight of the overlying crust. So, if the tensile strength of the country rocks is σ_T,

$$(P - P_L) = 2\sigma_T \qquad (4.2)$$

This failure condition is reached because the stress across the chamber walls increases as fresh magma arrives in the chamber from deeper levels within the volcanic system and increases the chamber pressure. The amount of magma that can be added to a chamber of given size before failure occurs is given by

$$\Delta V/V_c = [(\sigma_T/\beta)(1 + s)] + s \qquad (4.3)$$

where ΔV is the volume of magma added to the chamber, V_c is the initial chamber volume, β is the **bulk modulus** of the magma (the reciprocal of its compressibility) and s is the fractional increase in chamber volume which can be achieved inelastically, that is, without compressing the rocks surrounding the chamber (perhaps by the closure of any open pore spaces present in the surrounding rocks).

Figure 4.20 shows the relationship between chamber size and the volume which can be added to the chamber prior to failure. In general the larger the chamber, the larger the volume of magma which must be added before failure occurs. If the chamber is not allowed to expand in volume as the magma is added ($s = 0$, i.e., the chamber walls are rigid) then the additional volume is accommodated by compression of the magma and a corresponding increase in chamber pressure so that the failure criterion is rapidly reached. If, however, the chamber can expand inelastically to accommodate the added

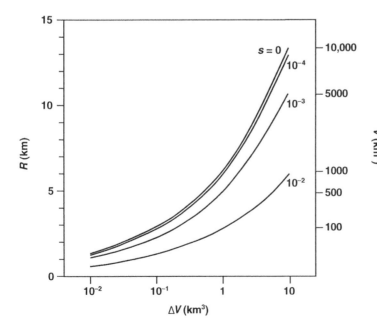

Fig. 4.20 The relationship between the size of a magma chamber, expressed as its volume (right side) and the radius of a sphere with the same volume (left side), and the volume ΔV of new magma which can be added to the chamber before it fails, i.e., its walls fracture. The curves are labeled by the amount of inelastic deformation that the chamber surroundings can undergo to relieve the rising pressure as magma is added. (Adapted by permission from Macmillan Publishers Ltd: *Nature*, Blake, S., Volcanism and the dynamics of open magma chambers, **289**, 783–785, copyright (1981).)

volume ($s > 0$) then the chamber pressure increases less for a given volume of magma added and so more magma must be added for the chamber failure criterion to be met. This simple model has two important implications for the patterns of activity at a given volcanic center.

First, as long as the chamber walls behave elastically, the volume added prior to failure is equal to the maximum volume that can be erupted from the chamber during a single eruption without the danger of caldera collapse, and this volume is ~0.1 to 1% of the chamber volume (see Fig. 4.20). Thus this model predicts that the larger the size of the magma chamber feeding an eruption, the larger the volume of the eruptions it can produce. Second, the frequency of eruptions can be thought of in terms of the repose time, R_T, between eruptions and expressed as:

$$R_T = \Delta V / M_m \qquad (4.4)$$

where, as above, ΔV is the volume of magma added to the magma chamber prior to failure and M_m is the volume rate at which magma is supplied to the chamber. If M_m does not vary too greatly between different magma chambers, it follows that for bigger chambers the repose time between eruptions will be longer. Or, put another way, the bigger the magma chamber, the less frequent its eruptions.

Taken together, these two implications of the model match the observed behavior of real volcanic systems, i.e., that larger magma chambers feed larger but less frequent volcanic eruptions. So magma chamber size plays a crucial role in regulating the scale and frequency of volcanic activity at any given volcanic center.

4.4.3 Volatiles and chamber failure

The model just discussed represents one extreme scenario for magma chamber failure: failure occurs because of the replenishment of the magma chamber with fresh magma from deeper levels. At the opposite extreme, models have been developed in which the failure of the magma chamber occurs solely as the result of exsolution of volatiles as magma cools and crystallizes within a closed chamber (i.e., one in which no fresh magma is being added). Just as in the replenishment model, failure occurs because the pressure in the chamber gradually increases to the point where the tensile strength of the chamber walls is exceeded and fracturing occurs. However, in this case the pressure increases because of the exsolution of volatiles from the magma. Volatile species such as carbon dioxide (CO_2) and water (H_2O) are contained dissolved in magmas as they form at great depths in the mantle. Chapter 5 describes in detail how the solubility of these gases decreases with decreasing pressure so that as magma rises it will eventually reach a pressure level at which the magma becomes saturated in one or more volatile species and gas will **exsolve** from the magma forming gas bubbles. The low solubility of CO_2 in magmas means that it is often present as a separate gas phase within shallow magma chambers. In addition, as magma cools and crystallizes within a magma chamber, the volatile species become increasingly concentrated in the residual melt and, by this mechanism, the melt may also eventually become saturated in more soluble volatiles, particularly water. The formation of gas bubbles within the magma requires space to be created. This space can be created by the compression of the magma, the deformation of the chamber walls and by the crystallization process itself. Just as with the addition of magma to the chamber, the exsolution of gas and formation of gas bubbles increases the pressure in the chamber. Modeling suggests that only a few percent crystallization is necessary to cause sufficient exsolution to raise the pressure to failure levels.

This mechanism of chamber failure can occur only if the gas stays trapped within the chamber. Observations in the summit regions of many volcanoes show that gas is constantly escaping upwards from the magma chamber. (Visitors to volcanoes such as Kilauea in Hawai'I will be immediately aware of this gas leakage in the summit region because of the sulfurous fumes, and caution is advised in spending too much time in certain localities because the gas release is so great that it is potentially lethal to someone with breathing problems!) The ability of gas to escape will depend primarily on whether the gas bubbles can move upwards through the magma in the chamber. In low viscosity magmas, such as basalts, bubbles can move with relative ease and thus have the greatest opportunity to rise to the chamber roof and escape through cracks in the overlying rocks. In high-viscosity magmas, movement of bubbles is inhibited and the gas is less liable to escape. Thus the importance of gas exsolution as a method of pressurization of a magma chamber is likely to be more significant in chambers containing chemically evolved, viscous magmas. Indeed, observations show that eruptions of evolved magmas frequently start with a highly explosive phase which appears to be caused by the exsolution of water from the magma while it is still in the magma chamber. The exsolution is thought to both trigger the chamber failure which leads to eruption and explain the highly explosive nature of the opening stage of such an eruption. Once this bubble-rich magma has been erupted, less gas-rich magma from deeper in the chamber may be erupted with a corresponding decline in the explosivity of the eruption.

The magma replenishment and gas exsolution models described here represent end-member cases for chamber failure. They are not, however, mutually exclusive, and in reality both magma replenishment and gas exsolution may contribute to the pressure rise within, and failure of, a particular magma chamber.

4.5 Summary

- Evidence from petrology, volcano morphology, geophysics and geology suggests that storage of magma in a crustal magma chamber prior to eruption is extremely common. The sizes and depths of these storage zones vary widely (Table 4.2).
- The development of magma chambers is not well understood, but study of eroded chambers suggests that they often develop from an initial sill-like body. As any intrusive body is susceptible to cooling and solidification, a chamber can develop only if it is frequently resupplied with fresh magma. For this reason, establishment of a deep plumbing system to feed the magma chamber is essential to its survival.

- Chamber shapes appear to vary widely from simple sheet-like and laccolithic structures to more equant shapes. The shape of a chamber is likely to evolve through time by a combination of processes, including deformation and fracturing of the surrounding rocks, cooling and stoping.
- The existence of a magma chamber has a profound effect on the chemistry of erupting magmas and hence also on the physical properties of the magmas. The storage of magma in a shallow magma chamber allows the magma to cool and crystallize and so lets its chemistry progressively evolve.
- The presence of a magma chamber within a volcanic system has a profound effect on the scale and frequency of volcanic eruptions and thus magma chambers are crucial regulators of volcanic activity. Small chambers give rise to small but frequent eruptions while larger chambers erupt less frequently but generate larger individual eruptions.
- Magma chamber failure, which is a prerequisite for eruption, can be triggered by both the influx of fresh magma or by the exsolution of gas within the magma chamber as a result of magma crystallization. Either process can raise the internal chamber pressure to the point where the tensile strength of the chamber walls is exceeded.

4.6 Further reading

Blake, S. (1981) Volcanism and the dynamics of open magma chambers. *Nature* **289**, 783–5.

Blake, S. (1984) Volatile oversaturation during the evolution of silicic magma chambers as an eruption trigger. *J. Geophys. Res.* **89**, 8237–44.

Marsh, B.D. (2000) Magma chambers. In *Encyclopedia of Volcanoes* (Ed. H. Sigurdsson), pp. 191–206. Academic Press, San Diego, CA.

Norton, D., Taylor, H.P. & Bird, D.K. (1984) The geometry and high-temperature brittle deformation of the Skaergaard intrusion. *J. Geophys. Res.* **89**, 10, 178–92.

Tait, S., Jaupart, C. & Vergniolle, S. (1989) Pressure, gas content and eruption periodicity of a shallow, crystallising magma chamber. *Earth Planet. Sci. Lett.* **92**, 107–23.

West, M., Menke, W., Tolstoy, M., Webb, S. & Sohn, R. (2001) Magma storage beneath Axial volcano on the Juan de Fuca mid-ocean ridge. *Nature* **413**, 833–6.

Yamamoto, M., Kawakatsu, H., Yomogida, K. & Koyama, J. (2002) Long-period (12 sec) volcanic tremor observed at Usu 2000 eruption: seismic detection of a deep magma plumbing system. *Geophys. Res. Lett.* **29**, 43-1-4.

4.7 Questions to think about

1 Describe four kinds of evidence that might be used to suggest that a currently active volcano has a region of magma storage at shallow depth below its summit.

2 What kinds of features would you look for in rocks now exposed at the Earth's surface by uplift and erosion as evidence for the presence of an ancient volcano?

3 What differences are there between the patterns of activity of basaltic volcanoes as a function of their tectonic setting? What effect does this have on the chemistry of the magma erupted?

4 For each of the eruptions described in Table 4.1, work out the area of the caldera (treating it as a circle or ellipse, as appropriate). Next divide the erupted volume by the caldera area to get the equivalent vertical depth of magma that was erupted from the shallow magma reservoir, assuming one exists. Finally assume that, as implied by the data in Table 4.2, most magma reservoirs extend vertically for ~5 km and work out what percentage of all the magma originally in the reservoir is represented by the erupted magma volume. Recalling that theory suggests that caldera collapse will occur if more than 0.1 to 1% of the magma in a reservoir is erupted, would you have expected collapse to occur in all of these eruptions?

5 The role of volatiles

5.1 Introduction

We saw in the previous chapter how magma moves through the mantle and crust. If this magma contains no dissolved **volatiles** then if it reaches the surface it will always be erupted effusively – simply pouring out of the vent to form lava flows or domes (depending on the chemistry and effusion rate – see Chapter 10). In practice, however, the majority of eruptions which occur subaerially involve some degree of explosiveness. As explained in section 1.2, in volcanology the term "explosive" is used to denote any eruption in which magma is fragmented and ejected from a vent within a stream of gas. In some cases volcanic explosions are transient events (these are described in Chapter 7) but often fragmentation can occur continuously during a steady eruption which might last hours or days (such eruptions are discussed in Chapter 6). So, in both a Hawaiian lava fountain (Fig. 1.1), in which clots of magma up to 1 m or more in diameter are carried up to heights of hundreds of meters above the vent, and a Plinian eruption, in which mainly tiny ash particles are carried to heights of several tens of kilometers, an explosive eruption involves fragmented magma being ejected from the vent within a stream of gas.

In some cases, eruptions are explosive because a volatile substance such as water is mixed with the magma as it approaches the surface. However, in many cases, eruptions are explosive because the rising magma has volatiles dissolved within it. As the magma rises towards the surface and the confining pressure decreases, the volatiles gradually **exsolve** from the magma forming the gas bubbles which are distributed throughout the liquid. It is the connecting together of a network of these bubbles that ultimately causes the continuous body of liquid to break apart or **fragment** into a spray of droplets or clots suspended in the gas (see Chapter 6). This process is similar to what happens when a bottle of any fizzy drink is opened. When these drinks are bottled they have carbon dioxide (CO_2) forced into them at high pressures. At high pressures the CO_2 dissolves in the liquid. When the bottle is opened the pressure is reduced to atmospheric. The solubility of CO_2 in the liquid is lower at lower pressures so when the bottle is opened not all of the CO_2 can remain dissolved and some of it exsolves and forms gas bubbles that expand and make the drink "fizz". In magmas, typically 95–99% of the "mass" of material erupted is liquid rock – at most the gas accounts for only a few percent of the weight; but that small amount of gas represents a very large "volume" as it expands to atmospheric pressure, and is fundamentally important in producing explosive eruptions.

This chapter reviews which gases are commonly dissolved in magmas, how the composition of the magmas influences the amount of dissolved gas, and how gases are released from magmas.

5.2 Volatiles in magma

When magma is in the mantle or the lower crust it contains a range of volatiles dissolved within it. The most common dissolved volatiles are H_2O (water) and CO_2 (carbon dioxide). However, anyone who has visited an active volcano is probably familiar

Fig. 5.1 Scientists R. Okamura and K. Honma making measurements at a fumarole on Kilauea volcano, Hawai'I, on August 27, 1973. (Photograph by R.L. Christiansen, Hawaiian Volcano Observatory, courtesy of the U.S. Geological Survey.)

with the fact that volcanoes also release a lot of sulfurous gases – the often-noticed smell of "rotten eggs" is caused by the release of hydrogen sulfide (H_2S), and frequently deposits of sulfur can be found near vents and **fumaroles** (Fig. 5.1). In fact the most common sulfur compound released by volcanoes is sulfur dioxide (SO_2). Sulfurous gases are particularly associated with basaltic magmas – a basaltic eruption will typically release about 10 times as much sulfur as a rhyolitic eruption of the same size. This is an important factor when considering the effect of volcanic eruptions on climate (see Chapter 12). A wide range of other volatiles can be found in varying amounts in magmas, including hydrogen chloride (HCl) and hydrogen fluoride (HF).

5.3 The solubility of volatiles in magma

The amount of a given volatile which can be dissolved in a magma depends on a number of factors such as the confining pressure, the composition of the magma, and the temperature of the magma. The pressure and the composition (see Table 2.1) are generally the most important. To understand the behavior of volatiles in rising magmas it is necessary first to know something about the solubility of volatiles in different magmas. By carrying out many laboratory experiments scientists have found empirical **solubility laws** for various magma-volatile combinations. Three examples are given below.

For H_2O dissolved in a basalt:

$$n = 0.1078 \, P^{0.7} \tag{5.1}$$

where n is the amount of dissolved gas given as a weight percentage (wt%) and P is the pressure in megapascals (MPa) acting on the magma. For H_2O dissolved in a rhyolite:

$$n = 0.4111 \, P^{0.5} \tag{5.2}$$

For CO_2 dissolved in most magmas:

$$n = 0.0023 \, P \tag{5.3}$$

The equations above apply only if a single volatile species is present in a magma. In fact magmas virtually always contain several volatiles, and the solubility functions are then more complex because the volatiles interact chemically with one another as well as with the magma. As soon as the least soluble volatile starts to exsolve and form gas bubbles, some small amounts of most of the other species present will also diffuse into those bubbles.

However, for simplicity, Fig. 5.2 shows the solubility of H_2O alone in rhyolite and basalt as a function of pressure and depth beneath the surface. This diagram illustrates several important points about the behavior of volatiles in magmas.

• For both basalt and rhyolite there is a general trend in which the amount of water which can be dissolved in the magma decreases as the pressure on the magma decreases (i.e., as the magma rises towards the surface).
• The solubility of water in rhyolite is considerably greater than that in basaltic magma. For instance, at a depth of 5 km beneath the surface the maximum amount of water which can be dissolved in a rhyolitic magma is ~4.8 wt% whereas in a basalt it is only 3.4 wt% (Fig. 5.2).
• The solubility curves show the maximum amount of water which can be dissolved within the magma at a given pressure. This does not mean that the magma will actually contain that amount of water dissolved within it. For instance, at a depth of

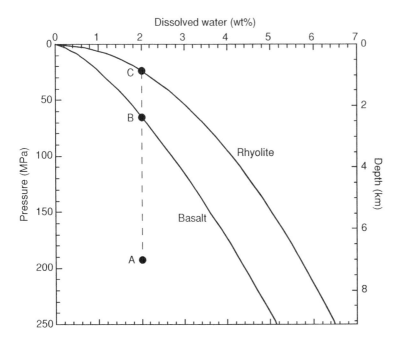

Fig. 5.2 The solubility of H_2O in rhyolite and basalt as a function of pressure and corresponding depth beneath the surface when no other volatiles are present in the magma. A magma ascending from a depth corresponding to point A would become saturated in H_2O at point B if it were a basalt but not until it had ascended to the lower pressure marked by point C if it were a rhyolite.

~8 km beneath the surface a rhyolitic magma is capable of containing about 6 wt% of water in solution (Fig. 5.2), but the magma might actually contain significantly less than this. The magma is then said to be **undersaturated** in water, i.e., more water could be dissolved in it if more water were available. In the deep crust and mantle, magmas will generally be very undersaturated in the common volatiles such as H_2O and CO_2 because there are not enough of these available to cause saturation.

If there are two magma bodies, one rhyolitic and one basaltic, each with say 2 wt% water dissolved within it at a depth of 7 km (point A in Fig. 5.2), we can illustrate what happens to the water within the magmas as they rise towards the surface. As the magmas rise nothing happens until point B (Fig. 5.2) is reached. Here the solubility curve for basalt is intersected. At this point the amount of water dissolved in the basaltic magma is 2 wt%, which is equal to the maximum amount of water which can be dissolved in such a magma. At this point the basaltic magma is said to be **saturated** in water. So as both magmas continue to rise the basaltic magma becomes **supersaturated** in water, i.e., it has more water dissolved within it than is allowed by the solubility laws, and water starts to

exsolve from the magma forming bubbles of water vapor. From here to the surface more and more water will exsolve from the basaltic magma as it constantly tries to accommodate the decreasing solubility of the water as the pressure decreases. Exactly the same process occurs for the rhyolite except that the higher solubility of water in rhyolite means that the rhyolitic magma does not reach saturation until point C (Fig. 5.2), i.e., at a shallower depth than that at which the basalt became saturated. From point C onwards the rhyolite also exsolves water and forms bubbles of water vapor just as the basalt does.

Figure 5.3 compares the solubility of H_2O and CO_2 in basaltic and rhyolitic magma. This diagram shows some other important aspects of gas behavior.

• The solubility of CO_2 in basaltic and rhyolitic magmas is very similar (eqn 5.3) and is considerably less than the solubility of H_2O in such magmas.
• The lower solubility of CO_2 means that it tends to exsolve from magmas at much greater depths beneath the surface than H_2O. For instance, if a basaltic magma contains 0.5 wt% CO_2 and 0.5 wt% H_2O, then the CO_2 begins to exsolve at a depth of ~8 km whereas the H_2O only begins to exsolve at a

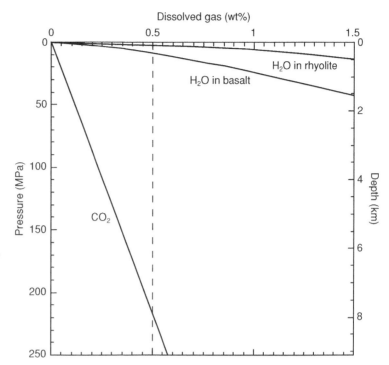

Fig. 5.3 A comparison of the solubility of H_2O in rhyolite and basalt and of CO_2 in either magma as a function of pressure and corresponding depth beneath the surface. The lower solubility of CO_2 means that it tends to exsolve from magmas at much greater depths beneath the surface than H_2O.

depth of ~0.35 km. For H_2O in rhyolite, exsolution begins at an even shallower depth (Fig. 5.3). The situation is actually more complicated when two volatiles are present because as soon as the less soluble volatile starts to exsolve, a very small amount of the more soluble volatile is also released. Fortunately, it is generally accurate enough to treat the solubilities of the two components as though they were independent.

The difference between the depths at which CO_2 and H_2O tend to exsolve can affect the behavior of the magmatic system as a whole. Say, for example, that beneath the summit caldera of a volcano there is a magma chamber with its top at a depth of ~2 km and its bottom at ~6 km, and that basaltic magma is fed to the chamber from below. By the time the magma enters the chamber, the low solubility of CO_2 (Fig. 5.3) will generally mean that the magma has already become supersaturated in this volatile, and so the magma will reach the chamber already containing some CO_2 bubbles.

Bubbles of CO_2 forming in the stored magma will rise through it to the roof of the magma chamber. Often this CO_2 will then seep out through cracks in

the overlying rocks causing considerable release of CO_2 in the summit caldera. The magma in the chamber, though, is still likely to be undersaturated in H_2O at all depths within the magma chamber, and so when the magma is eventually erupted it tends to have lost most of its CO_2 but none of its H_2O. As a result, H_2O will be the most important volatile in driving the eruption. This kind of behavior has been observed in real volcanic systems such as Kilauea volcano in Hawai'I. Other effects of exsolution of magma in magma chambers have been discussed in Chapter 4.

5.4 Bubble nucleation

In principle, bubbles should nucleate in magmas as soon as the least soluble volatile species becomes saturated in the melt. However, the process of bubble nucleation is not trivial. By definition it involves the coming together of a sufficiently large number of molecules to form a stable bubble: if the bubble that tries to form is too small, the force of **surface tension** acts to try to shrink the bubble which, at the molecular level, means pushing the volatile molecules back into the liquid. The spontaneous aggregation of molecules of a volatile species into bubbles in this way is called **homogeneous nucleation**.

The nucleation process is greatly aided if there is some, preferably irregular, surface onto which the volatile molecules can gather to minimize the effects of surface tension, in which case **heterogeneous** nucleation takes place. Thus, nucleation is helped by the presence of solid crystals, and such crystals are commonly present in many magmas, especially if the magma has been stored in a magma chamber before eruption for long enough to have cooled below its solidus temperature, so that at least one mineral has started to crystallize. This use of crystals in magmas as nucleation sites for gas bubbles has analogies with the way water vapor condenses onto dust motes in the atmosphere to form raindrops. Of course, any magma chamber or dike must have walls, and at first sight these are obvious potential sites for bubble nucleation (the next time you hold a glass of champagne look closely at the streams of bubbles nucleating at irregularities in the wall of the glass). However, the magma immediately adjacent to the wall will be relatively cool and viscous, and these factors reduce the ability of volatiles to migrate through the magma to the otherwise attractive nucleation sites.

If there are no aids to nucleation in a magma then there may be a nontrivial delay in the onset of bubble formation, and the magma may become very significantly supersaturated, by as much as ~100 MPa, before bubbles begin to form. The balance between supersaturation pressure ΔP and bubble radius r is

$$\Delta P = 2\sigma/r \qquad (5.4)$$

where σ is the surface tension, typically 0.05–0.1 N m^{-1}. With $\Delta P = 100$ MPa, the initial bubble sizes will be only a few nanometers (1 nm = 10^{-9} m). However, if heterogeneous nucleation on magma crystals is taking place at a small supersaturation, say 1 MPa, then the nucleating bubbles will be more like 1 μm in size.

If the magma has become very supersaturated before bubbles start to form, a large number of bubbles may nucleate at more or less the same time throughout the magma, and this will mean that the distances between the bubbles will be less than if nucleation has taken place in equilibrium with the decreasing pressure. Of course, there still has to be *some* movement of the volatile molecules to reach a nucleation site even if the magma is very supersaturated, and that movement, by diffusion through the liquid magma, takes a finite time. So the greater the rise speed of the magma toward the surface, the more out of equilibrium, and hence the more supersaturated, the magma is likely to become. At all stages in the rise of a magma that is exsolving volatiles there will be a competition between adding more molecules to existing bubbles and nucleating new bubbles. The distance between bubbles is a major factor in determining how efficiently volatile molecules can reach the nearest existing bubble or new bubble nucleation site. Thus the presence or absence of crystals (and also whether there are a large number of small crystals or a small number of large crystals) will have a major influence on how bubbles form and then grow.

5.5 Bubble growth

Once gas bubbles have formed within a rising magma, the bubbles grow progressively through some combination of three processes: **diffusion** of more gas into existing bubbles; **decompression** and expansion of the gas already in bubbles; and **coalescence** of bubbles.

5.5.1 Growth by diffusion

Growth by diffusion involves the migration into a bubble of molecules of the volatile compounds still dissolved in the surrounding magma. The main volatile entering the bubble will be the one that has first become supersaturated and so caused the nucleation of the bubble, but some molecules of any other volatiles present will also enter the bubble. The diffusion process of any one volatile is influenced by many factors, especially the magma composition, the magma temperature, and the mixture of other volatile species present.

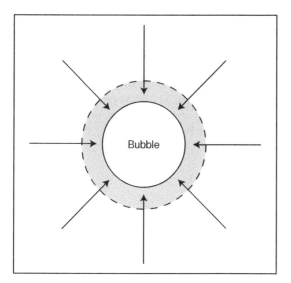

Fig. 5.4 Migration of the molecules of a volatile into a gas bubble from the surrounding liquid increases the concentration gradient in the nearby liquid (shaded area), driving more molecules towards the growing bubble.

When a bubble is very small the addition of even a relatively small amount of gas causes a relatively large increase in the size of the bubble. As the gas molecules in the magma closest to the bubble are the first to be added to the bubble, this sets up a concentration gradient around the bubble in which the number of gas molecules close to it is small compared with the number further away. This concentration gradient drives more molecules into the area of low concentration and thus drives more molecules towards the growing bubble, so further growth can occur (Fig. 5.4). This growth by diffusion is important when the bubble is small but becomes less important as the bubble grows bigger because the percentage increase in bubble volume for the addition of a given number of molecules becomes less as the bubble grows, and also because as the bubble grows there are less gas molecules left in the surrounding magma and so the concentration gradient decreases.

5.5.2 Growth by decompression

Boyle's Law (one of the **Gas Laws**) states that:

$$PV = \text{constant} \qquad (5.5)$$

or

$$P_1 V_1 = P_2 V_2 \qquad (5.6)$$

where P is the pressure in the gas and V is the volume occupied by the gas, and the subscripts 1 and 2 refer to conditions before and after some change takes place. This means that if the gas pressure decreases, the volume of the gas increases, i.e., the gas expands. So as magma rises and the pressure exerted on it by the surrounding rocks decreases, any gas bubbles within the magma also experience this decrease in pressure and expand in volume accordingly.

For example, if a bubble forms at a depth of ~200 m beneath the surface and grows by decompression until it reaches the surface then the initial pressure on the bubble, P_1, is

$$P_1 = \rho g h \qquad (5.7)$$

where ρ is the density of the surrounding rocks, g is the acceleration due to gravity and h is the depth beneath the surface. So for $\rho = 2800$ kg m^{-3}, $g = 9.81$ m s^{-2} and $h = 200$ m, the initial pressure P_1 is 5.5 MPa. The pressure at the surface, P_2, is 1 bar, i.e., 0.1 MPa. Therefore:

$$V_2 = (5.5/0.1)\, V_1 \qquad (5.8)$$

where V_2 is the final volume of the gas bubble and V_1 is the initial volume. The volume of a bubble is proportional to its radius cubed, and so the increase in the radius of the bubble in this example is found from

$$r_2 = 55^{1/3}\, r_1 = 3.8\, r_1 \qquad (5.9)$$

Figure 5.5 shows the relative importance of diffusion and decompression in bubble growth. Bubbles form at a depth of ~220 m beneath the surface and grow by both diffusion and decompression until they reach the surface. The bubble radius increases by a factor of ~1000 between the nucleation depth (220 m) and the surface (Fig. 5.5). Decompression over a distance of ~200 m can increase the bubble radius by only a factor of about four; thus the growth of bubbles in this case is

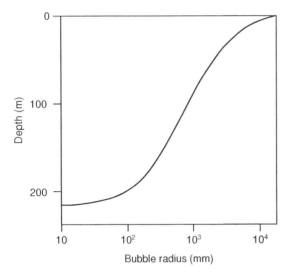

Fig. 5.5 The variation of gas bubble radius with depth beneath the surface in a rising magma. Over this range of depths the growth of the bubble is controlled mainly by the diffusion of volatile molecules from the surrounding liquid into the bubble. (Adapted from fig. 7(b) published in *Journal of Volcanology and Geothermal Research*, Vol 3, Sparks, R.S.J., The dynamics of bubble formation and growth in magmas: a review and analysis, 1–37, copyright Elsevier (1978).)

dominated by diffusion. However, if bubbles form at greater depths, the pressure change experienced by the bubbles is proportionately greater and so the importance of decompression growth is also greater. For example, for a bubble initially formed at a depth of 5 km beneath the surface, decompression alone will increase the bubble volume by a factor of more than 1000 and hence the bubble radius by a factor of 10 by the time the bubble reaches the surface.

5.5.3 Bubble coalescence

The final process by which bubbles can grow is through coalescence. This is only a significant process in certain circumstances, but it is important in transient explosive eruptions such as Strombolian and Vulcanian eruptions (see Chapter 7).

Bubbles within rising magma are always buoyant compared with the magma due to the low density of the gases they contain. This buoyancy means that

bubbles are always trying to rise through the overlying magma even as the magma itself rises. The buoyancy force which causes the bubbles to rise through the magma is counteracted by the drag force of friction exerted on the bubble as it moves through the magma. As long as bubbles do not rise too quickly they maintain a very nearly spherical shape and so the buoyancy force, F_B, can be defined as:

$$F_B = (4/3)\, \pi\, r^3\, (\rho_m - \rho_g)\, g \qquad (5.10)$$

where r is the bubble radius, and ρ_m and ρ_g are the magma and gas densities, respectively. As long as the bubble radius does not become extremely big, the drag force, F_D, is controlled just by the viscosity of the magma, η, and is given by:

$$F_D = 6\, \pi\, \eta\, r\, u \qquad (5.11)$$

where u is the rise speed of the bubble. An equilibrium is reached in which the drag force equals the buoyancy force and, thus, the rise speed, u, of the bubble is:

$$u = [(2/9)\, (\rho_m - \rho_g)\, g\, r^2]/\eta \qquad (5.12)$$

From this equation it is apparent that the rise speed of a bubble through the magma is proportional to the square of its radius, i.e., larger bubbles rise disproportionately faster than smaller ones. This difference in the rise speeds of bubbles of different size is a major factor in bubble coalescence. However, the absolute rise speeds, and hence the relative speeds of bubbles, are much greater in low-viscosity magmas than in melts that are very viscous (Table 2.1). Consider a 100 µm diameter bubble rising through a melt with density $\rho_m = 2500$ kg m^{-3} (we can neglect the gas density because it is much less than this). In a basaltic magma with viscosity 30 Pa s this bubble would rise at 1.8 µm s^{-1} whereas in a rhyolitic magma with viscosity 10^5 Pa s its rise speed would be more than 3000 times smaller at 0.54 nm s^{-1}. Now consider a bubble that has managed to grow to a diameter of 10 mm. This will rise at a speed of 18 mm s^{-1} in the basaltic magma and 5.4 µm s^{-1} in the rhyolite. The ratio of the speeds in the two magmas is the same, but the 10 mm bubble

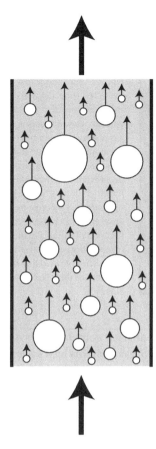

Fig. 5.6 The typical distribution of bubble sizes within magma rising toward the surface. The largest bubbles are those that formed earliest at the greatest depth beneath the surface and have grown by diffusion and decompression; the smallest bubbles are those that have most recently nucleated. The length of the arrows extending from each bubble reflects the relative rise speeds of the bubbles through the magma, largest bubbles having the greatest rise speeds.

rises by nearly twice its own diameter in one second in the basalt but by only about one-twentieth of one percent of its own diameter in the rhyolite.

Figure 5.6 shows schematically what might be seen if a "snapshot" could be taken of the bubbles within magma which is rising towards the surface. At any given time the magma will contain a population of bubbles of various sizes. Once bubbles start to form at the **exsolution level**, new bubbles continue to nucleate until the magma is finally erupted. However, as was seen earlier, the exact

way nucleation proceeds depends on the availability of sites for nucleation and the degree of supersaturation of the magma at any given time. Nevertheless, in general the magma will contain some relatively large bubbles that formed some time ago at deeper levels beneath the surface and which have been growing through diffusion and decompression, and also some very small bubbles which have just formed. A spectrum of bubble sizes will exist between these two extremes (Fig. 5.6).

All of these bubbles are rising relative to the magma because of their buoyancy but the larger bubbles are rising faster than the smaller ones (eqn 5.12). This means that the larger bubbles can overtake the slower, smaller bubbles. When this happens there are two possibilities. The small bubble may be swept around the larger one, effectively trapped in the liquid magma moving sideways and down to let the larger bubble pass, and thus may be left behind as the large bubble continues to move upward. But if the small bubble is close enough to the large one it may be swept into the wake of the large bubble. This wake consists of some liquid magma that is effectively trapped behind the large bubble and is moving up with it through the rest of the liquid. The small bubble is no longer left behind but instead rises slowly through the wake liquid and eventually collides with the bottom of the large bubble and coalesces with it, forming a single, larger bubble. This new, larger bubble then rises even faster, overtakes more bubbles and coalesces with some of them, growing larger and moving even faster and so on. This runaway process can lead to a situation where a single large bubble (called a **slug**) can fill the whole width of the dike or conduit, absorbing all of the smaller bubbles ahead of it as it continues to rise.

In basaltic magmas, a critical factor in determining whether bubble coalescence can occur is the rise speed of the magma. A simple example will illustrate why this is the case. Consider magma rising over a distance of 500 m. If the rise speed of the magma is 1 m s^{-1} then it takes 500 seconds for the magma to rise 500 m. If the magma rise speed is only 0.1 m s^{-1} then it takes 5000 seconds to move the same distance. If a bubble in the magma is rising relative to the magma at a speed of 0.01 m s^{-1} then in 500 seconds it will rise a distance of 5 m through

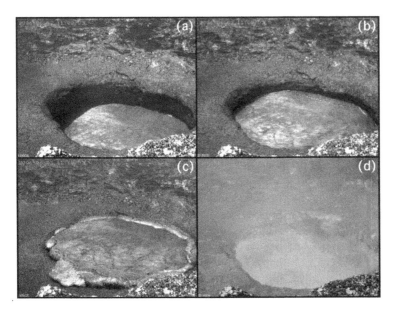

Fig. 5.7 A sequence of four frames extracted from a movie showing the rise of magma in a vent on Kilauea volcano, Hawai'I, during an episode of gas pistoning. In frames (a) to (c) the accumulation of gas beneath the lava crust causes the lava crust to rise progressively higher in the vent. In frame (d) the gas has escaped from beneath the lava crust by tearing the crust apart in a minor explosion and the level of the lava in the vent has fallen. (Photography by Tim Orr, Hawaiian Volcano Observatory, courtesy of the U.S. Geological Survey.)

the overlying magma, whereas in 5000 seconds it will rise 50 m. Thus, during rise of magma over a given distance, the slower rising magma allows the bubbles to travel further relative to their starting position in the magma. The further bubbles are able to rise through the magma, the greater the opportunity for collision with other bubbles and therefore for coalescence. In the extreme, the magma itself may be stationary and the bubbles rise up through it to reach the surface of a lava pond in the vent. Ascent of the bubbles through such magma gives the initially largest bubbles the greatest possible opportunity for overtaking smaller bubbles and reaching the runaway stage in which single, large bubbles form, filling the entire dike or conduit. In basaltic magmas this effect can manifest itself in vigorous **Strombolian** explosions or in more gentle "gas pistoning" (Fig. 5.7). These styles of eruption are discussed in detail in Chapter 7.

Figure 5.8 shows the influence of rise speed on bubble growth for two different magma gas contents (the amount of gas initially dissolved in the magma at depth) in a basaltic magma that does not become supersaturated. The graph shows that at magma rise speeds of greater than 1 m s^{-1} the final size reached by a bubble by the time it is erupted depends linearly on the rise speed and also on the total gas content. Bubbles grow larger in magmas

with higher initial gas contents and with lower rise speeds. Higher gas contents lead to larger bubbles because the bubbles start to form at deeper levels (supersaturation occurs at deeper levels) and thus the bubbles grow more by decompression during ascent. Smaller rise speeds lead to bigger bubbles because the bubbles have more time to grow by diffusion during ascent. For bubbles in basaltic magmas rising at speeds greater than 1 m s^{-1}, the maximum size that the bubbles reach is typically between 1 and 10 mm and essentially no bubble coalescence occurs. At rise speeds less than 1 m s^{-1}, although the total gas content still influences bubble size, coalescence is the dominant factor determining the final bubble size – the slower the rise speed the greater the final size of the bubble. When bubble growth is controlled by coalescence, bubbles in basaltic magmas can grow to sizes greater than 1 m.

Figure 5.8 shows one specific example of how bubble growth is influenced by rise speed in a basaltic magma. In the case illustrated, the critical rise speed determining whether or not coalescence occurs is about 1 m s^{-1}. The value of this critical rise speed varies as a function of magma gas content and magma viscosity. At larger gas contents the critical rise speed is greater. This is because, as long as supersaturation does not become important, in

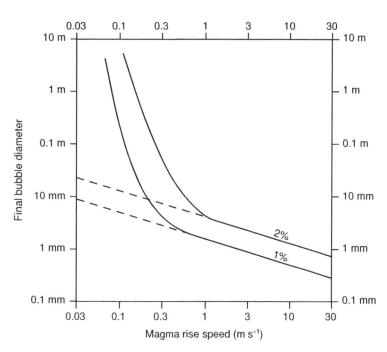

Fig. 5.8 The influence of magma rise speed on bubble growth for initial magma water contents of 1 and 2 wt% in a basaltic magma. At magma rise speeds greater than ~1 m s⁻¹ the bubbles grow by diffusion and decompression in the normal way, but at smaller rise speeds there is time for significant amounts of bubble coalescence and very much larger bubbles can be present in the magma reaching the surface. (Modified from fig. 8 in Wilson and Head (1981) Ascent and eruption of basaltic magma on the Earth and Moon. *J. Geophys. Res.* **86**, 2971–3001.)

more gas-rich magmas there is a greater number density of gas bubbles in the magma at any given depth, and thus bubble coalescence becomes more likely for a given rise speed. Thus the rise speed must be proportionately larger to prevent coalescence. The critical rise speed is smaller when the magma viscosity is greater. This is because the greater magma viscosity increases the drag on the bubbles and reduces their speed, so even if the rise speed of the magma is small the bubbles move so slowly that little coalesce can occur.

5.6 Magma fragmentation and the influence of volatiles on eruption styles

A critical consequence of the growth of gas bubbles in magmas is their ability to cause the magma to **fragment**, that is, to change from being a continuous liquid containing gas bubbles of various sizes to being a continuous body of gas in which pyroclasts – droplets or clots of the liquid – are carried along by the gas. Fragmentation may occur for any one of a number of reasons, and these tend to be linked to magma composition, but not exclusively so.

An early observation about all types of pyroclasts was that almost all of them contained a network of **vesicles** (see Fig. 5.9), these being the holes left behind by bubbles of volcanic gas that were trapped in the pyroclast as it was erupted. Commonly, but not universally, the bubbles are interconnected, so that the volcanic gas has been lost and replaced by air, and equally commonly the vesicularities of pyroclasts, in other words the volume fractions of the clasts that consist of the bubble spaces, are between 70 and 80%. This led to the idea that fragmentation occurs when the gas bubbles in a magma have grown so much that they become very closely packed, so that the liquid walls between the larger bubbles collapse, allowing the large bubbles to join together. Figure 5.9 shows the kind of bubble growth history that could lead to this. Clearly not all of the bubbles will be connected together, and so the pyroclasts that are formed would be expected to contain a lot of trapped bubbles, just as is observed.

However, there tends to be a difference between the size distributions of trapped bubbles in pyroclasts derived from magmas of different compositions. Pyroclasts of more evolved, highly viscous

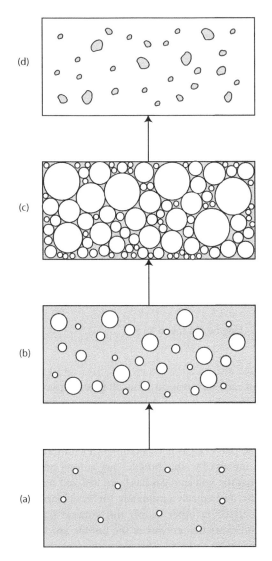

(d)

(c)

(b)

(a)

Fig. 5.9 Four stages in the growth of gas bubbles in a magma leading to magma fragmentation and an explosive eruption. Bubbles start to nucleate in (a); older bubbles have grown by diffusion and decompression and new bubbles have nucleated in (b); the bubble number density is so large in (c) that bubbles are extremely crowded and the walls between adjacent bubbles can start to collapse; in (d) so many bubble walls have collapsed that the magma has changed from a liquid containing bubbles to a gas containing liquid droplets.

magmas, called **pumice** clasts (see Fig. 5.10), generally contain a smaller range of vesicle sizes than more basaltic pyroclasts, usually called **scoria** (Fig. 5.11). This is due to the reduction in coalescence,

Fig. 5.10 Pumice clasts in the deposit from a pyroclastic density current erupted at Mount St Helens volcano, Washington State, USA, in August, 1980. Scale is in inches. (Photograph by James W. Head III.)

the greater likelihood of supersaturation and hence the nucleation of large numbers of small bubbles, and the reduced ability of volatile molecules to diffuse through viscous magmas as compared with basaltic ones. This suggests that in the more evolved magmas the cause of fragmentation is not just the close packing of large bubbles. Instead, it may often be the fact that it is difficult to force viscous liquid to flow through the narrow films of liquid separating the bubbles. On the time scale of the changing stresses to which the magma is subjected as it accelerates upward through a dike toward the surface (the subject of the next chapter) its rheology is no longer Newtonian, and it develops an effective strength. When the stresses exceed that strength, the magma fractures as though it were a brittle solid; a similar situation was described in Chapter 3, where rheology was a function of environmental stresses in connection with the behavior of rocks in the upper mantle.

There is a final very important factor connected with gas bubbles and magma fragmentation. When the fragmentation process occurs throughout the magma at some specific depth in a dike, every batch of magma rising through the system undergoes the

THE ROLE OF VOLATILES 75

Fig. 5.11 Scoria clasts in the Montana Colorada cone, Lanzarote. The cone formed during an eruption in April 1736. (Photograph by Elisabeth Parfitt.)

same process, and a relatively uniform and steady stream of gas and pyroclasts emerges through the vent. This is the commonest way in which viscous magmas are erupted. However, if a great deal of coalescence of bubbles occurs, especially to the point where giant bubbles fill nearly the whole width of the dike, then the magma rising though the dike is far from uniform. As the magma between the giant bubbles reaches the surface, it flows out of the vent as lava, containing whatever gas bubbles have avoided being swept up by the giant bubbles. As the giant bubbles reach the surface they up-dome the surface of the lava above them forming a skin. This skin may be plastic if lava is flowing out of the vent fairly quickly and will stretch and tear easily; alternately, if the lava is moving away only very slowly, the skin may have cooled significantly since the arrival of the previous giant bubble and will fracture in a brittle manner as it bursts. The gas

trapped inside the giant bubble is generally at a higher pressure than the atmosphere and so it expands, throwing clots of lava from the torn skin upward and outward in a Strombolian explosion. It was shown earlier that bubble coalescence occurs more easily in basaltic than more evolved magmas, and Fig. 5.8 demonstrates that, even for basaltic magmas, there is quite a sharp divide, in terms of magma rise speed and volatile content, between eruptions in which bubble coalescence is or is not important.

As a result of the above issues, explosive eruptions are split into two main classes. Where magma rise speed is high enough to prevent significant coalescence, gas bubbles stay coupled to the magma within which they form and this leads to "steady" eruptions – eruptions in which a continuous stream of gas and magma clasts is erupted (see Chapter 6). In basaltic eruptions with low enough rise speeds, bubble coalescence dominates and the bubbles decouple from the magma in which they form. Bubbles rising and coalescing lead to a situation in which the distribution of gas through the magma column is uneven. This can lead to a range of behaviors from relatively minor fluctuations in the intensity of the eruption (like a pulsing in the height of lava fountains in Hawaiian eruptions) through to intermittent, discrete explosions occurring as successive pockets of gas rise through the magma to be erupted. These eruptions are referred to as "transient" and they are discussed in Chapter 7.

5.7 Summary

• Most volcanic eruptions are to some extent **explosive**. In the context of volcanology this means that the magma is torn apart and ejected from the vent as clots or blobs within a stream of gas. Explosive eruptions can be transient events, occurring every few seconds to minutes or hours, or can be continuous, steady eruptions which last hours or days.

• Most explosive eruptions owe their explosive character to the presence of gases dissolved within the rising magma. The solubility of volcanic gases in magma decreases with decreasing pressure, so as magma rises towards the surface a point will be reached where the magma becomes **supersaturated** and gas starts to **exsolve** forming bubbles within the magma. The most common gases within magmas are H_2O, CO_2, SO_2 and H_2S.

- The solubility of gases in magmas is dependent on both the type of gas and the composition of the magma. The solubility of H_2O in rhyolite is, for instance, considerably greater than that in basalt, whereas the solubility of CO_2 is considerably less than the solubility of H_2O in either a basalt or a rhyolite. The solubility of volatiles is more complex when more than one species is present because they interact with one another.
- When bubbles first form in a magma they are typically extremely small. Bubbles can grow from their initial size as the result of three processes: diffusion, decompression and coalescence. The relative importance of each process depends on the amount of gas present in the magma, the magma composition and the magma rise speed.
- Bubble coalescence due to large bubbles overtaking smaller ones is important in low-viscosity magmas at low magma rise speeds. The rise speed, and its control on whether significant bubble coalescence does or does not occur, determines the fundamental character of the resulting eruption. If the magma rise speed is low and thus bubble coalescence is important, the gas segregates from the magma and rises through it, causing the kinds of discrete transient explosions described in Chapter 7. If the rise speed is higher then significant coalescence will not occur. In this case the magma and gas stay locked together and this leads to continuous, steady explosive eruptions in a wide range of magma compositions as discussed in Chapter 6.

5.8 Further reading

Cashman, K.V., Sturtevant, B., Papale, P. & Navon, O. (2000) Magma fragmentation. In *Encyclopedia of Volcanoes* (Ed. H. Sigurdsson), pp. 421–30. Academic Press, San Diego, CA.

Dixon, J.E. (1997) Degassing of alkali basalts. *Amer. Mineral.* **82**, 368–78.

Freundt, A. & Rosi, M. (Eds) (2001) *From Magma to Tephra*. Elsevier, 334 pp. ISBN: 0444507086.

Gerlach, T.M. (1986) Exsolution of H_2O, CO_2, and S during eruptive episodes at Kilauea Volcano, Hawai'I. *J. Geophys. Res.* **91**, 12,177–85.

Kaminski, E. & Jaupart, C. (1997) Expansion and quenching of vesicular pumice fragments in Plinian eruptions. *J. Geophys. Res.* **102**, 12,187–203.

Klug, C. & Cashman, K.V. (1996) Permeability development in vesiculating magmas: implications for fragmentation. *Bull. Volcanol.* **58**, 87–100.

Mader, H.H., Zhang, Y., Phillips, J.C., Sparks, R.S.J., Sturtevant, B. & Stolper, E. (1994) Experimental simulations of explosive degassing of magma. *Nature* **372**, 85–8.

Mangan, M.T. & Cashman, K.V. (1996) The structure of basaltic scoria and reticulite and inferences for vesiculation, foam formation, and fragmentation in lava fountains. *J. Volcanol. Geotherm. Res.* **73**, 1–18.

Sparks, R.S.J. (1978) The dynamics of bubble formation and growth in magmas: a review and analysis. *J. Volcanol. Geotherm. Res.* **3**, 1–37.

Sugioka, I. & Bursik, M. (1995) Explosive fragmentation of erupting magma. *Nature* **373**, 689–92.

5.9 Questions to think about

1 Why are magmas stored in magma reservoirs at depths of a few kilometers below the surface more likely to contain bubbles of carbon dioxide than bubbles of water vapor?

2 What is the main factor controlling whether volatiles exsolve in a way close to being in equilibrium with the decreasing pressure in a rising magma or out of equilibrium with the pressure?

3 What are the three processes by which the average size of the gas bubbles in a rising magma increases?

4 What property of the magma allows bubbles to coalesce more easily in basaltic magmas than more evolved magmas?

5 What is the main control on whether an explosive eruption involves the discharge of a steady or unsteady stream of gas and pyroclasts through the vent?

6 Steady explosive eruptions

6.1 Introduction

The previous chapter discussed how, as magma rises towards the surface, the volatiles dissolved within it begin to exsolve forming gas bubbles. Continued rise of the magma leads to further exsolution of gas and growth of gas bubbles through diffusion, decompression and bubble coalescence. This chapter considers what happens in eruptions in which the rise speed of the magma is sufficiently great to ensure that bubble coalescence is minor during rise of the magma (see sections 5.4 and 5.5), so that a more or less steady spray of gas and entrained magma clots emerges from a surface vent at a speed, an **exit velocity**, that can vary from several tens to several hundreds of meters per second. How gas in the magma behaves during the early stages of magma ascent is considered first, followed by consideration of how the gas affects the way the gas–magma mixture rises after fragmentation. Finally we look at what happens to the gas–magma mixture when it is erupted. The physical processes discussed in this chapter are fundamental to a wide range of styles of volcanic eruptions – from Hawaiian lava fountains (Fig. 1.1), through subPlinian and Plinian eruptions (Fig. 1.2), to ignimbrite-forming eruptions (Fig. 1.11).

6.2 Influence of gas bubbles prior to magma fragmentation

As soon as gas bubbles begin to nucleate in a rising magma they start to influence the magma rise speed. A simple way to illustrate this is by thinking

about the **mass flux** through the dike system. The mass flux is simply the mass of material passing any given point in the system in a given amount of time. The mass of material entering the dike system must be the same as the mass erupted at the vent (unless magma is intruded or stored somewhere on the way). This means that, if the mass flux could be measured at a whole series of points in the system, it would be found that its value was the same at every point. Figure 6.1 shows a schematic view of a

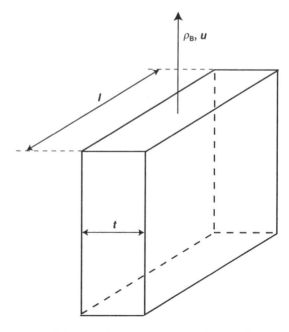

Fig. 6.1 Schematic view of a dike of length l along strike and width t in which magma of density ρ_B rises at an average speed u.

dike system. The dike has length l and width t, and the gas–magma mixture flowing through it has a bulk density ρ_B and rises at a speed u. The mass flux, M, through the dike is then given by

$$M_f = t \, l \, \rho_B \, u \tag{6.1}$$

The bulk density, ρ_B, of the gas–magma mixture is given by

$$\frac{1}{\rho_B} = \frac{n_f}{\rho_g} + \frac{(1-n_f)}{\rho_m} \tag{6.2}$$

where n_f is the exsolved weight fraction of gas, ρ_g is the gas density and ρ_m is the density of the magmatic liquid.

The gas density, ρ_g, varies with temperature and pressure and usually can be approximated with reasonable accuracy by the perfect gas law:

$$\rho_g = \frac{P \, m}{Q \, T} \tag{6.3}$$

where P is the pressure, m is the molecular weight of the gas, Q is the universal gas constant ($8310\,\mathrm{J\,kmol^{-1}}$) and T is the **absolute temperature** of the gas, its temperature relative to the absolute zero of temperature at about $-273.15°C$.

Before any bubbles form within the magma, the bulk density, ρ_B, is equal to the magma density, ρ_m. As bubbles start to form, the bulk density of the gas–magma mixture decreases because the density of the gas is much less than the density of the magma. Equation 6.1 shows that, if the mass flux, M_f, stays constant throughout the dike system, then as the bulk density of the gas–magma mixture decreases, the rise velocity, u, of the mixture must increase to compensate. Table 6.1 shows a simple example to illustrate this effect.

As long as the decrease in the pressure acting on the magma is not too great, the main effect of the gas expansion is simply the one illustrated in Table 6.1. However, as the magma approaches the surface, and the pressure decrease becomes very large, there are other factors that must be taken into account. These are related to the fact that if the

Table 6.1 Basaltic magma with a total gas content of 2 wt% water rises through a dike of width 2 m and horizontal length 2 km. The rise speed before any gas exsolves is $1\ \mathrm{m\ s^{-1}}$ and thus the mass flux through the dike system is $1.12 \times 10^7\ \mathrm{kg\ s^{-1}}$ (eqn 6.1). As the magma rises, gas exsolves from the magma according to the solubility law (eqn 5.1) and the bulk density of the gas–magma mixture decreases (eqn 6.2). The rise speed, u, therefore increases (eqn 6.1).

Depth (km)	P (Mpa)	ρ_B (kg m^{-3})	u (m s^{-1})
4	110.0	2800	1.0
2	55.0	2613	1.1
1	27.5	1758	1.6
0.5	13.7	150	18.7

speed of the magma is to increase, in other words if it is to accelerate, something must provide the increase in **kinetic energy** (energy of movement) that this represents. Also, the magma is being lifted in the Earth's gravitational field, so some **potential energy** must be supplied as well, and something must compensate for the energy that the magma is losing as a result of **friction** with the walls of the dike.

6.3 Acceleration of the gas–magma mixture

The energy for the acceleration is provided by the expansion of the gas which occurs as the gas–magma mixture rises and the pressure on it decreases. The effect can be illustrated by thinking about what happens when a tire on a bicycle is pumped up. The process forces air into the tire, compressing the air and raising the pressure inside the tire. To do that takes energy, and with a hand-pump the cyclist supplies that energy – it ultimately comes from the food the person eats! Anyone who has pumped up a tire in this way will recall that the body of the pump becomes slightly warm – some of the energy supplied goes into heating the gas. If, after pumping-up the tire, the end of the pump is released, the plunger is pushed out again by the gas in the tire. This happens because the pressure inside the tire is higher than outside. If there is nothing stopping the air inside from expanding, then the air will expand to try to equalize the pressure

STEADY EXPLOSIVE ERUPTIONS 79

difference and in doing so will provide energy to push the plunger out. Also, as the compressed gas expands, it will get a little cooler. This is not very noticeable in the case of the bicycle pump if only a little of the pressure is released, but if all of the air is let out of the tire it will be more obvious. Going back to our rising gas–magma mixture, as the mixture rises and the pressure decreases, the gas in the bubbles expands, cools a little, and releases energy. It is this energy that goes into raising and accelerating the gas–magma mixture through the dike, and the distribution of energy in the system is given by the **energy equation**:

$$0 = \frac{(1 - n_f)dP}{\rho_m} + g\,dh + u\,du + c_p\,dT \qquad (6.4)$$

Liquid internal   Potential Kinetic Gas internal

The zero on the left-hand side of this equation just means that the total energy of the system does not change. The terms on the right-hand side are labeled by the components of the energy that they represent. The rise speed of the mixture of gas and magmatic liquid is u and du is the change in the speed, so that $u\,du$ is the increase in the **kinetic energy** of the mixture. dh is the distance that the mixture is lifted against the gravitational field of the planet represented by the acceleration due to gravity, g (about 9.8 m s^{-2} on Earth), and so $g\,dh$ is the **potential energy** needed to do this. The two internal energy terms represent energy locked into the physical state of the materials; c_p is the specific heat at constant pressure of the mixture (the amount of heat each kilogram releases in cooling by one kelvin), and so $c_p\,dT$ represents the change in thermal energy of the system. Note that both $u\,du$ and $g\,dh$ increase as the magma rises toward the surface. However, the pressure P decreases, so dP is negative; it is not, therefore, trivial to predict what happens to the temperature T. In fact, except under very special circumstances, T decreases – the magma cools by some amount – as it ascends.

In parallel with the conservation of energy it is necessary to consider how the forces acting on a given batch of rising magma influence its motion, and this leads to the so-called **momentum equation**:

$$u\,du = -\frac{dP}{\rho_B} - g\,dh - \frac{f u^2\,dh}{t} \qquad (6.5)$$

Momentum Pressure Gravity Friction

Here account is taken explicitly of the friction between the magma and the wall of the dike. Both $g\,dh$ and $(f u^2\,dh)/t$ are positive quantities, and so the minus signs mean that they represent a **loss** of kinetic energy of the magma; however, the pressure is decreasing as the magma rises, so dP is negative, so $(-dP)$ is a **positive** number, and this makes it clear that the decrease in pressure is the key factor causing the eruption to happen.

The dependence of the friction factor, f, in eqn 6.5 on viscosity is important. Magmas are much more viscous than, say, water, which has a viscosity of about 10^{-3} Pa s. Basaltic magmas have typical viscosities of 10 to 100 Pa s, whereas magmas such as dacites and rhyolites can have viscosities as large as 10^5 to 10^{10} Pa s. Below the fragmentation level, the rising mixture consists of magma with some gas bubbles suspended within it, so the fluid in contact with the walls is magma (Fig. 5.9a–c). The magma has a significant viscosity, and so the friction factor, f, is relatively large and the energy needed to overcome wall friction is also large. After fragmentation, the rising mixture consists of a stream of gas with pyroclasts – clots of magma – suspended in it (Fig. 5.9d), and so the fluid in contact with the dike walls is now mostly gas, which typically has a viscosity of ~10^{-5} Pa s, which is ten million times smaller than the viscosity of a typical basaltic magma and ten billion times smaller than the viscosity of a typical rhyolite. Of course, from time to time clots of liquid magma will collide with the wall, and so the effective bulk viscosity of the gas–clot stream is greater than that of the gas alone. However, this bulk viscosity is so small compared with the viscosity of the liquid magma that after fragmentation the friction factor, f, becomes negligibly small. This means that the partitioning of energy in the energy equation (eqn 6.4) changes significantly. Prior to fragmentation, a lot of energy is used in doing work to overcome friction between the magma and dike walls, especially in narrow dikes. After fragmentation, the friction term becomes very much smaller, and so the energy which was used to overcome wall

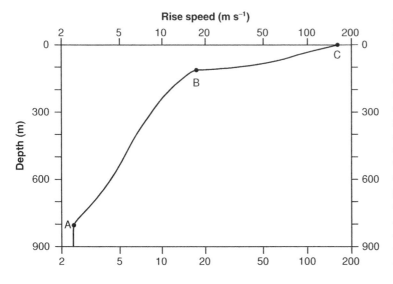

Fig. 6.2 The variation with depth beneath the surface of the rise speed of magma ascending in a dike. The magma rises at a constant speed from its source until gas bubbles first start to form at point A. Gas bubble expansion releases energy and the magma accelerates until bubbles become close-packed and fragmentation occurs at point B. Above this point wall friction becomes negligible and acceleration increases until the gas-pyroclast mixture erupts at the surface at point C. (Adapted from fig. 3 in Wilson, L. & Head, J.W. (1981) Ascent and eruption of basaltic magma on the Earth and Moon. *J. Geophys. Res.*, **86**, 2971-3001.)

friction becomes available for a different use. There is no change in the amount of potential energy required to raise the magma through a given distance toward the surface, and so the extra energy goes into the kinetic energy term and causes a large increase in the acceleration of the gas–pyroclast mixture.

The acceleration of the mixture during its ascent towards the surface can be simulated using computer programs. Figure 6.2 shows one such simulation. The ascending magma rises at a fixed speed of ~2.5 m s^{-1} until gas bubbles first start to form within it (point A). Once gas bubbles form and expand, energy is released and the magma mixture accelerates (between points A and B). At point B fragmentation occurs. Above this point wall friction becomes negligible and acceleration of the mixture becomes very much more pronounced. In this example, the gas–pyroclast mixture accelerates from less than 20 m s^{-1} at the fragmentation level (at ~100 m depth) to a velocity of ~170 m s^{-1} upon eruption at the surface (Fig. 6.2).

6.4 Controls on exit velocity

The speed with which gas and entrained small pyroclasts leave the vent in an explosive eruption is important for two reasons: it determines the rate at which the eruption products mix with the sur-

rounding atmosphere into which they emerge, and it determines the way large pyroclasts decouple from the stream of gas and small particles to fall to the surface. It is the measurement of the distribution of these large clasts around the vent that allows us to analyze the conditions during prehistoric eruptions.

6.4.1 Magmatic gas content and exit velocity

The exit velocity of the gas-pyroclast mixture in any particular eruption is sensitive to the gas content of the magma, with larger gas contents leading to higher exit velocities. There are three main reasons for this.

• The larger the gas content, the greater the total energy available for release during gas expansion and therefore the greater the energy available to accelerate the rising gas–pyroclast mixture.
• The larger the gas content of the magma, the greater the depth at which bubbles will first nucleate. The deeper the nucleation depth, the greater the total pressure decrease experienced by the gas as it ascends and, therefore, the greater the energy release through gas expansion.
• The larger the gas content, the deeper the fragmentation level. Deeper fragmentation means that the change from high to low wall friction also occurs deeper, the overall friction losses during

ascent are reduced, and so more of the energy released can be used in acceleration.

6.4.2 Dike shape, vent geometry and exit velocity

There is another factor which exerts a control on the speed at which fragmented magma erupts through vents, and also influences what happens just above the vent, and that is the pressure in the magma in the vent. Magma rising from a high-pressure source region beneath the surface must eventually reach a pressure equal to that of the atmosphere after it leaves the vent. In the simplest case, the pressure in the magma decreases smoothly as it rises and exactly equals atmospheric pressure at the vent. But this does not always happen. The reason is that every fluid (i.e., gas or liquid) has associated with it a natural speed at which pressure changes propagate through it. Pressure changes in air are what we experience as sound, and so this natural speed is called the speed of sound in the fluid. The speed of sound in magma is quite small, much less than in a pure gas or pure liquid. The exact value depends on the pressure and the gas/liquid volume ratio, but it is generally within a factor of two of 100 m s^{-1}. This is true both before and after fragmentation; in each case it is the interaction between the two components, one liquid and one gas, that causes the low speed. Before fragmentation the gas bubbles have to be deformed as the pressure in them changes when the sound wave goes by. After fragmentation magma clots are suspended in the gas, and these have to be pushed and pulled by the gas as the sound wave travels. It was shown above that it is quite common for the expansion of the gas exsolved by magmas to provide enough energy to accelerate the magma to speeds well in excess of 100 m s^{-1}. Thus, these magmas should erupt at speeds which, as far as the magma itself is concerned, are supersonic. However, it is well-established in fluid mechanics that a fluid flowing through a pipe can only become internally supersonic if, in the region where the transition occurs, the pipe first constricts slightly and then flares outward by a sufficiently great amount. The need to have this happen is what gives rise to the characteristic shape of the back end of a jet engine, and the flared shape is called a **de Lavalle nozzle**.

There is no guarantee that a dike will have the correct shape to allow the supersonic transition to occur. The dike width, t, is controlled by the distribution of stresses that caused the dike to propagate; in the early stages of the opening of a vent, the chances are that the dike will get narrower toward the surface, not wider. If the supersonic transition cannot occur, the best that the magma can do is to reach the surface traveling exactly at the local speed of sound. This condition is described as **choked flow**, and when the flow is choked the pressure in the magma in the vent can be much greater than the atmospheric pressure, and the eruption speed will then be much less than if the transition had occurred. There will be a violent expansion, both upward and sideways, of the magmatic gas just above the vent, with the pressure adjusting to atmospheric and the gas and pyroclasts accelerating to speeds similar to, but somewhat less than, those that they would have had if the whole acceleration had been a smooth process.

It is possible for quite complicated situations to develop if rocks break off from the wall of the dike as the eruption progresses. If the magma encounters a series of increases and decreases in width along the dike, a corresponding series of subsonic-to-supersonic-back-to-subsonic transitions can occur, and it is in these situations that, at least for a short vertical distance, the temperature of the magma can increase a little in contrast to its general decrease.

However, in most eruptions, significant erosion of the dike walls occurs mainly near the surface where the wall rocks tend to be weakest. Also, pyroclasts are deposited around the vent, building up a deposit which effectively increases the length of the dike to a small extent. In both cases, the stresses acting will preferentially cause the walls of the dike to adjust toward the de Lavalle nozzle shape which allows the supersonic transition to occur. Thus, what begins as a **choked** eruption can quite quickly evolve into a **pressure-balanced** eruption. Figure 6.3a shows a number of different dike shapes, and the effects of these differing shapes on the exit velocity of the gas–pyroclast mixture are illustrated in Fig. 6.3b. Note that model

(a)

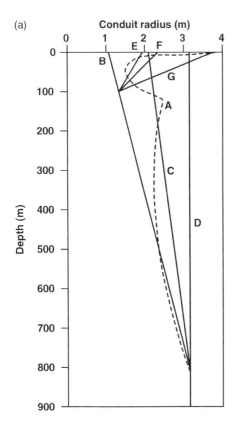

(b)

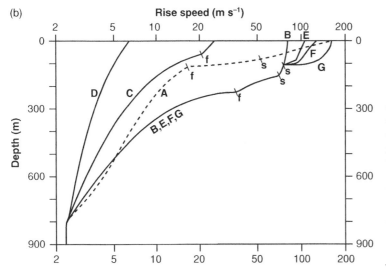

Fig. 6.3 (a) A number of possible shapes of dikes in the vicinity of the surface. (b) The effect of the dike shape on the exit velocity of the gas–pyroclast mixture. Model A in these figures is the same as the model shown in Fig. 6.2 and corresponds to the mixture leaving the vent at atmospheric pressure. (Adapted from figs 3 and 5 in Wilson, L. & Head, J.W. (1981) Ascent and eruption of basaltic magma on the Earth and Moon. *J. Geophys. Res.*, **86**, 2971–3001.)

A in Fig. 6.3 corresponds to the situation in which the mixture exits the vent at atmospheric pressure and is the same situation as that shown in Fig. 6.2.

6.5 Eruption plumes in steady eruptions

This section examines what happens to the erupting stream of gas and magma clasts as it exits from the vent. We are concerned here with eruptions which go on for hours or days. These include Hawaiian eruptions (Fig. 1.1), subPlinian and Plinian eruptions (Fig. 1.2) and ignimbrite-forming eruptions (Fig. 1.11). Although the products of these eruptions differ greatly from each other (Chapters 8 and 10), they are all dynamically similar. Here the basic physics which such eruptions have in common is considered; the differences between these eruption styles are discussed in Chapter 8. Initially we will look mainly at what happens to the stream or jet of gas once it exits the vent and largely ignore the magma clasts within it. The clasts are not critical to the overall dynamics and are discussed later.

6.5.1 Plume rise

The first thing that happens when the stream of gas and clasts exits the vent is that the jet starts to incorporate air from the surrounding atmosphere in a process known as **entrainment**. As the gas jet streams upward through the air it causes turbulent mixing between the air and the edge of the jet and so air is mixed in and added to the eruption jet or column. The giant, turbulent convection cells in which this entrainment occurs are what give the edges of rising eruption columns their characteristic "cauliflower cloud" appearance (Fig. 6.4). The further the column rises the more air becomes

mixed into it. The width of the column would have to increase as a result of the addition of air, but the expansion is enhanced by the fact that initially the volcanic gas and clasts in the column are much hotter than the entrained air and so they heat it up (cooling themselves in the process) and thus cause it to expand as it mixes (Fig. 6.5). The amount of entrainment which occurs depends primarily on the exit velocity of the gas–magma mixture and the radius of the vent (see sections 6.3 and 6.4).

One effect of entrainment is that as the material in the eruption column rises, the velocity at which it rises progressively decreases. This can be understood by considering the principle of **conservation of momentum** which states that

$$m_1 v_1 = m_2 v_2 \tag{6.6}$$

where m_1 is the initial mass in the eruption jet and v_1 is the initial upward velocity (i.e., the exit velocity) and m_2 and v_2 are the mass and upward velocity at some height above the vent. As the eruption column ascends and entrains air, the mass of the column increases and, therefore, conservation of momentum requires the rise velocity to decrease. Mention of momentum should make you recall eqns 6.4 and 6.5 for the motion of magma below the ground; all of the same considerations apply to motion above ground, with the added complication that instead of a fixed edge to the magma – the dike

Fig. 6.4 The turbulent convection cells at the edge of the eruption plume from the May 18, 1980, eruption of Mount St Helens forming the characteristic "cauliflower cloud" patterns. (Photograph by Donald A. Swanson, courtesy of Cascades Volcano Observatory, U.S. Geological Survey.)

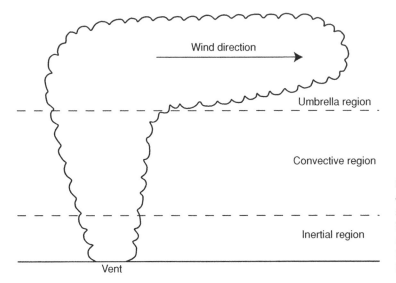

Fig. 6.5 The shape of an eruption column seen from a direction at right angles to the direction that the wind is blowing. The column expands as it rises and is eventually blown sideways by the wind when it reaches its neutral buoyancy height.

wall – there is now a flexible boundary where the volcanic material is mixing with the surrounding air.

A second major effect of entrainment is that the heat energy contained in the erupted gas–magma mixture is shared with the entrained air. Typically the erupting jet will have a temperature of ~900–1150°C. The temperature of the entrained air depends on the geographical location of the vent but typically will be ~0°C. The volume of air which is eventually entrained by the eruption column can vary widely depending on the mass flux and gas content of the erupting magma but, in general, will be between 100 and 10^5 times the volume of gas released in the eruption. Thus when the heat from the eruption jet and clasts is shared with the entrained air the temperature of the column as a whole ends up being only slightly greater than that of the surrounding air. This temperature contrast is, however, great enough to make the material in the eruption column less dense than the surrounding air, and this thermal buoyancy causes continued plume rise. This effect is the same as that which allows a hot air balloon to rise.

In thinking about this rising column or plume of material therefore we can define two main causes for its rise and define regions in which each is dominant. Immediately above the vent the plume rises because of the initial momentum that the

gas–magma mixture has as it exits from the vent. This lowest part of the plume is usually called either the **gas-thrust region** or the **inertial region** (Fig. 6.5) and generally dominates the rise of the plume for the first few kilometers above the vent. As the plume entrains air and the upward velocity of the plume decreases due to momentum sharing, the effects of the thermal buoyancy of the plume (usually) take over. The zone in which the plume rises as a result of buoyancy is referred to as the **convective region** (Fig. 6.5).

As the plume continues to rise in the convective region it entrains more air and the temperature of the plume decreases. The temperature decreases due to two effects: the sharing of the heat with progressively more air and the expansion of the plume as the atmospheric pressure decreases. As the plume rises and the atmospheric pressure acting on it decreases, the plume expands. Expansion of gases causes a decrease in the internal energy, and therefore the temperature, of the gases, and also of the pyroclasts which are still being carried in suspension in the plume. These pyroclasts are continually being lost from the plume as it rises, because the ability of the gas to support the clasts decreases as its speed and density decrease. Actually, the cooling of the gas increases its density somewhat, but this is counteracted by the decreasing atmospheric pressure. Thus as the plume rises, cools, and deposits

Fig. 6.6 The eruption plume from the March 22, 1915, eruption of Lassen Peak volcano, showing the classic umbrella shape of the upper part of the plume. (Photograph credit: R.I. Meyers, courtesy of the National Oceanic and Atmospheric Administration, National Geophysical Data Center.)

clasts, a point will be reached where the bulk density of the plume equals that of the surrounding air and the plume can rise no further through thermal buoyancy alone. The height at which this occurs is called the **level of neutral buoyancy**. However, the plume still has some inertia as it rises slowly through the neutral buoyancy level and so it overshoots this level somewhat, spreading sideways and downwind as it does so. This part of the plume is known as the **umbrella region** (Fig. 6.5) and an example can be seen in Fig. 6.6.

6.5.2 Controls on plume height

The heights of plumes in historic eruptions vary widely (Table 6.2). What controls the height to which a given plume rises? The biggest factor is the thermal buoyancy of the plume, and this is controlled by the thermal energy available, so the highest plumes are those with the most heat. The plumes with the greatest heat are those with the largest **mass flux**. Heat is provided to a rising plume by the clasts (and the volcanic gas) contained within it. The mass flux is a measure of the mass of magma erupted per unit time, so the higher the mass flux the greater the amount of heat supplied to the plume in a given time. Wilson et al. (1978) and Settle (1978) both showed that volcanic plumes follow very nearly the same relationship between mass flux and plume height that was pre-

Table 6.2 Plume heights generated during selected 20th century volcanic eruptions.

Eruption	Plume height (km)
Santa Maria (Guatemala), 1902	27–48
Soufriere (St Vincent), 1902	14.5–16
Hekla (Iceland), 1947	24
Bezymianny (Kamchatka), 1956	34–45
Hekla (Iceland), 1970	14
Fuego (Guatemala), 1971	10
Heimaey (Iceland), 1973	2–3
Ngauruhoe (New Zealand), 1974	1.5–3.7
Soufriere (St Vincent), 1979	18
Mount St Helens (USA), 1980	25
El Chichón (Mexico), 1982	20–24

Data taken from Wilson et al. (1978); Carey, S.N. and Sigurdsson, H. (1982) Influence of particle aggregation on deposition of distal tephra from the May 18, 1980, eruption of Mount St Helens Volcano. *J. Geophys. Res.*, **87**, 7061–7072; and Sparks, R.S.J., Moore, J.G. and Rice, C.J. (1986) The initial giant umbrella cloud of the May 18th, 1980, explosive eruption of Mount St Helens. *J. Volcanol. Geotherm. Res.*, **28**, 257–274.

dicted for pure gas plumes by Morton et al. (1956): the height of the plume is proportional to the fourth root of the eruption rate. For a "standard atmosphere" on Earth the relationship is

$$H = 0.236\, M_f^{1/4} \qquad (6.7)$$

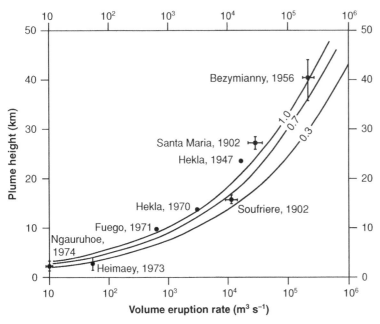

Fig. 6.7 The relationship between the heights of some observed eruption plumes and the estimated mass eruption rates in the eruptions producing them. The curves are the results of theoretical calculations and are labeled by the fraction of the heat contained in pyroclasts that is available to drive the plume upward. Not all of the pyroclast heat is available because some is removed by large clasts falling out of the plume. (Adapted from fig. 2 in Wilson, L., Sparks, R.S.J., Huang, T.-C. & Watkins, N.D. (1978) The control of volcanic column heights by eruption energetics and dynamics. *J. Geophys. Res.* **83**, 1829-1836.)

where H is the plume height in kilometers and M_f is the mass flux in kg s^{-1}.

A "standard atmosphere" is one that is typical of the whole of the Earth. In practice the thermal structure of the atmosphere varies as a function of latitude, longitude, elevation above sea level and season of the year, as well as with the local, short-term weather conditions, especially the humidity, i.e., the water vapor content. One systematic consequence of this is that, for the same eruption conditions, one expects to generate higher plumes at lower latitudes.

Figure 6.7 shows a plot of some observed eruption plume heights as a function of the corresponding estimated volume eruption rates, with three theoretical curves superimposed. One of these assumes that all of the available heat from the magma is used to drive the plume, as in eqn 6.7, and the others assume that either 70% or 30% of the heat is lost due to the early fallout of large clasts. There are inevitable uncertainties in the observations of plume heights and mass fluxes: both change with time during an eruption, and the eruption rate estimate involves measuring the amount of pyroclasts deposited around the vent over a finite time interval, and so it is only ever an average during part of the eruption. Given all these problems,

the match between theory and observation is surprisingly good.

6.6 Fallout of clasts from eruption plumes

Thus far we have largely ignored the fact that the rising eruption plume carries with it magma clasts or clots. This section looks at the fate of the clasts in an eruption plume.

6.6.1 Rise of clasts in an eruption plume

If you drop an object, it falls through the air under gravity and accelerates until the gravity force acting downward is balanced by the resisting frictional drag force of the surrounding air. The clasts in the eruption plume are no different from any other object: they are also trying to fall under gravity. The difference in the case of magma clasts is that they are also being dragged upwards within a rising gas stream. This is true of the clasts both when they are still within the dike system above the fragmentation level and when they are being carried upwards in the rising eruption plume above the vent. Thus a clast is subject to two competing forces – the force of gravity trying to making the clast fall and the drag

Table 6.3 The terminal velocity, U_T, is given for three clasts of different sizes. These values were calculated from eqn 6.9 assuming clast and gas densities of 1000 and 0.5 kg m^{-3} respectively and a drag coefficient of 1.0. If the gas stream in which the clasts are moving has an upward velocity of 160 m s^{-1} then the net upward velocity (relative to the ground surface) of the clasts is $(160 - U_T)$.

Radius (m)	U_T (m s^{-1})	Net upward velocity (m s^{-1})
0.02	30	130
0.15	90	70
0.48	160	0

force exerted on the clast by the gas stream through which it is moving. In practice a balance is reached between these two forces, and for a spherical clast this can be written:

$$\frac{1}{6} \pi d^3 \sigma g = \frac{1}{8} C_D \rho_g \pi d^2 U_T^2 \qquad (6.8)$$

Gravity   Drag

where d is the diameter of the clast, σ and ρ_g are the densities of the clast and gas respectively, g is the acceleration due to gravity, C_D is the drag coefficient and U_T is the terminal velocity of the clast relative to the gas. Thus the terminal velocity, U_T, of the clast in the gas stream is

$$U_T = \sqrt{\frac{4 \, d \, \sigma \, g}{3 \, C_D \, \rho_g}} \qquad (6.9)$$

For a given clast density, eqn 6.9 shows that the bigger the clast, the greater the speed at which it falls through the gas stream. The speed of the clast relative to the ground is, of course, its downward speed through the gas plus the upward speed of the gas relative to the ground. This just means that it is harder for the gas stream to drag a large clast up with it than it is to drag a smaller clast. This effect is shown by the example in Table 6.3, which demonstrates that the smallest clasts are carried upward at the greatest velocity relative to the ground surface. The largest clast in Table 6.3 has a terminal velocity which exactly equals the velocity of the gas stream and is thus suspended in the eruption plume at a

fixed height above the surface while smaller particles continue to stream upwards past it.

6.6.2 Fallout of clasts from eruption plumes

Section 6.5 showed that the upward velocity of the plume decreases with height. Figure 6.8 shows two examples of the variation in rise speed with height. In each case the velocity is highest as the gas stream exits the vent and then declines rapidly in the gas-thrust region. The velocity remains fairly constant through much of the convective region before declining rapidly near the top of the plume (Fig. 6.8). This variation in upward velocity of the gas in the plume has important implications for the fallout of clasts from it. Consider again the largest clast in Table 6.3. This clast has been carried upward through a region in which the gas speed exceeded 160 m s^{-1}. As the plume rises and its

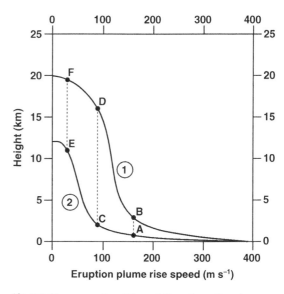

Fig. 6.8 Two examples of the variation of eruption plume rise speed with height. The velocity is always highest as the gas stream leaves the vent and then decreases rapidly through the gas-thrust region. The speed decreases slowly through much of the convective region before declining rapidly towards the top of the plume. (Adapted from fig. 5(b) in Wilson, L. & Walker, G.P.L. (1987) Explosive volcanic eruptions – VI. Ejecta dispersal in plinian eruptions: the control of eruption conditions and atmospheric properties. *Geophys. J. Roy. Astron. Soc.* **89**, 657–679, copyright Wiley-Blackwell Publishing Ltd.)

velocity decreased, the clast reached a height within the plume where the gas stream velocity has declined and become equal to the terminal velocity of the clast. The clast is then suspended at this height above the ground and cannot be carried any further upward (points A and B in Fig. 6.8). The two smaller clasts in Table 6.3 would continue to be carried higher in the plume because their terminal velocities are lower. As the plume rise speed continues to decrease, a point will be reached first where the terminal velocity of the intermediate size clast equals the rise speed of the plume and the clast becomes suspended at this height (points C and D in Fig. 6.8). Finally a height is reached where even the smallest clast can no longer rise (points E and F). As the two different cases in Fig. 6.8 illustrate, the height gained by a clast of a given size depends on the eruption conditions: the eruption with a larger mass flux (eruption 1) produces a higher plume and carries clasts of a given size higher above the vent than the smaller mass flux plume (eruption 2).

So what is the fate of these suspended clasts? The eruption plume is a highly turbulent place, so a clast that has reached its maximum height will not be passively suspended at this height but instead will be constantly moved around by eddies within the plume. Eventually, the clast is likely to find itself at the edge of the eruption plume. If it is nudged to the edge of the plume by the turbulence, it will leave the plume because it no longer has the support of the rising gas. The clast will then fall toward the ground. So for any particular plume speed, and therefore any particular height above the ground, there is a maximum size of clasts of any given density which can be supported at that height. The size of clast that can be supported decreases with height (because plume rise speed decreases – Fig. 6.8) so that large clasts will fall out from the plume at smaller heights above the vent than will small clasts. In practice, many particles are likely to fall out from the plume before they reach their maximum achievable heights because the turbulence of the plume causes them to reach the edge of the plume and be pushed out into the surrounding, still atmosphere prematurely. The deposition of tephra from eruption plumes is discussed in more detail in Chapter 8.

6.7 Unstable eruption columns

In a stable eruption plume, air entrained into the column is heated enough to be thermally buoyant despite the load of entrained pyroclasts that it is carrying (see section 6.5). This section looks at what happens to an eruption plume if it cannot achieve thermal buoyancy.

6.7.1 Plume density and column stability

We start by considering how the density of the eruption plume varies with height. Table 6.4 shows typical bulk densities for gas–magma mixtures as they leave an eruptive vent. From these values it is clear that, even in the most gas-rich eruptions, the gas–magma mixture is denser than air when it is erupted from the vent. It has been shown that rise of the plume in the lowest few kilometers occurs because of the initial momentum of the erupted material. During this initial rise, entrainment and heating of the air occurs and usually leads to a situation where the density of the plume material becomes slightly smaller than that of the surrounding air and hence the plume is able to rise due to thermal buoyancy. A situation may arise, however, in which entrainment causes the rise speed to decline to a very small value because of the momentum-sharing (eqn 6.6) before the plume has been able to entrain and heat enough air to become thermally buoyant. In other words, the plume reaches a point where its rise speed is negligible but the bulk density of the plume material (the overall density of the mixture of gas and pyroclasts) is still greater than that of the surrounding air. In this situation the plume can rise no further and the material in it

Table 6.4 The bulk density, ρ_B, of a gas–magma mixture as it is erupted for four different exsolved water contents, n. For comparison, the density of air at the Earth's surface is ~1.2 kg m^{-3}.

n (wt%)	ρ_B (kg m^{-3})
1	18
3	6
5	3.6
7	2.6

will fall back to the ground surface in a continuous stream forming a kind of enormous fountain over the vent. This effect is usually referred to as **column collapse**, although it would be much better described as column instability. The resulting system can be thought of as similar to an ornamental water fountain, in which the rise of water above the ground surface is driven purely by the initial high speed at which the water is ejected. The rise of the water droplets continues only until all their kinetic energy is converted into potential energy, at which point they must fall back to the ground. We refer to the volcanic version as a **pyroclastic fountain**.

6.7.2 Causes of column instability

It appears that, in many cases, eruption plumes in steady eruptions are initially stable but can, in certain circumstances, become unstable at a later stage. There are two main reasons why an eruption column might become unstable as the eruption continues: the mass flux may increase significantly or the exsolved gas content of the erupting magma may decrease.

Take the increasing mass flux case first. The mass flux, M_f, is given by

$$M_f = \pi r^2 \rho_B u \qquad (6.10)$$

where r is the vent radius, ρ_B is the bulk density of the gas–magma mixture and u is the exit velocity of the gas–magma mixture. Thus the mass flux is proportional to r^2. As the gas–magma mixture emerges from the vent it begins to entrain air around its margins. The surface area, A, over which air can be entrained is

$$A = 2 \pi r x \qquad (6.11)$$

where x is the vertical distance moved by the gas jet in one second (Fig. 6.9). Thus the surface area over which air can be entrained is proportional to the vent radius. Because the increasing mass flux of volcanic material gets larger in proportion to r^2 whereas the amount of surrounding atmosphere entrained only increases in proportion to r, entrainment does not keep pace with the increasing mass flux. So, a situation may be reached in which the

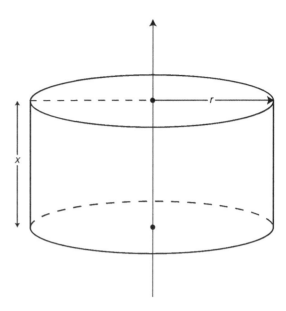

Fig. 6.9 Diagram of a "control volume" used to relate the flux of atmospheric gas entrained into an eruption plume to the flux of gas and pyroclasts already rising in the plume.

gas–magma mixture has not entrained enough air by the time it reaches the top of the gas-thrust region to be buoyant. In such a situation the eruption plume will cease to be stable and will collapse. This effect can be seen in the example in Fig. 6.10a. Here a Plinian eruption starts from a vent with a radius of ~12 m. As the eruption continues the mass flux progressively increases as erosion widens the vent, and the plume height also increases because more heat is being provided to the plume (see section 6.5.2). Eventually a point is reached, in this case when the vent radius reaches ~350 m, where the mass flux has become so great that not enough entrainment can occur to allow the plume to become thermally buoyant. The initially stable Plinian eruption plume then collapses from its peak height of ~40 km and a pyroclastic fountain forms over the vent. The fountain is driven solely by the initial momentum of the erupted material, and so is less than 10 km in height (Fig. 6.10a).

We saw in eqn 6.2 and Table 6.4 that the bulk density of the gas–magma mixture depends on the amount of gas exsolved from the magma and, therefore, the initial gas content of the magma. Thus, if the gas content of the erupting magma decreases

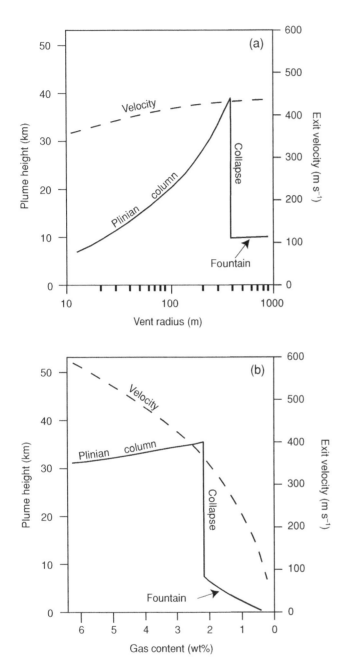

Fig. 6.10 Diagrams showing the variations of the velocity of gas and pyroclasts exiting a vent, and also of the height of the resulting Plinian eruption column or pyroclastic fountain, with (a) changing vent radius and (b) changing magma gas content. In each of the cases shown conditions evolve from a stable Plinian column though a collapse event to a stable but much lower fountain. (Adapted from fig. 15 in Wilson, L., Sparks, R.S.J. & Walker, G.P.L. (1980) Explosive volcanic eruptions – IV. The control of magma chamber and conduit geometry on eruption column behaviour. *Geophys. J. Roy. Astron. Soc.*, **63**, 117–148, copyright Wiley-Blackwell Publishing Ltd.)

progressively during an eruption, the bulk density of the erupted gas–magma mixture increases and more air must be entrained to ensure that the plume will be able to rise by thermal buoyancy. Such an erupting jet, though, will entrain **less** not **more** air. This is because, with less gas exsolution, the exit velocity of the erupting jet will be smaller, and a smaller exit velocity leads to less entrainment (section 6.4.1). So, if the gas content decreases sufficiently during an eruption, this may also cause the eruption plume to collapse. An example of this effect is shown in Fig. 6.10b. Here the mass flux of the eruption is constant but the gas content of the magma decreases progressively. Initially the eruption generates a stable Plinian column greater than 30 km in height (Fig. 6.10b). As the gas content decreases a point is eventually reached, in this case when the gas content is ~2.3 wt%, where the plume can no longer entrain enough air to become thermally buoyant. From this time onward, the new material emerging from the vent feeds a pyroclastic fountain. The gas and pyroclasts already in the eruption cloud continue to climb convectively into the atmosphere, although their motion, and the shape of the eruption cloud, change as the supply of new heat from below is cut off, and eventually disperse downwind. As the gas content declines further, the height of the fountain will also decline because its height is critically dependent on the exit velocity of the material, and, therefore, on the gas content as described in section 6.4.1. Many magma chambers, especially those containing rhyolitic magma, evolve in a way that causes the volatiles to be concentrated into the shallowest part of the system. It is this part of the system that erupts first, and so a decrease in volatile content during an eruption will be more common than an increase.

The pyroclastic fountains produced by convective instability in eruption columns feed various kinds of pyroclastic density currents, the largest of which generate deposits called **ignimbrites**, as described in Chapter 8.

6.8 Summary

- As gas bubbles form and grow within a rising magma they become increasingly close-packed. With continued growth the bubble walls become so thin that they collapse causing the individual bubbles to coalesce into a continuous gas stream. This process is called **fragmentation** as it causes the tearing apart of the magma into individual clots and clasts. Below the **fragmentation level** the rising fluid consists of a continuous stream of magmatic liquid which contains individual gas bubbles within it. After fragmentation the rising fluid is a continuous stream of gas with individual clots and clasts of magma within it. The clots and clasts may still contain trapped – and even interconnected – gas bubbles.

- As the gas–magma mixture rises towards the surface its pressure decreases. This causes an expansion and cooling of the gas and a resulting release of energy. This released energy causes acceleration of the rising material and does work against gravity and friction.

- The speed at which the gas–magma mixture emerges from the vent is called the **exit velocity** and is typically tens to hundreds of meters per second. The exit velocity is dependent on a number of factors but is most strongly controlled by the gas content of the magma. The higher the initial gas content of the magma, the higher the exit velocity.

- As the gas stream exits from the vent, air is mixed or entrained into the rising eruption plume. Entrainment causes two important effects. It causes the rise speed of the material in the plume to decline progressively above the vent. This happens because the initial rise of the material is due to the momentum the material has on eruption. As air is entrained the mass of material in the plume increases, and so **conservation of momentum** requires that the rise speed decreases (eqn 6.6). The erupted material has an eruption temperature of ~900–1150°C whereas the entrained air has a temperature of about 0°C. This means that as the air is entrained it is heated. The volume of air which is entrained is large compared with the volume of magmatic material erupted and this means that the temperature of the plume is eventually only slightly higher than that of the surrounding air. This heating is, however, sufficient to cause the plume to be thermally buoyant compared with the surrounding air.

- In the lowest few kilometers of the eruption plume rise is due to the initial momentum of the erupted material. This section of the eruption plume is called the **inertial** or **gas-thrust region**. As entrainment proceeds, the rise speed due to the initial momentum becomes minimal. By the time this point is reached, however, the entrainment and heating of air will mean that the plume material is of slightly lower density than the surrounding air and so rise continues through thermal buoyancy. The section of the plume in which rise is due to thermal buoyancy is known as the **convective region**. With continued rise the temperature of the material in the plume declines and a point will be reached where the density of the plume material is equal to that of the surrounding atmosphere. Above this point, known as the **level of neutral buoyancy**, plume rise rapidly ceases and the plume spreads out. This uppermost part of the plume is known as the **umbrella region**.

- The height to which an eruption plume can rise within the atmosphere depends most strongly on the **mass flux** of the eruption. The larger the mass flux, the more heat is supplied to the plume in a given time and so the higher the plume can rise.
- Fragmentation generates a range of sizes of clasts and these clasts are carried upwards in the gas stream both within the dike system and then above the vent in the eruption plume. Two competing forces act on the clasts: the force of gravity which tends to make the clasts fall and the drag of the rising gas stream which tends to make the clasts rise. A balance is reached in which the clasts will fall at their terminal velocity through the gas stream. As long as the terminal velocity of the clast is smaller than the rise speed of the gas, the clast will be carried upwards relative to the ground surface. Rise speed in an eruption plume decreases with height, and so a clast of given size will be carried to a height at which the rise speed becomes equal to the clast's terminal velocity and no higher. Large clasts have large terminal velocities and are, therefore, carried only to relatively small heights within the plume. Smaller clasts have smaller terminal velocities and will be carried higher above the vent before falling out from the eruption plume.
- Most steady eruptions initially generate stable, convecting eruption plumes. In some cases, however, an eruption plume may become unstable, i.e., it cannot achieve thermal buoyancy. In such eruptions the plume will rise until it no longer has any upward momentum and will then collapse. The resulting pyroclastic fountain can give rise to a pyroclastic density current ultimately depositing an ignimbrite. Eruption plumes may become unstable due to an increase in mass flux at the vent, a decrease in the gas content of the magma, or a combination of both effects.

6.9 Further reading

GAS RELEASE AND FRAGMENTATION

Gardner, J.E., Thomas, R.M.E., Jaupart, C. & Tait, S. (1996) Fragmentation of magma during Plinian volcanic eruptions. *Bull. Volcanol.* **58**, 144–62.

Sparks, R.S.J. (1978) The dynamics of bubble formation and growth in magmas: a review and analysis. *J. Volcanol. Geotherm. Res.* **3**, 1-37.

Thomas, N., Jaupart, C. & Vergniolle, S. (1994) On the vesicularity of pumice. *J. Geophys. Res.* **99**, 15633-44.

ACCELERATION IN THE DIKE SYSTEM

Buresti, G. & Casarosa, C. (1989) One-dimensional adiabatic flow of equilibrium gas-particle mixtures in long vertical ducts with friction. *J. Fluid Mech.* **203**, 251-72.

Wilson, L. & Head, J.W. (1981) Ascent and eruption of basaltic magma on the Earth and Moon. *J. Geophys. Res.* **86**, 2971-3001.

Wilson, L., Sparks, R.S.J. & Walker, G.P.L. (1980) Explosive volcanic eruptions – IV. The control of magma properties and conduit geometry on eruption column behavior. *Geophys. J. Roy. Astron. Soc.* **63**, 117-48.

PLUME RISE AND RISE HEIGHTS

Carey, S.N. & Sparks, R.S.J. (1986) Quantitative models of the fallout and dispersal of tephra from volcanic eruption columns. *Bull. Volcanol.* **48**, 109-25.

Morton, B.R., Taylor, G. & Turner, J.S. (1956) Turbulent gravitational convection from maintained and instantaneous sources. *Proc. Roy. Soc. Ser. A* **234**, 1-23.

Settle, M. (1978) Volcanic eruption clouds and the thermal output of explosive eruptions. *J. Volcanol. Geotherm. Res.* **3**, 309-24.

Sparks, R.S.J., Bursik, M.I., Carey, S.N., Gilbert, J.S., Glaze, L.S., Sigurdsson, H. & Woods, A.W. (1997) *Volcanic Plumes* (Chapters 1-4). Wiley, Chichester, 574 pp.

Wilson, L. & Walker, G.P.L. (1987) Explosive volcanic eruptions – VI. Ejecta dispersal in Plinian eruptions: the control of eruption conditions and atmospheric properties. *Geophys. J. Roy. Astron. Soc.*, **89**, 657-79.

Wilson, L., Sparks, R.S.J., Huang, T.C. & Watkins, N.D. (1978) The control of volcanic column heights by eruption energetics and dynamics. *J. Geophys. Res.* **83**, 1829-36.

OBSERVATIONS OF ERUPTION PLUMES

Sparks, R.S.J. & Wilson, L. (1982) Explosive volcanic eruptions V: observations of plume dynamics during the 1979 Soufriere eruption, St. Vincent. *Geophys. J. Roy. Astron. Soc.* **69**, 551-70.

Sparks, R.S.J., Bursik, M.I., Carey, S.N., Gilbert, J.S., Glaze, L.S., Sigurdsson, H. & Woods, A.W. (1997) *Volcanic Plumes* (Chapter 5). Wiley, Chichester, 574 pp.

Wilson, L. & Self, S. (1980) Volcanic eruption plumes: density, temperature and particle content estimates from plume motion. *J. Geophys. Res.* **85**, 2567-72.

PLUME INSTABILITY

Sparks, R.S.J. & Wilson, L. (1976) A model of the formation of ignimbrite by gravitational column collapse. *J. Geol. Soc. London* **132**, 441-51.

Sparks, R.S.J., Wilson, L. & Hulme, G. (1978) Theoretical modelling of the generation, movement, and emplacement of pyroclastic flows by column collapse. *J. Geophys. Res.* **83**, 1727-39.

Sparks, R.S.J., Bursik, M.I., Carey, S.N., Gilbert, J.S., Glaze, L.S., Sigurdsson, H. & Woods, A.W. (1997) *Volcanic Plumes* (Chapter 6). Wiley, Chichester, 574 pp.

6.10 Questions to think about

1 What two aspects of the expansion of gas bubbles in magmas rising toward the surface cause an increase in the magma rise speed?

2 Why does magma fragmentation make a big contribution to increasing the rise speed of magma in a dike?

3 In what three ways does a large volatile content in a magma contribute to a high eruption speed?

4 Why do some magmas reach the surface at pressures greater than atmospheric pressure?

5 Why does the material in an eruption cloud slow down at first after leaving the vent but then increase its speed for a while before eventually slowing down again?

6 What controls the maximum height to which an eruption cloud can rise in the atmosphere?

7 Why are large pyroclasts always found close to the vent in air fall eruption deposits whereas small clasts are found at all distances from the vent?

8 Why is it likely that an eruption may evolve from being Plinian to being ignimbrite-forming in nature but is much less likely to evolve in the other direction?

7 Transient volcanic eruptions

7.1 Introduction

In Chapter 5 we saw that gas bubbles forming in rising magma behave differently depending on the rise speed of the magma. At high rise speeds the gas bubbles are locked to the magma and steady fragmentation of the magma results in the sustained style of eruption described in Chapter 6. If, however, the magma is rising slowly or is stationary, gas bubbles can rise through a low-viscosity magma resulting in transient, explosive magmatic eruptions. Transient explosions can also occur when external surface or subsurface water (often called **meteoric** water since it derives ultimately from rainfall from the atmosphere) comes into contact with magma. Such events are best referred to as hydromagmatic eruptions, although the term **phreatomagmatic** is used when the water involved is groundwater. This chapter discusses the styles of these transient eruptions and current ideas about the mechanisms which cause them.

7.2 Magmatic explosions

Initially the two main types of transient explosions which can result from the segregation of magmatic gas from its parent magma are considered: Strombolian and Vulcanian eruptions. These types of eruption share many similarities:

- each explosion is short-lived (seconds to minutes in duration);
- repose times between individual explosions are usually short (seconds to hours);
- explosions usually occur in sequences (although the lengths of the sequences vary widely – from days to years);
- both types of eruption are characterized by the ballistic ejection of large blocks of country rock and volcanic bombs;
- both types of explosions generate eruption columns, column height being relatively small except when the explosions are closely spaced in time.

The eruptions differ primarily in the violence of the events and the composition of the magmatic material involved. Strombolian explosions are mild events involving basaltic magmas in which almost all of the erupted material is magmatic and is ejected at relatively low velocities (less than 200 m s^{-1}) and deposited close to the vent. By contrast, Vulcanian explosions generally involve intermediate to evolved magmas and generate some of the highest eruption velocities of any type of volcanic eruption ($200–400 \text{ m s}^{-1}$) and eject large blocks of magmatic material and country rock to distances of up to ~5 km from the vent (see Chapter 1). The basic mechanism behind both types of eruption appears to be very similar, the key difference between them being the pressures which are generated in the vent prior to eruption: Vulcanian eruptions are associated with much higher vent pressures than Strombolian eruptions.

The basic mechanism proposed to explain both types of eruption is that they occur within open magmatic systems in which the vent or magma column is capped with relatively cool magma. The rise speed of magma within the system is low or negligible which means that magmatic gas bubbles rise upwards through the overlying magma and accumulate beneath the vent "plug". In Strombolian eruptions it is common for this accumulating gas to cause an updoming of the cooled magma skin on the top of the magma column (Fig. 1.14). This "skin", although cooled, is still plastic and can stretch and deform just like the rubber of a balloon being blown up. Eventually the bubble "skin" tears, releasing the gas within and ejecting the gas and fragmented "skin" upwards and outwards (Fig. 1.14). If the magma forming the "skin" is cooled to a greater degree then it may form a brittle plug within the vent which has much greater strength than a plastic "skin". In this case more pressure will need to build-up before the plug fails and an eruption occurs. In the case of Vulcanian eruptions the plug material may consist of cool magma and/or material from previous explosions which has fallen back into the vent blocking it. The higher velocities in Vulcanian explosions suggest that the strength of the plug is considerably greater in Vulcanian events than in Strombolian ones, so that a much higher pressure is reached before failure. The fact that the two types of eruption are associated with different magma compositions suggests that a link exists between plug strength and composition. Such a link may arise because the higher viscosity of more evolved magmas means that bubbles rise more slowly through them. Slow rise and accumulation of bubbles means that more cooling of the skin/plug will occur prior to an explosion. In the extreme case, bubble movement in high viscosity magma is so slow that segregation and rise of bubbles is likely to be negligible. This fact leads to the suggestion that meteoric water is involved in many Vulcanian events. Cooling of the magma by contact with external water stops it rising, and at the same time that water is boiled to steam and trapped, thus providing a lot of the gas to drive the explosion (see below).

In purely magmatic events, the upward and outward movement of gas and fragmented magmatic material is driven by the expansion of the magmatic gas which occurs when the plug fails, exposing the compressed trapped gas to the lower pressure environment of the surrounding atmosphere.

7.2.1 Modeling transient magmatic explosions

RELATING VENT PRESSURES TO ERUPTION VELOCITIES

We saw in the last chapter (section 6.3) that in steady-state eruptions it is the expansion of gas which provides the energy to accelerate and erupt the gas–magma mixture. The energy use in an eruption can be defined by the **energy equation** (eqn 6.4). A similar **energy equation** likewise forms the basis for any treatment of the dynamics of transient explosions. During a transient explosion a certain mass of gas is trapped at a certain starting pressure, P_i, behind the vent "plug". This pressure is higher than atmospheric pressure, P_a, so when the plug fails the gas will expand in order to reduce its pressure to atmospheric. A finite amount of energy will be released, the actual amount being a function of the initial mass of gas and the initial pressure it was under. The released energy is shared in the explosion by the expanding gas, the mass of plug material that it expels, and the mass of atmosphere that is displaced as the expansion takes place.

There are two extreme ways in which gases can expand: adiabatically or isothermally. As they expand, their temperature will decline unless heat is supplied to them. If no heat is supplied and the temperature declines, then the expansion is said to be adiabatic. If an infinitely large amount of heat can be supplied to the gas there is no temperature change and the expansion is said to be isothermal. In the steady-state case treated in Chapter 6 it was assumed that expansion occurred isothermally. This assumption is valid in the steady-state case because the gas represents only a small part of the total mass being erupted, and the much greater mass of magma forms a heat reservoir which is able to maintain the gas temperature almost constant as long as the contact between the gas and magma is sufficiently good. In the case of transient explosions, however, the contact between the gas and magma is not as good – the short time scale of the explosion means that not much heat can travel from the inside of a pyroclast to its surface to reach the gas. Also, because gas has been accumulating under the trapping lid, the gas to magma mass ratio in the mixture that is expelled is larger. Taken together, these factors mean that it is unlikely that the magma can buffer the gas temperature to any significant extent. Instead the expansion is likely to occur nearly adiabatically. In this case the **energy equation** can be written as

$$\frac{n \, Q \, T_i}{m} \frac{\gamma}{\gamma - 1} \left[1 - \left(\frac{P_f}{P_i} \right)^{(\gamma-1)/\gamma} \right] + \frac{(1-n)}{\rho_m} (P_i - P_f)$$

$$= \frac{1}{2} (U_f^2 - U_i^2) + g \, h + \text{friction} \qquad (7.1)$$

where n is the gas mass fraction, Q is the universal gas constant, m is molecular weight of the gas, T_i is the initial temperature of the gas and magma, γ is the ratio of the specific heats of the gas, P_i is the initial gas pressure, P_f is the final pressure (equal to the atmospheric pressure, P_a), ρ_m is density of the magmatic material, U_i is the initial velocity of the gas (which is essentially zero since the magma rise speed is very small), U_f is the final velocity of the gas at the end of the expansion phase, g is the acceleration due to gravity, and h is the vertical distance risen by the magmatic material during the expansion. Just as in eqn 6.4, the left-hand side of the equation represents the energy released by expansion of the gas (in this case adiabatically) while the right-hand side represents the three ways in which this released energy is used in the eruption. The first term represents the change in the kinetic energy of the system, in other words the energy used to accelerate the gas and magma in the explosion. The second term represents the energy used to raise the magmatic material within the gravity field while it is being accelerated. The final term is concerned with the friction between the gas and magmatic material and the air through which it is passing, i.e., the **air drag**.

The treatment of the air drag is a critical issue in modeling these eruptions. When an explosion occurs the gas and magmatic materials move upwards and outwards from the point of the explosion. In doing so they must push the surrounding air out of the way, and the resistance of the air to this pushing is the air drag. Initial models of tran-

sient explosions assumed that the air ahead of the expanding gas and magma was stationary and thus that a large amount of energy from the explosion would be used in overcoming the air drag acting on individual pieces of solid material. However, observations of transient explosions show that they commonly send a shock wave outwards from the explosion point slightly ahead of the expanding cloud of gas and solids. This shock wave pushes the surrounding air out of the way and so, in the early stages of the explosion, instead of thinking in terms of drag forces acting on individual solid fragments we have to evaluate the energy used to displace the atmosphere as a whole.

When all of these factors are taken into account, the results can be summarized as in Fig. 7.1, which shows the maximum speed of the ejected solid material as a function of the pressure which built up in the trapped gas before the explosion and the weight fraction which the gas represents of all of the materials ejected. Recall that the typical dissolved volatile contents of basaltic magmas likely to be involved in Strombolian explosions are less than 1 wt% and the volatile contents of the more evolved magmas commonly involved in Vulcanian explosions are up to a few weight percent. The curves in Fig. 7.1 are given for much larger volatile contents because of the expected accumulation of gas before the explosion. The range of pressures used in the calculations is cut off at 10 MPa because no rocks are strong enough to allow pressures this large to accumulate before they break.

Figure 7.1 shows the expected basic relationship between skin/plug strength and eruption velocity: i.e., that the highest velocities will correspond to the greatest strengths. Observations of Strombolian eruptions at Heimaey and Stromboli give clast velocities of ~ 150 m s^{-1} as typical for Heimaey (with a maximum of 230 m s^{-1}) and 50–100 m s^{-1} as typical for Stromboli. Estimates have been made in both cases of the weight percentage of gas in these eruptions: the minimum gas content in both cases is ~ 11 wt% and ranges up to values of 36–38 wt%. These combinations of ejecta velocity and gas content suggest that the strength of the skin prior to rupture is < 0.3 MPa. Such low "skin" strengths are consistent with the idea that the skin is still

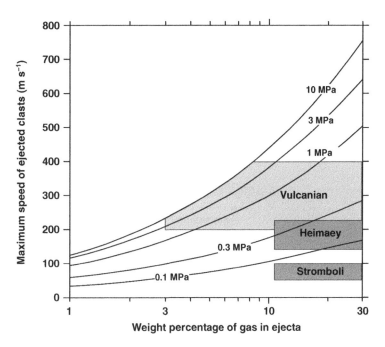

Fig. 7.1 The maximum speed of the solid rock fragments ejected in a transient explosion as a function of the pressure which built up in the trapped gas before the explosion (curves labeled with pressure in MPa) and the weight fraction which the gas represents of all of the materials ejected. The ranges of conditions observed in explosions at Stromboli, Heimaey, and in various Vulcanian explosions are indicated.

plastic and deformable prior to the explosion. Where the skin is still plastic the limiting factor on its deformation and ultimate failure is the **yield strength** of the magma. Observations on lava flows and on lava cooling in lava lakes show that the yield strength increases as the lava cools, with maximum recorded yield strengths for basaltic lavas being ~0.23 MPa. Thus the strengths estimated from Fig. 7.1 are consistent with the idea that it is the yield strength developed in a cooling but still plastic skin of lava which controls the pressure developed prior to eruption and hence the velocities of the ejected clasts.

In the case of Vulcanian explosions observations show that typical ejecta velocities range from 200 to 400 m s^{-1}. The weight percentage of gas in these explosions is unknown but, assuming some segregation of gas occurs, it must exceed the initial magmatic gas content which is likely to be a few weight percent. These constraints show, as expected, that the plug strength in these eruptions is greater than in Strombolian eruptions, but a wide range of strengths is possible, from ~0.3 MPa up to the maximum value used in the calculations of 10 MPa. In the case of Vulcanian explosions it is

the strength of solidified magma which is thought to be the controlling factor on the pressure generated prior to eruption. The **tensile strength** of solidified magma varies greatly depending on how fractured the rock is. The 10 MPa value used as an upper limit in these calculations represents the strength of pristine unfractured igneous rocks. In practice the vent plug is likely to contain fractures or, indeed, to be composed of unconsolidated material which has slumped back into the vent. In either case the tensile strength is likely to be significantly less than the 10 MPa upper limit. The range of strengths estimated from Fig. 7.1 is thus consistent with the likely strength of the vent-plugging material.

PREDICTING THE RANGE OF BALLISTIC CLASTS

Thus far the theoretical model has been used to look at the relationship between vent pressures and ejecta velocities in transient explosions. The ejection of large blocks in both types of transient explosion represents a significant hazard. In the case of Vulcanian explosions these blocks may land as much as 5 km from the vent! So from a hazard

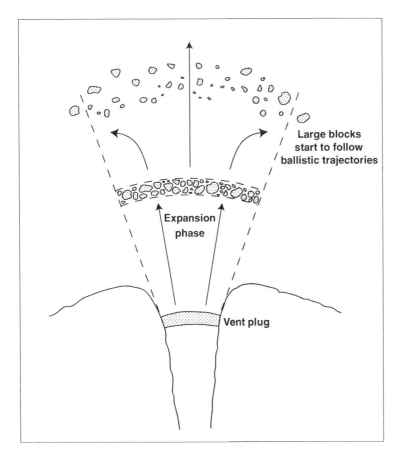

Fig. 7.2 Diagram showing the progressive fragmentation and dispersal of the material originally forming a plug in the vent as the compressed trapped gas is released by the explosion and expands and accelerates. (Adapted from fig. 1 published in *Icarus*, Vol. 123, Fagents, S.A. and Wilson, L., Numerical modelling of ejecta dispersal from transient volcanic explosions on Mars, 284–295, copyright Elsevier (1996).)

point of view it is useful to extend the model to predict the range of clasts ejected in transient explosions. To do this we need first to visualize what happens during these explosions.

During the initial phase of the explosion the gas is expanding and the gas and clasts making up the ejected material move upwards and outwards from the explosion point as a single mass (Fig. 7.2). Towards the end of the gas expansion process, however, the sharp boundary between the atmosphere and the ejected material breaks down. Gas from the explosion mixes with air and slows down rapidly, and the smallest solid particles, which suffer a very large drag force from the gas, also slow down. These particles share their heat with the air and warm it up, so that a convecting eruption column forms – this is discussed in the next section. Meanwhile the larger blocks of solid material carry on moving at high speed.

These blocks now experience air drag in the normal way and this has to be taken into account in order to work out the maximum distance to which they can be thrown. We make the analysis a bit easier by imagining that the blocks are spheres. The drag force, F_d, acting on a sphere of radius R moving at a high speed U through a gas of density ρ_a is

$$F_d = 0.5\, \rho_a\, C_d\, \pi\, R^2\, U^2 \tag{7.2}$$

where C_d is a dimensionless number called the drag coefficient and has a value between about 0.5 and 1.3 for real blocks depending on their shape and roughness (and even on whether they are spinning or not!) – an average value of 0.7 is used here. A second force acts on the block, its weight – gravity pulls it downward toward the Earth's surface. The gravitational force F_g acting on a block of density ρ_b and radius R is given by

$$F_g = (4/3) \, \pi \, R^3 \, \rho_b \, g \qquad (7.3)$$

where g is the acceleration due to gravity, about 9.8 m s^{-2} on Earth. These forces each produce an acceleration in the block given by Newton's third law of motion, that the acceleration is the force divided by the mass of the block. The mass of a spherical block is $[(4/3)\pi R^3 \rho_b]$, so eqn 7.3 is just a statement of Newton's law. Equation 7.2, however, shows that the air drag acceleration, A, is given by

$$A = (0.5 \, \rho_a \, C_d \, \pi \, R^2 \, U^2)/[(4/3) \, \pi \, R^3 \, \rho_b] \qquad (7.4a)$$

which simplifies to

$$A = (3 \, \rho_a \, C_d \, U^2)/(8 \, R \, \rho_b) \qquad (7.4b)$$

The drag force F_d always acts in exactly the opposite direction to that in which the block is moving, but the gravity force F_g always acts straight down toward the center of the Earth, and this makes calculating the path of a block quite complicated. But as an illustration consider two blocks of radii 1 m and 0.1 m, both with density $\rho_b = 2000$ kg m^{-3}, traveling straight upward at a speed of 100 m s^{-1} relative to the air. The air density near the Earth's surface is about 1 kg m^{-3}, and so eqn 7.4b shows that the accelerations (actually decelerations since the blocks are slowing down) are about 1.3 m s^{-2} for the larger block and about 13 m s^{-2} for the smaller one.

These numbers have to be compared with the downward accelerations due to gravity, $g = 9.8$ m s^{-2}, that the blocks will each be feeling at the same time. Thus the large block notices only a 13% difference from the situation if there were no atmosphere around it, whereas the smaller block experiences a 130% difference! Of course, for a block launched at an angle to the vertical, the drag force will change in size and direction as the speed of the block changes in both the upward and sideways directions, but this simple calculation gives a good idea of why meter-sized blocks travel on nearly ballistic trajectories, similar to those they would have if there were no atmosphere around them, while ejected blocks less than a fraction of a meter in size travel only to much smaller distances.

The above ideas can be used to simulate a given set of eruption conditions and predict the maximum

ranges of blocks of different sizes. For a given block size, initial speed, and launch angle from the horizontal, eqn 7.4b is used to find the total acceleration caused by drag. This total acceleration is broken down into its vertical and horizontal components and the acceleration due to gravity is added to the vertical component. Next, the vertical and horizontal accelerations are used to find how much the vertical and horizontal speeds change in some small time interval, and the average speeds multiplied by the time interval give the vertical and horizontal distances traveled. Thus we have a new block position, speed and direction, and therefore can repeat the calculation for another small time interval. This procedure is repeated a large number of times until the block finally reaches the ground. Here this process is illustrated by looking at the ranges of large blocks in four well-documented transient explosions and inferring what the eruption conditions must have been to allow these blocks to be ejected to their observed final positions (Table 7.1).

Consider the ejection of large blocks in the two Vulcanian explosions (Arenal and Ngauruhoe) first. In both cases the vent pressures and ejecta velocities are inferred to be high and are consistent with the range inferred for Vulcanian explosions from Fig. 7.1. In both cases the inferred gas content is moderate (4–6 wt%) but, as these are andesitic eruptions, it probably represents a small amount of gas segregation prior to eruption. These results contrast, as expected, with those for the two Strombolian eruptions (Heimaey and Stromboli) in which inferred eruption speeds and vent pressures are lower. In both Strombolian eruptions the inferred gas content is ~20 wt%, implying considerable gas concentration prior to eruption. These values are consistent with the idea that gas segregation and concentration is easier and greater in the basaltic magmas of Strombolian eruptions than in the more evolved magmas commonly involved in Vulcanian explosions.

The final three columns in Table 7.1 allow a comparison to be made between the amount of energy used in accelerating the ejected material, in raising the material in the gravitational field and in overcoming air drag, i.e., the partitioning of energy depicted in the energy equation (eqn 7.1). Note

Table 7.1 Parameters obtained for a number of transient explosive eruptions (S: Strombolian; V: Vulcanian) using the model represented by eqn 7.1: D, diameter of the largest volcanic bomb measured; R, maximum range to which bombs were observed to be thrown; U_f, calculated maximum speed of the ejecta at the end of gas expansion; P_i, inferred pressure in the gas at the start of the explosion; n, implied weight fraction of the explosion products that consisted of gas; KE, fraction of all of the explosion energy that appears as kinetic energy; PE, fraction that appears as potential energy; DE, fraction that appears as energy used to displace the atmosphere.

Volcano	Type	D (m)	R (m)	U_f (m s^{-1})	P_i (MPa)	n (%)	KE (%)	PE (%)	DE (%)
Arenal (1968)	V	1.3	5000	300	10	6	39	4	57
Ngauruhoe (1975)	V	0.8	2800	250	5	4	47	2	51
Stromboli (1975)	S	?	25	150	0.1	20	20	5	75
Heimaey (1973)	S	0.2	500	200	0.35	20	20	1	79

that in all cases more than half the available energy goes into pushing the atmosphere out of the way during the expansion phase of the explosion.

PLUME HEIGHTS IN TRANSIENT ERUPTIONS

We saw in Chapter 6 that the height of the plume formed in a steady eruption depends primarily on the rate at which heat is supplied to it and thus on the mass flux of the eruption. In a transient explosion it is the total amount of heat released, rather than the release rate, that matters, and so it is the total mass of erupted material that controls the plume height. The relationship for the Earth's standard atmosphere is

$$H = 0.042 \, M_e^{1/4} \tag{7.5}$$

where H is the plume height in kilometers and M_e is the total mass of solids and gas ejected in kilograms.

However, another factor also influences the heights of the eruption plumes generated in transient eruptions: the time gaps between individual explosions. For instance, individual explosions occur at time intervals of ~10 minutes at Stromboli which allows the plume from any one explosion to disperse before the next event takes place. Typical masses of ejecta are up to 500 kg and these generate plumes with maximum heights of ~200 m. By contrast, Strombolian activity at Heimaey in 1973 consisted of individual explosions which occurred every 0.5–2 s. The short time gap between these individual explosions meant that the plume generated by any one explosion would hardly have had

time to complete its formation, let alone disperse, before the next ejection of gas and clasts took place. As a result, the plume was able to be maintained in a way more analogous to that of a steady-state eruption, and the resulting plume heights averaged 6–10 km. Typical masses of material ejected at Heimaey were about 5×10^5 kg, so that if the explosions had been widely spaced in time the expected plume height would have been just over 1 km. Instead, the ejection of 5×10^5 kg typically once every second corresponded to an average release rate of 5×10^5 kg s^{-1}, and eqn 6.7 shows that this should have led to a plume height of 6.3 km, in reasonable agreement with what was observed. Needless to say, the fact that the plume rose six times higher than if the explosions had been less frequent means that the smaller ejected clasts were deposited from the plume over a much greater area around the vent.

7.3 Transient eruptions involving external water

Some Vulcanian explosions are thought to result from the interaction of magma with groundwater rather than from the exsolution and segregation of magmatic volatiles, and these explosions are thus hydromagmatic eruptions. Interactions between magma and an external source of water are common, and show a wide variety of eruption styles of which Vulcanian explosions are just one example (see Chapter 1). Many of these types of eruption are transient in character like the magmatic eruptions just discussed (although sustained hydromagmatic eruptions can also occur). The wide variety of styles of hydromagmatic eruptions reflects the complexity involved in the way magma and water can mix and the diversity of settings in which interactions occur. Vulcanian explosions fueled by external water are thought to result from interactions between magma and groundwater on land, but interactions can also occur in deep and shallow submarine settings, for example where lava enters the sea, and during eruptions through lakes or beneath glaciers, etc. (see Chapter 1).

7.3.1 Types of hydromagmatic eruption

The type of interaction which occurs in a hydromagmatic eruption depends upon the relative volumes of water and magma involved and on the extent to which the two can mix. When water is abundant, for example in a submarine setting, the interaction between magma and water may not even be explosive. When lava erupts under water it commonly forms flow lobes known as pillows. The buoyancy force of the water supports some of the weight of the lava and so pillows are thicker and more rounded in cross-section than subaerial flow lobes (Fig. 7.3). The surface of the growing pillow is covered by a layer of steam about 1 mm thick which acts to control the heat transfer from the lava

Fig. 7.3 Lobes of pillow lava erupted on the ocean floor. Displacement of the surrounding water effectively reduces the weight of the lava and makes pillows less flattened than pahoehoe lobes erupted in air. (Image courtesy of the Monterey Bay Aquarium Research Institute. (c) 2001 MBARI.)

to the water (gases are bad conductors of heat). Outside the thin steam film the water does not boil, and the film is quite stable, only being broken over a small area for a brief moment if a bubble of magmatic gas escapes from inside the pillow.

When the volume of water available is small compared with the magma volume, the water is thoroughly heated by the magma and is all converted to highly compressed steam. If this subsequently expands, the large volume increase (by a factor of nearly 3000) causes an explosion, but the small volume of water available restricts the size of the resulting event. For example, when lava flows over wet ground it sometimes happens that pockets of water trapped as the front of the flow advances over them boil faster than the steam can escape along the contact between the flow and the ground. The steam bursts through the flow, throwing out clots of lava and creating a crater, but this crater is never much wider than the flow is thick. A crater formed in a lava flow in this way is one example of a **rootless vent**, this term being used because, although lava is thrown out from this location, it is not a place where fresh lava is coming directly out of the ground. Another example of water being trapped and unable to escape fast enough occurs when lava flows which have formed tubes reach the sea. Waves break over the front of the flow and water is driven into the tube. The steam formed as the water boils is trapped by the inertia of the inrushing seawater and again a relatively small explosion can occur, blowing the top off the tube.

Between these two extremes, it sometimes happens that more nearly equal volumes of magma and water will mix together rapidly with all of the water being converted to steam. The magma cools as the water is heated and the final temperature of the mixture is at least one-third of the initial magma temperature, say 500 K. The density of water is very close to 1000 kg m^{-3}, but the density of steam at 500 K is 0.43 kg m^{-3}. Thus the conversion of water to steam in this way causes a more than 2000-fold increase in volume, and this expansion is what drives a hydromagmatic explosion. This kind of interaction gives rise to major hydromagmatic explosions that produce craters called **maars**. A well-documented example of this occurred at

Ukinrek in Alaska in 1977, when violent explosions due to interactions between magma and groundwater created two nearby craters with diameters of 170 and 300 m (see Chapter 1).

We can illustrate the importance of the relative amounts of water and magma in controlling the violence of a hydromagmatic explosion by calculating the equilibrium temperature, T_e, of the resulting steam and chilled magma fragments when a given mixture of water and hot magma mix together. The heat lost by the magma, H_m, is

$$H_m = V_m \, \rho_m \, c_m \, (T_m - T_e) \qquad (7.6)$$

where V_m, ρ_m, c_m, and T_m are the volume, density, specific heat, and initial temperature, respectively, of the magma. The heat gained by the water, H_w, consists of the amounts of heat needed to warm the water to its boiling point, then to boil it, and finally to heat up the steam produced, and is given by

$$H_w = V_w \, \rho_w \, [c_w \, (373 - T_w) + L_w + c_{st} \, (T_e - 373)] \qquad (7.7)$$

where V_w, ρ_w, c_w, and T_w are the volume, density, specific heat, and initial temperature, respectively, of the water, L_w is the amount of heat needed to boil the water, and c_{st} is the latent heat of steam. Since

these two amounts of heat must be equal, we can find the final temperature as a function of the volume fraction of water in the mixture [$V_w/(V_w + V_m)$]. Assuming that the initial temperatures of the water and magma are 300 K and 1450 K, respectively, that the magma is somewhat vesicular with a density of 2000 kg m^{-3}, and taking values of the thermal parameters for water from standard tables, the final temperatures are found to vary with the water content as shown in Fig. 7.4. Also shown in Fig. 7.4 is the maximum speed of the chilled magma and steam thrown out in the explosion. The speed is calculated using the equivalent of eqn 7.1, with the peak pressure during the explosion being found from the fact that the water is turned into steam so fast that there is essentially no change in volume. Notice that the most energetic (high velocity) explosions occur when water makes up about 20–25% of the volume of the exploding mixture. If the magma has a very low vesicularity the optimum water volume fraction is about 30%.

7.3.2 Mechanisms of violent magma–water interactions

The physical interaction between magma and water which occurs in a very violent hydromagmatic

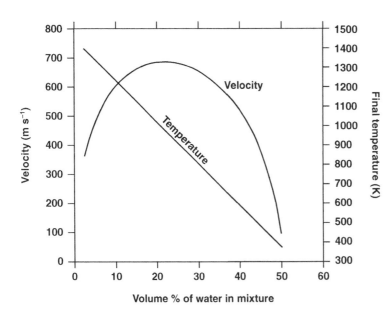

Fig. 7.4 The final velocity and temperature of fragments ejected in hydromagmatic explosions driven by the intimate mixing of magma and water. The most violent explosions occur when the water forms 20–30% of the volume of the mixture.

eruption when water and magma mix together efficiently is analogous to what in an industrial setting is called a fuel-coolant interaction (FCI). Fuel-coolant interactions are events in which the rapid transfer of heat from a hot fluid (the fuel) to a cold volatile fluid (the coolant) results in the rapid conversion of the thermal energy into kinetic energy, i.e., an explosion. A good example of where such an FCI might occur is when molten steel is poured into a crucible for transfer elsewhere within a steelworks but rainwater has been allowed to accumulate in the bottom of the crucible. The resulting explosion shatters the crucible into high-speed shrapnel. Fortunately such accidents are rare. Because FCIs have been studied extensively, this wealth of engineering literature can provide insights into the processes which are important in the magma–water interactions of hydromagmatic eruptions.

Whether a mixing event between magma and water is explosive or nonexplosive depends critically on the rate of transfer of heat between the fuel (magma) and coolant (water), and this in turn depends on the surface area over which the two components are in contact. The initial step in any FCI is one in which the fuel and coolant first come into intimate contact with each other. If this interaction is sufficiently gentle the boiling of the water will create a stable vapor layer between the fuel and coolant which limits the transfer of heat between the two, as in the case of pillow lavas. However, if the initial interaction is sufficiently vigorous the vapor layer may collapse or never form properly. This could happen for a number of reasons, for instance because of the passage of a seismic wave through the mixture, because a local implosion occurs as vapor condenses or because a magmatic explosion (e.g., a Strombolian bubble burst) suddenly accelerates the magma into the water.

Once the vapor layer is disrupted the fuel and coolant come into good thermal and mechanical contact and rapid conversion of a small volume of water to a large volume of steam occurs. If the expansion of the steam forces the interface between the water and magma to deform and become wavy, the surface area over which they are in contact increases and this leads to even more steam formation. This is a positive feedback situation. The removal of heat from the magma may be

so great that near the contact surface it becomes solid. If it cools fast enough, the stress caused by the large temperature gradient near the surface may shatter the solid. This liberates fragments of solid magma that are stirred into the water and may still be hot enough to boil some more water. Probably more important, however, the removal of the fragments exposes yet more hot magma to come into close contact with water. A runaway situation can develop in which each new pulse of steam generation provides more than enough energy to drive the next cycle of deformation, cooling, shattering, and stirring. Once this critical state is reached, the explosive interaction between the water and magma must continue violently until one or other of the two components is completely used up.

7.3.3 Tephra from hydromagmatic eruptions

The tephra from these eruptions often differs significantly from that produced in purely magmatic eruptions. For instance, in general hydromagmatic eruptions produce finer grained material than magmatic eruptions. The tephra usually consists of magmatic glass, crystals, and lithic material (wall rock and material which has slumped into the vent system), often with the lithic component dominating. Examination of the juvenile glass under an optical or electron microscope further shows that the glass typically takes on a number of distinct forms: blocky, moss-like, platey, and spherical, which experimental studies have suggested are linked to the details of how the magma–water interaction occurs in a particular eruption.

7.4 Summary

- There are two main types of purely magmatic transient explosions: Strombolian and Vulcanian. These share a number of features in common, including: short duration; short time between individual explosions; they often occur in long sequences; they are characterized by the ejection of large blocks and volcanic bombs; and they generate eruption plumes which are usually moderate in height but with the height being strongly influenced by how rapidly one explosion follows on from the previous one.

- The main differences between the two types of explosion are the difference in their violence and in the chemistry of the erupting magma. Strombolian events are generally very mild and are associated with basaltic magmas while Vulcanian events are more powerful and associated with more evolved magmas.
- The basic mechanism in both types of explosion is thought to be similar with gas rising through a near stationary magma column to accumulate beneath a vent cap or plug. The difference in violence between the two eruption types seems to be linked to the strength of this plug, and hence the amount of pressure build-up.
- Modeling work shows that the eruption velocities of the two types of eruption are linked to the vent pressures developed prior to each explosion and hence to the strength of the vent plug, the typical velocities in Strombolian explosions of < 200 m s^{-1} being linked to the weak vent plug formed between explosions, the higher velocities (200–400 m s^{-1}) of Vulcanian explosions reflecting the greater strength of vent plugs involved in these eruptions.
- Modeling can be used to predict the range of large blocks thrown out in transient magmatic explosions and, conversely, the position of blocks can be used to infer the initial eruption conditions. Study of a number of eruptions shows that the vent conditions thus inferred are consistent with the link between vent pressure and plug strength just described.
- Modeling of plumes generated in transient explosions shows that the plumes are usually only of limited height but that when explosions occur in close succession the eruption becomes more like a sustained event (from the point of view of the plume) and so much higher plumes can be generated.
- Transient explosions are also common in hydromagmatic eruptions. In these events the relative quantities of magma and water involved and the details of how the magma and water are brought into contact with each other have a profound influence on the character of the eruption which results. The most explosive interactions are those in which water forms 25–30% of the mixture volume.

7.5 Further reading

Fagents, S.A. & Wilson, L. (1993) Explosive volcanic eruptions – VII. The range of pyroclasts ejected in transient volcanic explosions. *Geophys. J. Int.* **113**, 359–70.

Morrissey, M. & Mastin, L.G. (2000) Vulcanian eruptions. In *Encyclopedia of Volcanoes* (Ed. H. Sigurdsson), pp. 463–75. Academic Press, San Diego, CA.

Morrissey, M., Zimanowski, B., Wohletz, K. & Buettner, R. (2000) Phreatomagmatic fragmentation. In *Encyclopedia of Volcanoes* (Ed. H. Sigurdsson), pp. 431–45. Academic Press, San Diego, CA.

Wilson, L. (1980) Relationships between pressure, volatile content and ejecta velocity in three types of volcanic explosion. *J. Volcanol. Geotherm. Res.* **8**, 297–313.

7.6 Questions to think about

1 In what two main ways can transient volcanic explosions occur?

2 What determines the violence of a transient volcanic explosion as judged by the eruption speed of pyroclastic fragments?

3 How would you expect an explosive eruption under water to differ from one in air?

4 Imagine two eruptions, both of which release magma at an average rate of 2×10^5 kg s^{-1} (which corresponds to about 80 m^3 of bubble-free magma per second). The first eruption proceeds steadily forming a Hawaiian lava fountain with a constant plume above it. The second eruption is Strombolian and has one explosion every ten seconds, forming a series of transient plumes. Use eqns 6.7 and 7.5 to deduce the height of the plume in each case.

5 What is the difference between vents and rootless vents?

6 What is it that makes maar-forming explosions so violent?

8 Pyroclastic falls and pyroclastic density currents

8.1 Introduction

Chapter 6 discussed the upward transport of clasts in eruption columns, their eventual release from such columns, and the factors that may cause columns to become unstable, leading to pyroclastic fountaining that feeds pyroclastic density currents. This chapter looks in more detail at all of these processes and at the mechanisms of emplacement of the resulting deposits.

8.2 Fallout of clasts from eruption columns

Chapter 6 showed how the terminal velocity of a clast within the eruption column, which is a function of its radius, density, and shape, largely determines the height above the vent to which it can be carried. The general pattern is for the largest clasts to fall out from lower levels with smaller clasts being carried to greater heights. The exact height reached by a clast of given size and type, however, depends on the eruption conditions, as discussed further below, and the location at which it lands depends on the state of the atmosphere, especially the wind. The distribution of fallen pyroclasts on the ground can be used to infer at least some of the properties of the eruption that produced them. This is particularly important for volcanoes that erupt very explosively, but do so only very rarely (perhaps at intervals of tens to hundreds of thousands of years), because it means that the products of prehistoric eruptions can be analyzed in order to gain some idea of what to expect in the future.

8.2.1 Fallout from the rising eruption column

Chapter 6 explained that an eruption column can be divided into three regions: the gas-thrust region, the convective region and the umbrella region. The width of an eruption column increases with height in the gas-thrust and convective regions, becoming even wider in the umbrella region, where it commonly spreads preferentially downwind (Fig. 6.5). Figure 8.1 shows the same eruption column seen from directly upwind of the vent. In the gas-thrust and convective regions the expansion in the eruption column is caused by the entrainment and heating of air from the surrounding atmosphere. The

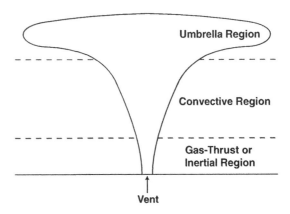

Fig. 8.1 A Plinian eruption cloud viewed from upwind of the vent. The progressively greater lateral spreading of the cloud continues for most of its rise, but eventually incorporation of air that brings with it the momentum of the wind causes the top of the cloud to be carried away in the downwind direction and spreading becomes negligible.

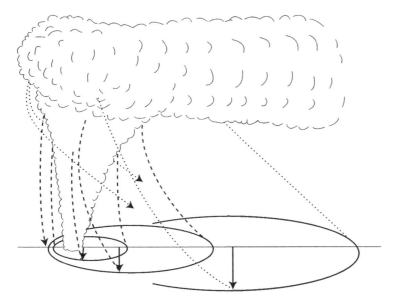

Fig. 8.2 A Plinian eruption cloud viewed obliquely from a direction at right-angles to the direction of the wind. A clast released from the edge of the cloud is carried downwind from its release point and travels a distance that depends on the wind speed and its terminal fall velocity.

entrainment in the gas-thrust region is dominated by momentum transfer, and an analysis by a pioneer of fluid mechanics named Prandtl showed that this should cause the edge of the eruption column to expand at a rate of one unit sideways for every eight units of height gained. This pattern is close to what has been seen in historic eruptions, especially if an allowance is made for the fact that the bulk density of the erupting jet of clasts and volcanic gas is greater than the density of the surrounding atmosphere. Higher up the eruption column, as discussed in Chapter 6, things become more complicated due to the expansion of entrained air, as a result of both heating by pyroclasts and the decreasing atmospheric pressure. The spreading rate increases, and as a general approximation one can think of the lateral expansion rate as being 1 in 8 for the first quarter of the plume height, 1 in 4 for the second quarter, 1 in 2 for the third quarter, and 1 in 1, i.e., a spreading angle of 45° from the vertical, near the start of the umbrella region (Fig. 8.1).

A clast will only fall out from the eruption column if it is carried to the edge of the column. Once it is released, it will be carried along by the wind, but this will only move it in the direction of the wind. Using the theoretical models of eruption cloud development described in Chapter 6, the eruption conditions (magma gas content and erupted

mass flux) can be used to predict the shape of the eruption column and the height to which a clast of a given size and density will be carried. Then if the speed at which the clast will fall and the speed of the wind are known, it is possible to predict where on the ground any given clast should land (Fig. 8.2). The smaller the clast, the greater the height to which it is carried above the vent and thus, because the column expands with height, the greater the cross-wind distance it will land from the vent; also, the greater the wind speed, the greater the downwind distance it will land. This description suggests that there should be a unique relationship between clast size and position in the final deposit. Unfortunately things are a little more complicated, because at any given height in an eruption column a range of particle sizes will be released. The largest clast will, of course, be the one that is just supported in the column by the rising gases at that height; but there will also be many smaller clasts that would normally be carried to greater heights but that happen to be moved by turbulence to the edge of the column where the upward velocity is much smaller as the column grades into the bulk of the atmosphere, and some of these smaller clasts will be released as well. This means that at every location a pyroclastic fall deposit will contain a range of grain sizes.

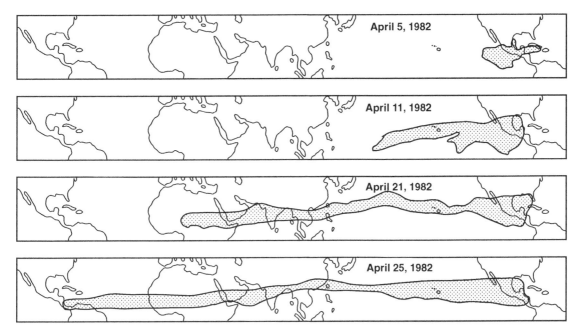

Fig. 8.3 The progressive westward drift of the eruption cloud produced by the eruption of El Chichón volcano in Mexico in 1982 was tracked by satellites. Note that dispersal of the cloud north and southward is limited. (Redrawn from fig. 1 in Robock, A. and Matson, M. (1983) Circumglobal transport of the El Chichón volcanic dust cloud. *Science*, **221**, 195-197. Reprinted with permission from the American Association for the Advancement of Science.)

8.2.2 Fallout from the umbrella region

The smaller clasts within the eruption column are carried up into the umbrella region, the region in which the column is neutrally buoyant and stops rising (Figs 6.5, 6.6 & 8.1). In this region the column spreads laterally (Fig. 8.2) and becomes what might be better called an eruption cloud. This spread is caused by the constant addition of material from below. In the absence of any wind, the umbrella cloud would spread out symmetrically so that if viewed from above it would appear circular. In most cases, however, the cloud is affected by the wind and moves downwind as it spreads laterally. This effect can be seen in images of active eruption clouds (Fig. 1.2) and is evident in the distribution pattern of deposition of the deposited fallout clasts (Fig. 8.2).

In the laterally spreading umbrella region the average upward velocity is essentially zero. Without any net support from a rising gas stream, the clasts that are still in this region will fall out progressively. They do not all do so at once because, although the average upward speed is zero, there is still a great deal of turbulence in the cloud, so some particles are swept up in updrafts while others are being helped to move down. As discussed below, the small particles carried up into umbrella clouds only fall very slowly when released, and may spread in the atmosphere for days to weeks (Fig. 8.3). The very smallest particles are the aerosols formed when droplets of water containing dissolved gases such as SO_2 collect around tiny silicate grains. Turbulence in upper atmosphere winds can carry these particles around the planet for months to years.

8.2.3 Fall speeds of pyroclasts

A clast released from an eruption column or umbrella cloud accelerates downward until the air drag retarding it is just balanced by its gravitational weight, at which point it reaches a steady final speed, its **terminal velocity**. The terminal velocity of a clast is dependent mainly on its radius and density and also, to some extent, its shape. In the early 1970s experiments were carried out to investigate the terminal velocities of clasts of various densities, sizes, and shapes by dropping them and filming their motion with high-speed cameras. For very small clasts, normal laboratories were high enough to allow clasts to reach their terminal velocity, but larger clasts required the use of the stairwells or tall buildings!

The experiments showed that the normal rules of fluid mechanics applied to volcanic particles provided due account was taken of their often very irregular shapes. We already used the relevant relationships in eqns 6.8 and 6.9 in Chapter 6 to describe how pyroclasts of a given size, density, and shape are suspended in the gas stream inside an eruption column. For a clast falling through air, the equivalent of eqn 6.9, now written in terms of the average diameter, d, of the clast instead of the radius, is

$$U_\mathrm{T} = \sqrt{\frac{4\,d\,\sigma\,g}{3\,C_\mathrm{D}\,\rho_\mathrm{a}}} \qquad (8.1)$$

where U_T is still the terminal velocity of a clast of density σ, but now it is the density of the atmospheric air, ρ_a, that controls the clast fall speed. The influence of the shape of the clast is represented by the value of the drag coefficient C_D, which has to be determined from experiments on actual volcanic particles. Furthermore the value of C_D for a given clast shape will depend on how its "average" diameter d is defined: d could be the arithmetic mean of the longest dimension of the clast and the two dimensions at right angles to this, or it could be defined as the diameter of a sphere with the same volume as the clast, and so on – various alternates have been used, and care must be taken to use consistent values of C_D.

In describing the support of clasts inside an eruption column, Chapter 6 was mainly concerned with the relatively large clasts in the lower part of the column. The gas flow around these pyroclasts is turbulent, and so it was possible to assume that the drag coefficient, C_D, in eqn 8.1 was nearly constant for a given clast shape. Here we also need to be concerned about the fall of very small particles released from the upper part of the column and the umbrella cloud. The air flow past these pyroclasts is laminar, i.e., smooth, not turbulent, and under these con-

ditions the drag coefficient is not a constant but instead is inversely proportional to a dimensionless number called the Reynolds number, Re. This number represents the ratio between the inertial forces and the viscous forces acting on the clast and is defined by

$$Re = (d\,U\,\rho_\mathrm{a})/\eta_\mathrm{a} \qquad (8.2)$$

where U is the speed of the clast through the gas and η_a is the viscosity of the gas. What is observed is that for spheres, C_D is equal to $24/Re$, and combining this with eqn 8.2 shows that in laminar flow conditions eqn 8.1 has to be replaced by

$$U_\mathrm{T} = (d^2\,\sigma\,g)/(18\,\eta_\mathrm{a}) \qquad (8.3)$$

For irregularly shaped particles the constants differ from 24 and 18, and again have to be determined from experiments, but the basic relationships are the same. Some examples of the experimental determinations of clast terminal fall speeds consistent with these theoretical equations are shown in Fig. 8.4. These values all correspond to conditions at ground level under average temperature and pressure conditions. Equation 8.1 shows that the terminal velocities of large clasts depend on the reciprocal of the density of the atmosphere, and the density, like the pressure, of the atmosphere decreases with height. This means that the terminal velocity of a large clast must be greater when it is high in the atmosphere than when it nears the ground. Similarly eqn 8.3 shows that the terminal velocities of small clasts depend on the reciprocal of the viscosity of the atmospheric gas. The viscosities of gases are mainly a function of temperature, decreasing as the temperature decreases, and since the atmospheric temperature generally decreases with height under normal conditions, the terminal velocities of small clasts, like those of large clasts, are greatest when they are high in the atmosphere.

8.2.4 Other factors affecting fallout from eruption columns

The physical processes described above and in Chapter 6, which determine how clasts are carried up in an eruption column and how they fall from

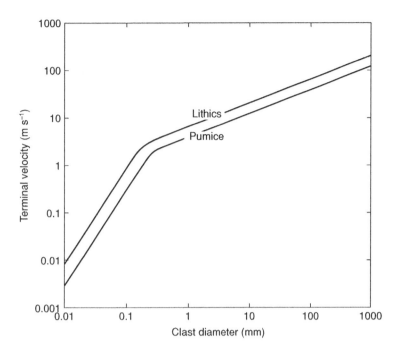

Fig. 8.4 Terminal velocities in the atmosphere at sea level of typical dense lithic clasts and vesicular pumice clasts.

the column, are quite complex. So far it has been shown how they depend on the eruption conditions (magma gas content and mass flux from the vent) and the general atmospheric properties, especially the decrease in atmospheric pressure with height. However, the properties of the atmosphere vary with latitude and season, and the water content of the lower atmosphere – the troposphere – varies enormously with local weather conditions. All of these factors influence the details of eruption column behavior and particle fallout, and so to simulate a particular eruption of a specific volcano one needs to use appropriate values for the properties of the local atmosphere. If it is a historic eruption in a populated area the weather conditions may be known extremely well, but if it is a prehistoric eruption and there is no information on the time of year when the event took place, the best we can do is to use the appropriate annual average conditions.

8.2.5 Common features of fall deposits

Study of fall deposits from recent steady eruptions that produce eruption plumes, including Hawaiian eruptions as well as subPlinian and Plinian events,

shows that certain characteristics are common to them all.

• The size of the largest clast found at any location in the deposit decreases with increasing distance from the vent. This is because the largest clasts fall out from the lowest heights in the eruption column where the column is narrowest, and therefore land closest to the vent (Figs 8.1 & 8.2).
• At any given site a fall deposit exhibits a range of clast sizes. This is mainly the consequence of turbulence carrying clasts to the column edge, allowing them to fall out prematurely, before reaching the maximum height that they could have attained in the column, and so land closer to the vent than expected. There are other causes. Large clasts have a high terminal velocity in the atmosphere and so may land at such a great speed that they break. Hot clasts may cool so much that thermal stresses cause them to crack. Both of these processes give a false impression of the maximum clast size, and this is why volcanologists always measure size and density of clasts. It is very common for a fall deposit to contain not only clasts derived from the erupting magma but also pieces of dense rock torn from the walls of the dike and vent system feeding the erup-

tion. These are usually called **lithic clasts**, and are far less likely to break on landing than fragile pieces of pumice. Although at any one site the lithic clasts are generally smaller than the lower density pumice clasts (recall from eqn 8.1 that it is ($d\sigma$), the product of size and density, that controls where clasts land), it is common to find that, near the vent, the value of ($d\sigma$) for the largest lithic clast is greater than the values from any of the pumices, due to pumice breakage, and it is this largest value of ($d\sigma$) that is used in any analysis of the deposit.

• The thickness of the deposit progressively decreases with distance from the vent. While this is generally true there are exceptions. Thus, if the deposit accumulates on a steep slope (greater than ~30°) slumping may occur; and at any location there may be erosion of the deposit by nonvolcanic processes after it is emplaced. Furthermore, soil formation by weathering can change the thickness of a deposit. There may also be effects related to the dynamics of the eruption itself. Thus, if there is an unusually large proportion of medium-sized pyroclasts leaving the vent, relative to the larger and smaller sizes, these will fall mainly at intermediate distances from the vent, and so a situation may occur where the downwind thickness of the deposit increases away from the vent for some distance before eventually decreasing in the more common way. Another complication can occur when an eruption column contains a lot of condensing water vapor, perhaps because the eruption occurs during a rain storm so that both air and water droplets are being entrained in the gas-thrust region. When this happens, a water film forms on the small pyroclasts and they can stick together progressively to form larger aggregates called **ash clusters** or **ash pellets**. In dry eruption clouds, frictional effects can cause particles to become electrostatically charged, and this too can cause aggregates to form. Being larger, aggregates fall to the ground faster than the small clasts forming them would have done, and again this can cause unusual patterns in the thickness of the deposit.

• A deposit may exhibit vertical **grading**. This means that the range of grain sizes at any given location in the deposit changes with height in the deposit. If the average grain size decreases upward in the deposit this is called **normal** grading; if the

average grain size increases upward the deposit is **inversely** graded. Grading can occur for two reasons. First, the wind speed may change during an eruption. An increase in wind speed will carry all clasts further from the vent, so at any one location increasingly larger clasts will be deposited with time and inverse grading will develop. Second, there may be a change in the mass flux of the eruption. An increase in mass flux causes an increase in the eruption cloud height (see eqn 6.7) and this too causes the dispersal of clasts of a given size to increase with time and so leads to inverse vertical grading. Both of these processes will act over most or all of a deposit. However, local grading can occur in regions of slumping on steep slopes due to the differential movement of clasts of differing sizes.

8.3 The application of eruption column models

A model is of value only if it actually reproduces behaviors that we see in reality. Theoretical eruption models predict how pyroclast dispersal is related to the eruption conditions and the atmospheric conditions. Since the development of models of this kind in the 1970s and 1980s there have been several well-observed eruptions for which the models have been shown to work quite well, certainly well enough to justify their use in diagnosing conditions during ancient eruptions whose products are preserved. For prehistoric eruptions, which include the largest eruptions in the geological record, events vastly greater than anything experienced thus far by humans (see Chapter 10), such techniques provide us with the only way of determining the mass fluxes and eruption cloud heights. This is very important in assessing the possible hazards presented by active volcanoes that have little or no historical record of eruption (see Chapter 11). Modeling has also enabled volcanologists to develop a long-term view of the effects that large eruptions might have had in the past and could have in the future on our climate and environment. This issue is discussed further in Chapter 12.

Both the testing and the application of the models require the same steps to be taken: the fall deposit produced by the eruption must be characterized in a way that makes it possible to compare the actual dispersal pattern with the theoretical predictions. We now describe how geologists go about doing this. Needless to say, it helps if the deposit is exposed and well-preserved at as many locations as possible.

8.3.1 Analyzing a fall deposit

At each location the thickness of the deposit is measured and the sizes and densities of the largest clasts (up to 10 of each type present) are obtained. This is done for each recognizable horizon within the deposit, and wherever possible these horizons are traced from location to location based on any distinctive properties (color, unusual grain size, etc.). For very large clasts the sizes are found in the field using a measuring scale, whereas smaller clasts are taken back to a laboratory and passed through standard-sized sieves. Note that the dimension that determines if a clast can pass through a sieve is neither its longest nor shortest dimension, so that typical shapes of clasts need to be recorded as well as size. In many cases a sample of the clasts will also be taken so that the overall size distribution of clasts in the whole deposit can be determined. The clast densities can be found in various ways: sawing clasts into cubes to measure their volumes and weighing them for their masses; or coating their surfaces with a waterproof layer and weighing them first in air and then under water. Recall that many pyroclasts will be very vesicular, and the **vesicles** (the holes previously occupied by volcanic gases) may be interconnected and open at the surface with many sharp edges or they may be closed off and trapped within the relatively smooth surface of a clast. The fact that there can be a range of clast types, even within a single eruption, can make measuring densities a complicated and often inaccurate procedure.

When all of the information on clast size and density and on deposit thickness has been collected, two maps are typically drawn for a deposit – an **isopach** map and an **isopleth** map. An isopach map shows the contours of deposit thickness. For example, Fig. 8.5a shows the isopachs for the main part of the pumice deposit from the 1875 eruption

of Askja volcano in Iceland. Notice how the isopachs converge towards a point, in this case just inside the north end of a lake called Öskjuvatn. This convergence point is the approximate location of the vent. The lake formed soon after the eruption finished when the ground above the magma reservoir that fed the eruption collapsed to form a small caldera.

An isopleth is a line joining points of equal clast size. An isopleth map will typically be drawn using the maximum clast sizes in a deposit and will therefore show contours of the maximum distance from the vent reached by clasts of a given size. Figure 8.5b shows the isopleth map for the pumice clasts from the 1875 Askja eruption. The isopleths also converge on the location of the vent. Note how the isopachs and isopleths indicate the direction of the wind; the more elongate the contours the stronger the wind that was blowing.

We can now discuss how the isopleth and isopach maps are used to quantify the eruption conditions.

8.3.2 Estimating the eruption rate and the eruption speed

Recall the comments made earlier about Figs 8.1 & 8.2: pyroclasts leaving the column at a given height and hence at a given distance from the vent are blown downwind, but the wind does not greatly change their distance from the vent measured at right angles to the wind direction, in other words their cross-wind range. This is the key to using fall deposits to analyze eruptions, and the critical steps in the whole process can be stated as follows. For a given set of eruption conditions (magma gas content and erupted mass flux) there is a specific height that the top of the resulting eruption column will reach, and there is a specific variation of gas ascent velocity and gas density with height in the eruption column. This means that there is a maximum height to which a pyroclast of a given size and density can be carried in the column. But this also means that there is a well-defined maximum lateral distance from the vent at which that pyroclast can be released and hence an equally well-defined maximum cross-wind distance at which that pyroclast will be found. There is thus a direct relationship

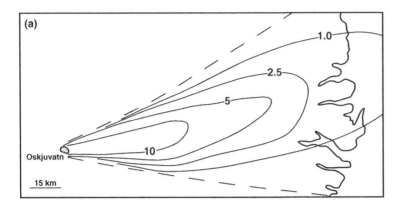

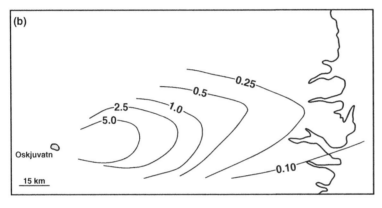

Fig. 8.5 (a) Isopach and (b) isopleth data for the main Plinian fall deposit from the 1875 eruption of Askja volcano, Iceland. (Redrawn from fig. 10 in Sparks, R.S.J., Wilson, L. and Sigurdsson, H. (1981) The pyroclastic deposits of the 1875 eruption of Askja, Iceland. *Phil. Trans. Roy. Soc.*, A **299**, 242–273.)

between the maximum cross-wind ranges of clasts and the eruption conditions that produced the eruption column that dispersed them. This means that the isopleth map can be used to find the maximum cross-wind ranges, b, of clasts of a given diameter, d; the diameter d is then multiplied by the corresponding clast density σ and the product $(d\sigma)$ is plotted as a function of b.

If this is done for a wide range of different eruption conditions, the results shown in Fig. 8.6 are found. At small distances from the vent, pyroclasts fall from the gas-thrust region of the eruption column, where the support for the clasts is mainly controlled by the upward speed of the gases in the column. We saw in Chapter 6 that the speed of materials coming out of the vent is mainly determined by the amount of gas exsolved from the magma; thus, not surprisingly, as the distance from the vent goes to zero, the curves in Fig. 8.6 converge on values of $(d\,\sigma)$ that are more or less independent of the mass eruption rate and are just a function of the magma gas content (the values quoted are the

water contents assuming that no other gases are exsolved). However, pyroclasts landing further from the vent were released from the convective thrust part of the column or the umbrella cloud. In these regions the support for clasts is controlled mainly by the gas speed corresponding to the momentum generated by buoyancy, and since the buoyancy is derived from the heat flux into the column, the gas speed is controlled by the mass eruption rate, independently of the magmatic gas content. Thus at large distances from the vent the curves in Fig. 8.6 are just labeled by the mass eruption rate.

Figure 8.7 contains the theoretical curves of Fig. 8.6 but also shows data points derived from field measurements on various named fall deposits. Notice how few data are available from near the vent in these eruptions. This is because it is common for the fall deposit near the vent to be hidden by material from later, smaller scale eruptions or destroyed when caldera collapse occurs, as happened after the 1875 eruption at Askja shown

Fig. 8.6 The theoretically predicted variation of the product of clast diameter and density ($d\,\sigma$) with cross-wind range b (see Fig. 8.2) for various values of the mass eruption rate and the magma gas content, expressed as the equivalent mass fraction of water exsolved from the magma. Values of mass eruption rate range from 10^6 to 10^9 kg s^{-1}. The short-dashed, long-dashed and solid curves are for 5, 3 and 1 wt.% water, respectively. (After fig. 8 in Wilson, L. and Walker, G.P.L. (1987) Explosive volcanic eruptions – VI. Ejecta dispersal in plinian eruptions: the control of eruption conditions and atmospheric properties. *Geophys. J. Roy. Astron. Soc.*, **89**, 657–679, copyright Wiley-Blackwell Publishing Ltd.)

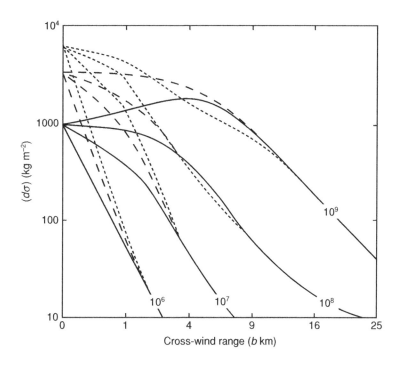

Fig. 8.7 Data for four well-studied eruptions plotted on the theoretical diagram of Fig. 8.6. (Data for Fogo 1563, Pompeii AD 79 and Taupo were provided by G.P.L. Walker; data for Askja 1875 are from Sparks, R.S.J., Wilson, L. and Sigurdsson, H. (1981) The pyroclastic deposits of the 1875 eruption of Askja, Iceland. *Philos. Trans. Roy. Soc. Ser. A*, **299**, 242–273.)

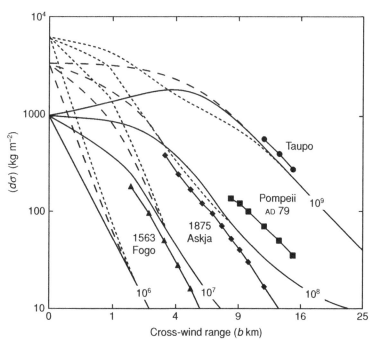

in Fig. 8.5. Also, notice that in some cases the observed data do not follow the trends of the theoretical curves very well. There is no doubt that in some cases this is a result of problems in the field where only a few poorly preserved exposures are available. In other cases erratic changes may have taken place in the mass eruption rate and exsolved magma gas content during the course of the eruption.

Table 8.1 Values of the magma eruption speed, the corresponding magma water content, the mass eruption rate, the eruption cloud height, and the factor by which the wind speed during the eruption exceeded the current annual average wind speed for the vent location, for a series of prehistoric (Avellino, Toluca and Fogo A) and historic (Pompeii, Askja and Fogo 1563) eruptions.

Eruption name and location	Gas eruption speed (m s^{-1})	Magma water content (wt%)	Mass eruption rate (kg s^{-1})	Eruption cloud height (km)	Wind speed factor
Avellino (Vesuvius)	210	1.1	2.9×10^8	31	1.4
Pompeii (Vesuvius, AD 79)	230	1.3	5.8×10^8	37	1.1
Askja (Iceland, AD 1875)	400	2.9	1.7×10^8	27	0.6
Toluca (Mexico)	500	4.3	2.3×10^8	29	2.3
Fogo A (Azores)	520	4.6	1.1×10^8	24	0.8
Fogo (Azores, AD 1563)	415	3.2	6.5×10^6	12	1.9

Data for Avellino, Fogo A, Pompeii and Fogo 1563 provided by G.P.L. Walker; data for Askja taken from Sparks, R.S.J., Wilson, L. and Sigurdsson, H. (1981) The pyroclastic deposits of the 1875 eruption of Askja, Iceland. *Philos. Trans. Roy. Soc. Ser. A*, **299**, 242–273; and data for Toluca taken from Bloomfield, K., Sanchez Rubio, G. & Wilson, L. (1977) The plinian pumice-fall eruptions of Nevado de Toluca Volcano, Central Mexico. *Geol. Rundsch.*, **66**, 120–146.

A third possibility is the failure of the model to take account of the detailed weather conditions prevailing during the eruption. For various eruptions, Table 8.1 gives values that have been deduced in this way for the magma eruption speed, the corresponding magma water content, the mass eruption rate, and the eruption cloud height. The cloud heights were found from the mass eruption rates using eqn 6.7.

8.3.3 Finding the wind speed

Clearly the elongation of the isopleths and isopachs in Fig. 8.5 must be an indicator of the wind speed during the eruption. In fact the wind speed – and even the wind direction – commonly varies with height under normal conditions on Earth, and so whatever is deduced from the deposit will represent some kind of average of conditions between the ground and the level of the top of the eruption cloud.

Application of the analysis described in the previous section will provide an estimate of the eruption cloud height and also the maximum height above the ground from which any given size of pyroclast has fallen. This means that the terminal velocities discussed in section 8.2.3 (taking account of the way they vary with height) can be used to find the times taken by each clast size to reach the ground from its release height. The next step is to find the differences between the maximum downwind and cross-wind ranges of clasts of a given size and density. This difference provides a good approximation to the downwind distance that the clast was transported by the wind while falling, and so dividing the transport distance by the fall time gives the average wind speed. A separate average wind speed is obtained from each clast size used in the analysis, and so some idea of the variation of wind speed with height can be deduced from these values. The last column of Table 8.1 shows some examples of average wind speeds deduced in this way. The values are given in terms of the amount by which the value deduced for the average wind speed exceeds the current annual average for the location of the vent. This serves as a check on whether the analysis is sensible. There is no particular reason to expect eruptions to occur in unusually windy or unusually calm conditions, so we might expect the average value of these wind speed factors to be unity; in fact for the six eruptions given it is 1.35, which seems a reasonable result given the small sample and all of the potential errors involved.

8.3.4 Finding the fall deposit volume and the eruption duration

The isopachs in Fig. 8.5a are the key to finding the total volume of magma erupted. Imagine a deposit that is a perfect cone. If you take a series of closely spaced horizontal slices through the cone, then the average of the areas of the top and bottom of a slice, multiplied by the slice thickness, is a good approximation to the volume of the slice. In the same way, the average of the areas of any two isopachs, multiplied by the difference between the deposit thicknesses that they represent, is a good approximation to the volume of the deposit between the isopachs. Adding up the contributions from all pairs of isopachs gives the total deposit volume. Of course, it is necessary to estimate the location of the zero-thickness isopach; and the more closely spaced the isopachs the more accurate the volume.

The resulting volume is the volume of the deposit as it lies on the ground. It is normal to convert this to the equivalent volume that it would occupy if all of the pore space within the pyroclasts and all of the vacant space between the rather loosely packed clasts were removed – the so-called dense rock equivalent (DRE) volume. Multiplying the DRE volume by the density of the solid version of the erupted magma gives the total mass of the deposit. These are important adjustments to make, because they allow the duration of the eruption to be estimated. This is done by dividing the mass of the deposit by the mass eruption rate derived from fitting the field data to the curves in Fig. 8.6. The estimate of the duration may not be very accurate because, as mentioned earlier, the mass flux may have changed during the eruption, and so the fit to the theoretical curves may not give a good estimate of the average eruption rate.

8.3.5 Fall deposits: summary

When used on prehistoric eruptions, the kinds of analyses described above, from which we can estimate magma volatile content, mass eruption rate, eruption cloud height, and wind speed and direction during the eruption, are best applied to the deposits from large Plinian eruptions. This is because the large dispersal area of such eruptions maximizes the chances of significant amounts of the deposit being preserved in the geological record and exposed for us to examine. There is clearly the potential to apply the same methods

to fall deposits from the convecting eruption clouds above Hawaiian eruptions, but little work has been done on this yet.

We now turn our attention to the second family of pyroclastic deposits, those emplaced not by fallout from a high eruption column but rather by flow of a mixture of clasts and gas as density currents close to the ground.

8.4 Pyroclastic density currents and their deposits

The deposits from pyroclastic density currents include some of the smallest volcanic layers preserved in the geological record and also the deposits from what are undoubtedly the largest-volume and most destructive volcanic phenomena on Earth. However, they have certain key properties in common which is why they are discussed together.

8.4.1 Nature of the deposits

The three main terms used to describe these deposits are **ignimbrite**, **pyroclastic surge deposit** and **block-and-ash flow deposit**, and the magmas giving rise to them are almost always of evolved composition. The deposits are dominated by **juvenile** material but commonly also contain **lithic** fragments torn from the walls of the dike and vent system feeding the eruption. The juvenile clasts include intact lapilli and blocks of vesicular pumice, glassy shards which represent fragments of pumice clasts which have been broken up, and broken pieces of crystals which grew in the magma prior to eruption and which were released when pumices shattered. Ignimbrites are generally ash-rich and very poorly sorted, forming extensive sheets or fans (Fig. 8.8) that cover large areas, up to tens of thousands of square kilometers. They bury or drape pre-existing topography and thicken in local depressions. Pyroclastic surge deposits are somewhat better-sorted than ignimbrites, although still poorly sorted, and show distinct internal stratification (Fig. 8.9). They may occur within or immediately adjacent to ignimbrite deposits. Block-and-ash flow deposits differ from the other two types in generally containing clasts that are less vesicular.

Fig. 8.8 A thin (~1 m thick) nonwelded ignimbrite flow unit, part of a deposit on the island of Terceira in the Azores. This unit shows a good example of a basal layer which is depleted in coarse clasts and an upward concentration in larger pumice clasts. (Image courtesy of Stephen Self.)

Fig. 8.9 Cross-bedding in surge deposits from hydromagmatic eruptions at the prehistoric Hana Uma vent, O'ahu, Hawai'I. (Photograph by Elisabeth Parfitt.)

They tend to be of small volume and are mainly associated with events in which lava domes or lava flow fronts collapse.

The term **flow unit** is used to describe the deposit from a single pyroclastic density current at a particular location. Any given eruption may produce a number of separate pyroclastic density currents resulting in a compound deposit consisting of several flow units. Many ignimbrites are emplaced hot enough that some **welding** between clasts occurs. Also there may be some compaction of the deposit under its own weight. This causes clasts that are hot enough to deform in a plastic fashion to be flattened to form structures called **fiamme** (Italian for flames). The gaps in time between emplacement events at a given location may allow one or more flow units to be laid down in quick succession and to cool significantly before another emplacement event occurs. This process produces **cooling units** which may be recognized by changes in color of the deposit, reflecting differences in oxidation rate as the mass cools, or differences in the amount of flattening or welding of clasts.

The morphological features of the deposits from pyroclastic density currents that do not completely bury the pre-existing topography suggest that the bulk of the flowing material has a density greater than that of the surrounding atmosphere. Thus the distributions of the deposits indicate that the cur-

Fig. 8.10 A pyroclastic density current erupted from the crater of Mount St Helens volcano on August 7, 1980. The dense, ground-hugging current is almost completely hidden by the low-density convecting cloud formed above it as hot gases lift small particles out of the body of the current. (Photograph by Peter W. Lipman, courtesy of U.S. Geological Survey, Cascades Volcano Observatory.)

rents moved downslope at right-angles to contours in a ground-hugging fashion. Observations of pyroclastic density currents that have been seen in historical eruptions confirm this tendency (Fig. 8.10), but also show a cloud of gas and fine material forming a convecting cloud above the basal part of the current. This strongly implies that there is a vertical variation of properties within the density current, i.e., there is density stratification. Other features of pyroclastic density current deposits support this idea. Thus in places where the bulk of a deposit lies in a valley, there may be a veneer of pyroclastic material on top of ridges on either side of the valley or in adjacent valleys. And in some cases it has been observed that pyroclastic density currents reaching an ocean or lake shore split into a component that travels under the water, following the topography, and a component that travels for some distance across the water surface.

In some instances, ignimbrites are found in adjacent valleys separated by ridges up to 1500 m high. At one time this was taken to indicate that the ridge height could be used as a measure of the speed of the pyroclastic density current forming the deposit.

The method was to equate the potential energy needed to raise material to the height h of the ridge ($g\,h$, where g is the acceleration due to gravity) to the kinetic energy of material approaching the ridge at speed v ($0.5\,v^2$). For a ridge 1500 m high this would imply a speed of ~170 m s^{-1}. Speeds approaching 100 m s^{-1} have been observed for historic pyroclastic density currents, so this was assumed to be plausible. However, the realization that such currents have a vertical density stratification casts doubt on the reliability of this kind of calculation. It may be that only the upper parts of a vertically extensive current cross the ridge, implying less vertical rise of material and hence a smaller speed. Nevertheless, it is clear that speeds up to ~100 m s^{-1} are a common feature of all types of pyroclastic density current, and this must be linked to the ways in which they form.

8.4.2 Origins of pyroclastic density currents

Chapter 6 introduced one mechanism for forming these features: the collapse of Plinian and sub-Plinian eruption columns to form pyroclastic fountains. Other mechanisms that have been observed in historic eruptions include the formation of "directed blasts" by the partial disruption of viscous lava domes growing over vents, the complete collapse of viscous lava domes, and the explosive disintegration of the fronts of viscous lava flows.

COLUMN COLLAPSE AND PYROCLASTIC FOUNTAINING

Considerable potential for confusion exists in the literature in connection with the term "column collapse" as a way of forming pyroclastic density currents. The actual collapse of what had previously been a stable eruption column is a discrete, fairly short-lived event. If an eruption column ceases to be stable because the bulk density of the material within it becomes greater than that of the surrounding air, the material at the top of the column will take a similar length of time to reach the ground as a stone dropped from that height. The time needed to fall a distance s when the acceleration due to gravity is g and there is no drag force from the surrounding atmosphere is $[(2s)/g]^{1/2}$, so the collapse times for eruption columns 20, 30 and 40 km high are ~63, 77 and 89 seconds. In practice there will be some interaction of the outer part of the collapsing column with the surrounding air, but it is still clear that the process will take only a few to at most several minutes.

There is no reason why the collapse of an eruption column should cause the eruption to stop; and it is clear from well-studied deposits that post-collapse eruptions may continue for hours or even days. We have no observational evidence to tell us what happens at the vent during and after a large-scale column collapse event because no one who has ever been close to one has lived to describe it (Chapter 11). A mixture of pyroclasts and volcanic gas is still emerging from the vent at high speed, but the mixture is no longer able to entrain enough air to become buoyant. Theory suggests that a fountain would form over the vent (Fig. 8.11). On reaching the ground, gas and pyroclasts falling down the outer edge of this fountain would move away as a pyroclastic density current. We can get some idea of the height of the fountain because, although the material in its outer part would interact with the atmosphere, the material in the core would mainly

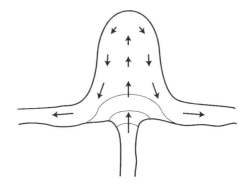

Fig. 8.11 Diagram showing the formation of a pyroclastic fountain over a vent when insufficient air is entrained into the eruption column to cause convection. The arrows show the direction in which gas and particles are moving, and the thin solid lines show contours of constant pressure.

be influenced just by gravity. The height, h, reached by an object thrown vertically upward at speed u when there is no interaction with the atmosphere is given by $[u^2/(2g)]$. Table 8.1 showed that magma eruption speeds in Plinian eruptions range from 200 to ~500 m s^{-1} for magma water contents in the range 1 to 5 wt%. If this range of values is used, the corresponding fountain heights are predicted to be ~2 km to ~13 km!

In fact this analysis is too simple. The material falling down the outer part of the fountain exerts a pressure on the ground surface where it lands (imagine standing under a waterfall), and in fact it is the distribution of high pressure around the vent that is responsible for deflecting the fountain material from its near-vertical fall into a ground-hugging flow (Fig. 8.11). The high pressure at the level of the vent has another effect. It reduces the speed with which the mixture of pyroclasts and magmatic gas emerges from the vent because it reduces the amount of gas expansion in the shallow part of the dike system, as discussed in section 6.4.2. The erupting mixture speeds up as the gas expands above the vent, but some of the energy released by the expansion must be used to give the mixture a lateral velocity as well as a vertical velocity, and so the effective upward speed that we should use to find the fountain height is less than it would be in an eruption column that has not collapsed. Numerical simulations of this process (Table 8.2) suggest that

Table 8.2 Parameters of pyroclastic fountains forming pyroclastic density currents. For a range of pre-eruption magma water contents dissolved in the magma chamber, n_{dis}, values are given for the amount exsolved from the magma emerging from the vent, n_{exs}; the pressure at which the gas emerges, P_v; the speed at which the gas and entrained pyroclasts emerge, U_v; the bulk density of the emerging mixture, β_v; and the bulk density β_e in the gas–clast mixture after the gas has decompressed to reach equilibrium with the atmospheric pressure.

n_{dis} (%)	n_{exs} (%)	P_v (MPa)	U_v (m s^{-1})	β_v (kg m^{-3})	β_e (kg m^{-3})
1.57	1.00	1.93	89.4	360	13.7
2.81	2.00	3.91	126.4	365	7.7
4.00	3.00	5.93	154.8	369	5.4
5.16	4.00	7.99	178.7	373	4.2
6.31	5.00	10.08	199.8	376	3.4

a pressure of up to 10 MPa must exist at the vent, the value increasing with the amount of gas in the magma. The consequence is to reduce the upward speed into the fountain to values in the range 100 to 200 m s^{-1}, leading to revised fountain heights of 1 to 8 km for magma water contents in the range 1 to 5 wt%. The smallest of these fountain heights, especially for even smaller magma gas contents, probably corresponds to what has been described as "magma boiling over from the vent" in the few very small-scale eruptions of this kind that have been observed.

DIRECTED BLASTS AND COLLAPSES FROM LAVA DOMES AND FLOWS

The common factor in these three processes seems to be that a body of viscous magma is being erupted, or has recently been erupted, to form a lava dome or a short lava flow in a way that has allowed a continuous, stable, cooled surface layer to form on the lava. For a while, this cooled shell or carapace has enough strength to resist the outward pressure exerted on it by gas trapped in vesicles within the lava body. Then, something disturbs the stability of the shell. This may be a sudden increase in the rise rate of magma in the dike beneath the lava body, the increased deformation rate causing the rheological response to change from plastic to brittle; it may be the fact that some part of the edge of the lava body becomes too steep and collapses under its own weight – this might happen where the dome edge or flow front advanced onto a steeper slope. A third option is the one that caused the start

of the 1980 eruption of Mount St Helens. In that case magma was intruded beneath the surface of the steep flank of the volcano. This caused the flank to bulge, oversteepening it until it collapsed as a landslide, thus uncovering the intrusion and almost instantly relieving the pressure exerted on it. Bubbles of gas that had already exsolved from the magma abruptly expanded and the magma exploded.

No matter what the triggering mechanism, the subsequent events are probably very similar in all of these types of eruption. The pressure on a layer of trapped gas bubbles is released and the outer walls of these bubbles break, freeing the pressurized trapped gas which starts to expand and carry with it the bubble-wall fragments that have been formed. In other words, this first layer of bubbles explodes in a kind of Vulcanian explosion. But this releases the pressure on the next exposed layer of trapped gas bubbles, and these too explode. A wave of decompression, called an **expansion wave**, travels into the exposed lava body, and this feeds the expanding mixture of gas and fragments until, in the extreme case, the entire lava body is destroyed. The speed of the expansion wave will be a large fraction of the speed of sound in the lava which, because of the presence of the gas bubbles, will be a few to several tens of meters per second. The speed reached by the expanding gas and pyroclasts will depend, as we have seen for all explosive eruptions, on the gas mass fraction and the difference between the pressure in the trapped gas and the atmospheric pressure. The description of Vulcanian activity in Chapter 7 provides the closest approximation to what is happening, with the strength of the confining layer defining the pressure in the trapped gases. The equivalent of the calculations given in Chapter 7 suggests that for the magmatic volatile contents of order 1–3 wt% that commonly apply to these evolved magmas, and for carapace strengths of a few megapascals, the initial speeds in these kinds of explosions could easily be at least 100 m s^{-1}.

The fact that this process causes the development of a pyroclastic density current, rather than a vertical eruption column, must be strongly influenced by the bulk density of the mixture of gas and pyroclasts that is produced by the lava disintegration; it will also be at least partly dictated by the detailed geometry of the lava body and the location where the disintegration starts. Of course, if the erupted lava is connected to hot magma in the underlying dike system, it may well be that, as soon as the overlying erupted lava is removed, the subjacent magma explosively erupts upward and forms a Plinian eruption column (as at Mount St Helens in 1980) or, if this would not be stable, forms a pyroclastic fountain. Certainly, numerous examples have been observed of vertical eruption columns being produced very shortly after dome collapse events.

8.4.3 Ignimbrite emplacement mechanisms

Our understanding of the way pyroclastic density currents emplace their characteristic deposits has advanced enormously over the past two decades, although there are still many details that are not fully understood. One of the most important issues relates to the density of the currents while they are moving and the motions within them – especially the extent to which those motions are turbulent or laminar. These factors control the way clasts are deposited on the ground as the current passes a given location.

Because of the way they are formed, the currents are likely to start out as homogeneous mixtures of gas and pyroclasts, but they may have a wide range of bulk densities. If a current is formed from a pyroclastic fountain, the analysis described in the previous section shows that the pressure in the gas emerging from the vent will be 1 to 10 MPa (Table 8.2) and as a result the bulk density of the gas–pyroclast mixture will be in the range 360–370 kg m^{-3}. This is much greater than the density of the Earth's atmosphere, ~1 kg m^{-3}. However, as the mixture accelerates laterally away from the vent, its bulk density will rapidly decrease as the gas decompresses to atmospheric pressure. The bulk density will become similar to that expected from the disintegration of a lava dome or lava flow; the values, obtained from eqn 6.2, range between ~3 and ~20 kg m^{-3} (Table 8.2), depending on the magma volatile content. Even the smallest of these densities, corresponding to a very volatile-rich magma, is greater than that of the atmosphere. This ensures that a ground-hugging density current is always formed, whether from a pyroclastic fountain or the disintegration of a dome or lava flow, but clearly there is the potential for pyroclastic density currents to have a wide range of initial particle concentrations.

As soon as a pyroclastic density current moves away from its source, its top begins to entrain the air above it in the same way that the edge of an eruption column entrains adjacent air. This process starts even sooner if the current is fed by a fountain, since some mixing with the surrounding atmosphere must happen on the outer surface of the fountain as material descends toward the ground. This inclusion of air must decrease the bulk density of the part of the current involved, and so it begins to cause a density stratification within the current. As part of the interaction between the density current and the overlying air, a mixture of heated air and volcanic gas **elutriates** some fraction of the smallest clasts from the main body of the current to form a growing convection cloud above it. This is called a **co-ignimbrite cloud**, or **phoenix cloud** (see Fig. 8.10). These clouds may be carried by the wind in a quite different direction from that of the density current, eventually depositing their small clasts to form a characteristic **co-ignimbrite ash fall deposit**.

There is another important source of stratification within the main body of the density current: the tendency of clasts to fall through the gas phase. In principle every pyroclast will try to reach its equilibrium terminal velocity in the gas, which eqns 8.1 and 8.3 show to be a function of the clast size and density together with the properties of the gas. However, several processes act to prevent this happening. The separations between the moving particles may be small enough that collisions between particles moving at different speeds are common. As clasts settle downward, i.e., undergo **sedimentation** through the gas–clast mixture, they displace upward the gas and the very small particles effectively locked to the gas by their extremely small terminal velocities – we speak of a **dusty gas**. These processes collectively lead to strongly **hindered settling**. When particle-particle collisions dominate the interaction the term **granular flow** is used to describe the motion. In contrast, at least in the upper parts of the flow, collisions between particles may be much less frequent, and the bulk motion of the gas–clast mixture is turbulent, which leads to a constant stirring of the mixture, also reducing particle settling.

The nature of the deposit formed by a pyroclastic density current appears to be determined by the relationship between the upper turbulent zone and the lower hindered-settling zone. The deposit consists of all of the clasts which have segregated to the base of the current and ceased to have any lateral movement. Thus the boundary between deposit and current is constantly moving upward relative to the pre-eruption ground surface. In some currents, it appears that the turbulent dilute zone extends essentially all the way to the base of the current. There is a very large gradient in the horizontal velocity between the current and the deposit, and **tractional** forces cause clasts at the boundary to be rolled, dragged or bounced (**saltated**) along the interface before they come to rest. This appears to be what produces the internal stratification in pyroclastic surge deposits (Fig. 8.9). In contrast, if the base of the current is dominated by a laminar zone of hindered settling and granular flow, the gradient in the horizontal velocity between the current and the deposit is small, traction is minimal, and stratification is essentially absent.

Figure 8.12 shows these two extreme velocity distributions near the base of a pyroclastic density current. They represent end-members of a continuum of possible configurations. Which type of deposition will dominate at a given location in the deposit can change with time if conditions at the vent (mass flux, magma volatile content) are changing; thus vertical, as well as lateral, variations in the texture and grain size of the deposit are readily explained. Furthermore, to the extent that irregular topography will have its greatest influence on the motion of the densest part of the current, it is easy to understand how lateral gradations between ignimbrites and surge deposits can occur.

A final comment concerns deposits from very small pyroclastic density currents. Many of the deposits formed by currents produced in the dome-collapse events at Mount St Helens in 1980 were somewhat reminiscent of lava flows insofar as they had distinctive levées, flow fronts, and central channels (Fig. 8.13). The levées and flow front deposits were dominated by coarse pumice clasts and the central channels contained particles that were very much finer and supported occasional much larger pumice clasts at the surface. Examination of these deposits with instruments very shortly after their emplacement showed that the central channel materials had extremely nonNewtonian rheological properties, and attempts were made to link the structures of the deposits with the way the moving density currents had interacted with the atmosphere at their flow fronts and margins. On the scale of these small deposits (lengths of a few kilometers, widths of ~30 m, thicknesses of ~2 m) this was probably valid, but it seems unlikely that incorporation of atmospheric air has much influence on the depositional processes at the bases of large-scale pyroclastic density currents; the only significant effect is the production of phoenix clouds.

8.5 Summary

- Eruption columns contain a wide range of clast sizes, and some clasts of all sizes can fall out of the column at all heights. However, whereas small clasts can be carried right to the top of the column, there is a maximum height to which large clasts can be carried, and the larger the clast the smaller the maximum height.
- The well-defined shapes of eruption columns (Fig. 8.1) means that the lateral distance at which a clast of a given size and density is released can

(a)

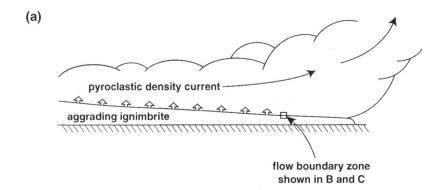

(b)

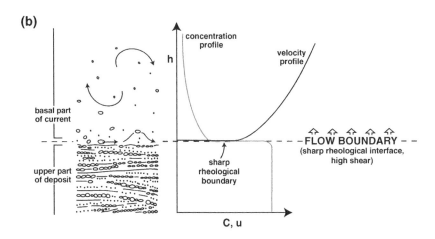

(c)

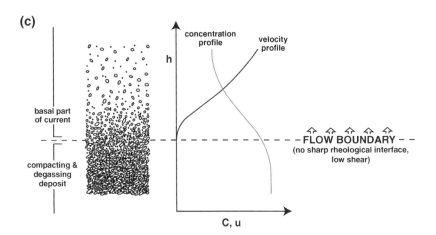

Fig. 8.12 Possible vertical variations in particle concentration, *C*, and horizontal velocity, *u*, in pyroclastic density currents. (Taken from fig.1.2 in Branney, M.J. and Kokelaar, P. (2002) *Pyroclastic Density Currents and the Sedimentation of Ignimbrites*. Geol. Soc. Memoir No. 27, Geol. Soc. Lond.)

Fig. 8.13 The deposit from a pyroclastic density current erupted from Mount St Helens volcano in August, 1980. The current bifurcated just before coming to rest, forming two lobes, each about 10 m wide. The coarse, rubbly levées are about 2 m high and the mainly fine-grained central channel deposit is ~1 m deep. (Photograph by Lionel Wilson.)

be related to the height from which it is released. The wind blows all clasts downwind, but if their positions on the ground are measured at right-angles to the wind direction we obtain a record of their release distances and hence the release heights.

- The release height of a clast of a given size and density is an indicator of the rise speed of the column at that height. Low down in the column, the rise speed depends mainly on the speed of the gas coming out of the vent and hence on the magma volatile content. Higher in the column, the rise speed depends mainly on the heat flux driving the column and hence on the mass flux of magma being erupted from the vent (Fig. 8.6).

- Observations of pyroclast dispersal patterns (Fig. 8.5) can be compared with theoretical predictions (Fig. 8.7) to deduce erupted mass fluxes, which also define the column heights,

and magma volatile contents (Table 8.1). If we add up the deposit volume from the isopachs, convert this to a corresponding total mass, and then divide the mass by the mass flux, the duration of the eruption can be calculated. Finally, the asymmetry of the deposit, together with the cloud height, can give us an estimate of the wind speed during the eruption.

- When a mixture of gas and pyroclasts forming an eruption column in a steady explosive eruption cannot obtain enough buoyancy from mixing with the surrounding air as it emerges from a vent, the lower part of the column collapses to form a fountain over the vent. The upper part of the column continues to convect and drifts away in the direction of the wind. The fountain now being fed by the eruption leads to the formation of one or more pyroclastic density currents.

- Pyroclastic density currents are fast-moving, ground-hugging clouds of very hot clasts and gas that flow like liquids. They contain a very wide range of clast sizes. Their speeds can be at least 100 m s^{-1} and their travel distance can be many tens of kilometers. Although they mainly flow along valley floors, these speeds imply that they can climb over topographic obstacles in their paths a few hundred meters high if there is no way to flow around.

- Smaller scale pyroclastic density currents can occur when only part of an eruption column collapses, or when a volcanic dome or lava flow becomes unstable and disintegrates into pyroclasts and released gas. These travel to smaller distances, although still at high speeds. A pyroclastic surge can form when the upper, more dilute, part of a pyroclastic density current becomes detached from the lower, denser part when the flow meets an obstacle.

- As they travel, pyroclastic density currents deposit pyroclasts on the ground. The deposits from large pyroclastic density currents dominated by grain flow, called ignimbrites, tend to be massive, whereas surge deposits formed from more dilute currents exhibit internal stratification.

- The fronts and tops of pyroclastic density currents can incorporate air which is strongly heated and rises above the body of the density current, carrying small clasts up with it to form a co-ignimbrite or phoenix cloud which can itself generate a fine-grained fall deposit.

8.6 Further reading

TEPHRA DISPERSAL IN STEADY ERUPTIONS

Carey, S. & Sparks, R.S.J. (1986) Quantitative models of the fallout and dispersal of tephra from volcanic eruption columns. *Bull. Volcanol.* **48**, 109-25.

Sparks, R.S.J. (1986) The dimensions and dynamics of volcanic eruption columns. *Bull. Volcanol.* **48**, 3-15.

Sparks, R.S.J., Bursik, M.I., Ablay, G., Thomas, R.M.E. & Carey, S.N. (1992) Sedimentation of tephra by volcanic plumes. Part 2: controls on thickness and grain size variations of tephra fall deposits. *Bull. Volcanol.* **54**, 685-95.

Walker, G.P.L. & Croasdale, R. (1971) Two Plinian-type eruptions in the Azores. *J. Geol. Soc. London* **127**, 17-55.

Wilson, L. & Walker, G.P.L. (1987) Explosive volcanic eruptions – VI. Ejecta dispersal in Plinian eruptions: controls of eruption conditions and atmospheric properties. *Geophys. J. Roy. Astron. Soc.* **89**, 657-79.

PYROCLASTIC DENSITY CURRENTS

Branney, M.J. & Kokelaar, P. (1992) A re-appraisal of ignimbrite emplacement: progressive aggradation and changes from particulate to nonparticulate flow during emplacement of high-grade ignimbrite. *Bull. Volcanol.* **54**, 504-20.

Branney, M.J. & Kokelaar, P. (2002) *Pyroclastic Density Currents and the Sedimentation of Ignimbrites.* Memoir 27, Geological Society Publishing House, Bath, 143 pp.

Burgisser, A. & Bergantz, G.W. (2002) Reconciling pyroclastic flow and surge: the multiphase physics of pyroclastic density currents. *Earth Planet. Sci. Lett.* **202**, 405-18.

Bursik, M.I. & Woods, A.W. (1996) The dynamics and thermodynamics of large ash flows. *Bull. Volcanol.* **58**, 175-93.

Druitt, T.H. (1998) Pyroclastic density currents. In *The Physics of Explosive Volcanic Eruptions* (Eds J.S. Gilbert & R.S.J. Sparks), pp. 145-92. Special Publication 145, Geological Society Publishing House, Bath.

Levine, A.H. & Kieffer, S.W. (1991) Hydraulics of the August 7, 1980, pyroclastic flow at Mount St. Helens, Washington. *Geology* **19**, 1121-4.

Neri, A. & Macedonio, G. (1998) Numerical simulation of collapsing volcanic columns with particles of two sizes. *J. Geophys. Res.* **101**, 8153-74.

Sparks, R.S.J., Wilson, L. & Hulme, G. (1978) Theoretical modelling of the generation, movement and emplacement of pyroclastic flows by column collapse. *J. Geophys. Res.* **83**, 1727-39.

Wilson, C.J.N. (1984) The role of fluidization in the emplacement of pyroclastic flows, 2: experimental results and their interpretation. *J. Volcanol. Geotherm. Res.* **20**, 55-84.

Wilson, L. & Head, J.W. (1981) Morphology and rheology of pyroclastic flows and their deposits, and guidelines for future observations. In *The 1980 Eruptions of Mount St. Helens, Washington. U.S.Geol. Surv. Prof. Pap.* **1250**, 513-24.

Wilson, L. & Heslop, S.E. (1990) Clast sizes in terrestrial and martian ignimbrite lag deposits. *J. Geophys. Res.* **95**, 17,309–314.

Woods, A.W., Bursik, M.I. & Kurbatov, A.V. (1998) The interaction of ash flows with ridges. *Bull. Volcanol.* **60**, 38–51.

8.7 Questions to think about

1 If you were able to find only one exposure of a fall deposit but knew the likely vent it came from, what could you infer about the eruption?

2 What should be the main difference between a fall deposit produced by an eruption on a calm day and a fall deposit produced by the same eruption conditions on a windy day?

3 A volcano 1000 m high with steep slopes has a volatile-poor eruption that ejects gas and pyroclasts at 100 m s^{-1}. The erupted mixture collapses down the side of the volcano to produce a pyroclastic density current. What is the maximum speed of the current when it reaches the foot of the volcano? Assume that the acceleration due to gravity is 10 m s^{-2}.

4 What do you think might happen to a pyroclastic density current if it reaches a shore line and starts to flow over the water surface?

9 Lava flows

9.1 Introduction

A lava flow is probably the first kind of feature that comes to mind when most people think about volcanic activity. The kinds of lava flows most commonly illustrated in documentary films are the basaltic flows that form on the shield volcanoes of Hawai'I (Fig. 9.1). There are, however, other kinds of lava flow, especially those formed by magmas of more evolved composition (Figs 9.2 & 9.3), and flows can differ greatly in terms of length (from several meters to literally a few hundred kilometers), volume, geometry, advance rate (from less than millimeters per second to several meters per second), and potential threat (see Chapter 11). This chapter first describes the general morphological features of lava flows and then considers the factors that control the features of both individual lava flows, called **flow units** after they have come to rest, and of collections of flow units, called **lava flow fields**.

Fig. 9.1 A pahoehoe toe lobe advancing over older lava. Note the rope-within-rope structure of the central part of the ~2 m wide lobe and the smoother surface of the small break-outs. (Photograph by Pete Mouginis-Mark, University of Hawai'I.)

Fig. 9.2 A dacite dome formed during the eruption of Unzen volcano, Japan, that began in May 1991. (Photograph by K. Scott in 1994, courtesy U.S. Geological Survey.)

Fig. 9.3 A rhyolite flow forming the ~80 m high Novarupta dome marking the site of the vent for the June 1912 eruption in Alaska that emplaced the Valley of Ten Thousand Smokes ignimbrite. (Photograph taken by T. Miller, courtesy U.S. Geological Survey.)

9.2 Origin of lava flows

The simplest way that a lava flow can form is when magma overflows from a volcanic vent onto the surrounding surface and moves downhill under gravity. We have seen that most magmas exsolve volatiles as they approach the surface, and thus should not be surprised to find that most active lava flows contain significant numbers of volcanic gas bubbles which, after the flow has cooled and atmospheric air has replaced the original gas in the interconnected bubbles, are quite vesicular. We know from the evidence in Chapter 5 that the high juvenile volatile content of many magmas has the potential to make them erupt explosively. If the magmas erupt on the deep ocean floor on Earth, the pressure of the overlying water greatly reduces the amount of gas that can exsolve, and so many of these magmas behave as though they had a low volatile content and do not erupt explosively.

However, even when volatile-rich magmas erupt subaerially, there is still a way that lava flows can be formed. All that is required is that at least some of the clots of fragmented magma ejected through the vent into a fire-fountain or into the base of an eruption column should fall back to the ground near the vent sufficiently quickly and in sufficient numbers. "Sufficiently quickly" is important so that the clots do not have time to cool much while they are in flight. Short travel times are encouraged by the magma having a relatively low volatile content

(although still high enough to cause the explosive activity of course) so that the speeds of the clots leaving the vent are not too large. "In sufficient numbers" implies that the mass eruption rate from the vent is high; this has two effects. First, if there is a large mass flux there will be a large number of clots in flight at any one time and so they form a dense cloud, with clasts on the outside shielding those near the middle from being able to "see" the sky and radiate away their heat. Second, with a large mass flux each clot that lands on the ground will be buried by the later-arriving clots more quickly, again conserving its heat. When the accumulation rate of hot clasts is large enough they completely merge together, the slightly cooled outer skins of the clots being reheated by the hot interiors as heat is shared by conduction, and the whole assemblage moves downhill away from the vent as a **rootless lava flow** (Fig. 9.4). The temperature of such a flow may be only a few kelvins to a very few tens of kelvins less than that of a flow formed directly by effusion from a vent.

It is very common for basaltic magmas erupting as fire-fountains to form rootless lava flows as part of a spectrum of features. If clots lose almost no heat, and coalescence is complete, a flow forms. If clots lose rather more heat, their skins may reach a high enough temperature, after heat sharing with the interiors, to weld together, but the interiors may then be too cool to deform significantly and so no flow occurs; instead a welded **spatter rampart** forms (Fig. 9.5). Finally, if too much heat is lost in flight, the clots accumulate as a warm but unwelded **scoria** deposit (Fig. 9.6). Figure 9.7 shows qualitatively how the combinations of magma volatile content and mass flux through the vent control this range of possible features.

In principle the same range of processes can occur in more evolved magmas. However, in dacitic magmas it is not common for the combinations of volatile content and eruption rate to favor this. Rhyolitic magmas, on the other hand, are much more likely to erupt at high mass fluxes and, because of high volatile contents, very explosively. The resulting great efficiency of magma fragmentation causes the average sizes of the pumice clasts that are formed to be small which, together with the high mass flux, makes the jet of clasts and gas

Fig. 9.4 A rootless lava flow being formed from a lava fountain at the Pu'u 'O'o vent on Kilauea volcano, Hawai'I, in 1984. (Photograph by J.D. Griggs, courtesy U.S. Geological Survey, Hawaiian Volcano Observatory.)

Fig. 9.5 View of one side of a welded spatter rampart on Kilauea volcano, Hawai'I. (Photograph by Lionel Wilson.)

Fig. 9.6 An unwelded deposit of scoria clasts. The scale is in centimeters. (Photograph courtesy of Rebecca Horne.)

emerging from the vent very opaque, thus conserving heat. The same mass of material broken up into a large number of small particles, rather than a small number of large particles, forms a much more opaque cloud: compare walking through mist or cloud, consisting of tiny water droplets, to walking through a normal rain shower. Thus in some explosive rhyolitic eruptions complete welding and some flowage of pumice takes place, forming a **rheomorphic** deposit. This can happen in the proximal parts of plinian fall deposits, but is much more likely to happen in ignimbrites (Fig. 9.8). The great viscosity of rhyolitic magmas ensures that such rheomorphic flows do not normally travel for more than a few meters to tens of meters.

9.3 Types of lava flow

Empirically, the morphological shapes of lava flows can be classified into a few general categories.

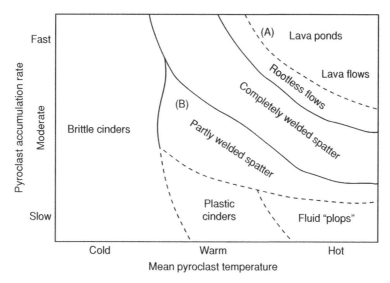

Fig. 9.7 Diagram showing how the mass flux erupted from the vent and the exsolved volatile content of the magma control the type of deposit produced near the vent in an explosive eruption. The pyroclast accumulation rate is directly proportional to the erupted mass flux. A high volatile content causes pyroclasts to be erupted at high speeds and so to travel further before landing, giving them a greater opportunity to cool. Hot deposits form preferentially from low volatile-content, low-speed eruptions. (Adapted from fig. 5 published in *Journal of Volcanology and Geothermal Research*, Vol. 37, Head, J.W. and Wilson, L., Basaltic pyroclastic eruptions: influence of gas-release patterns and volume fluxes on fountain structure, and the formation of cinder cones, spatter cones, rootless flows, lava ponds and lava flows, 261–271, copyright Elsevier (1989).)

Fig. 9.8 A rheomorphic ignimbrite deposit in the Trans-Pecos volcanic province of southwest Texas, USA. Note the extreme flattening and stretching of dense, highly alkaline juvenile (pumice) clasts caused by the lateral flow of the hot, welded deposit. (Image courtesy of Stephen Self.)

These morphologies are intimately connected with the rheological properties of the flow materials, as will be seen shortly, and also depend on the mass or volume eruption rate of the lava and the slope of the ground over which the lava is flowing, but it is convenient to describe the general features first.

Many lava flows, especially those of basaltic composition but also some more intermediate flows, consist of lava moving in a central zone bounded by a stationary accumulation of rock on either side which consists of lava that has been carried down the channel to the flow front and then pushed aside to make way for the liquid following it. Flows with these properties are called **channelized lava flows**, and the stationary banks bordering the central channel are called **levées** (French for "raised" features) because the depth of lava in the channel is often (but not always) less than the height of the levée (see Fig. 9.1). In other cases, the boundary between channel and levée is much less distinct, and all of the lava over almost the entire width of the flow is moving. In this case we have a **sheet flow** (Fig. 9.9). There is a strong tendency for sheet flows to be wider, for a given thickness, than channelized flows, and there is some indication that they are commoner on the ocean floor than on land. This may be at least partly related to the fact that sheet

Fig. 9.9 A basaltic sheet flow forming from a breakout from a pahoehoe lava flow. (Photograph by Richard Hoblitt, courtesy U.S. Geological Survey, Hawaiian Volcano Observatory.)

flows are more commonly of basaltic, rather than any other, composition and, as shown in Chapter 1, most eruptions on the ocean floor are of basaltic lava.

An immediately obvious feature of all lava flows is that as they spread away from a vent they initially move downhill and grow in width, but soon the lateral spreading ceases and the flow maintains a fairly constant width thereafter, unless significant changes in ground slope occur. This is one consequence of the characteristic cooling-induced rheology of the lava and is in marked contrast to the spreading of liquids such as water: a water flood will continue to spread sideways as well as downhill more or less indefinitely, at least until it becomes so thin that surface tension forces become important. Lava too has a surface tension, but its effects are confined to influencing the sizes of the gas bubbles it contains, as was seen in Chapter 5, and are quite irrelevant at the scale of even small lava flows.

Another striking feature of lava flows is that any one flow will not continue to lengthen indefinitely but, for a given set of eruption conditions, will reach a well defined maximum length. This maximum length is ultimately controlled by the cooling of the flow, as discussed later, and flows which cease to advance just because they have cooled too much are called **cooling-limited** flows. Of course

it may happen that the magma reservoir feeding the flow is small enough that it becomes depleted of eruptible magma before a flow has reached its cooling-limited length. In that case the flow stops when the available magma is used up, and has a smaller length and volume than it might have had – this is a **volume-limited** flow.

If magma continues to emerge from a vent after the first cooling-limited flow unit has stopped growing, one of three things can happen. First, a new flow lobe may begin to grow from the original vent along the side of the first flow. Presumably the first lava flow will have flowed down the steepest topographic gradient leading away from the vent, and the new flow will therefore take whatever is now the line of least resistance over the new terrain. Second, a new flow may begin to form on top of the previous flow unit. Normally this only happens if the pre-eruption topography around the vent is such as to confine the new flow and prevent it finding a path alongside the older unit. Third, some part of the boundary of the original flow unit may give way and allow lava from its interior to spill out to start to form a new flow.

This third process, called a **lava breakout**, can happen because all lava flows are constantly cooling along all their boundaries; the base cools by conduction into the cold ground under the flow and the top and sides lose heat both by radiation and as a result of convection currents in the surrounding air. However, liquid rock is a very poor conductor of heat, and so the information that the outside of the flow has become cool does not reach the center of the flow very quickly. It can be shown that the wave of cooling penetrates the flow in such a way that the greatest depth below the surface or above the base that experiences significant cooling after the flow has been traveling for a time t is λ given by

$$\lambda = \sim 2.3 \, (\kappa \, t)^{1/2} \tag{9.1}$$

where κ is the thermal diffusivity of the lava, commonly about $10^{-6} \, \text{m}^2 \, \text{s}^{-1}$ for all magma compositions. Consider two flows, one of which has been growing for an hour and the other for a day; cooling will have penetrated a distance of only $\sim$0.14 m into the first and 0.68 m into the second. Thus although the outside of a flow may be very cool, as long as it is a meter or two thick the lava in the core of the flow can still be as hot as when that lava left the vent many hours earlier. In this case we talk about the hot lava flowing within a **lava tube**. The lava may drain out of a lava tube at the end of an eruption leaving a cave-like tunnel; "tide-marks" are commonly seen on the walls of such drained lava tubes marking successive near-constant levels of the lava.

Hot, fluid lava transmits pressure changes with perfect efficiency, and so if the front of a lava flow stops moving, the rising pressure as the vent continues to supply lava is transmitted throughout the core of the flow and can rupture the cooled skin at its weakest point and let a new flow begin to form as a breakout. Breakouts from lava flows can occur for other reasons. Sometimes the eruption rate from the vent fluctuates, and if there is an increase in flow rate the level of lava in the central channel of the flow rises and the lava overflows the existing levées. However, the pressure due to the extra lava depth may even be enough to push part of one of the levées aside so that the original flow effectively splits into two. Both the new and the old lava flow may continue to be fed at the same time, although of course the level of the lava in the original channel downstream of the breakout will decrease due to the diversion of part of the supply from the vent into the new flow lobe. Sometimes the reason for the overflow and breakout is that a section of the inner part of one of the levées breaks away from the main mass of stationary material and dams the existing channel. When this occurs we talk about an **accidentally breached** flow.

When the front of a lava flow stops moving while the vent is still supplying lava, it is not always the case that a breakout occurs at one specific place around the margin of the flow to create a new flow unit. Instead, large numbers of small fractures may form around the margins of the flow, each of which allows the edge of the flow to expand sideways a little but also, more importantly, raises the top surface of the flow slightly. Lava from the hot core oozes into each of these fractures and seals it again, even as new fractures are forming elsewhere. The net effect is that the flow increases its area slightly, but much more importantly it also gets thicker as the top surface is raised. This occurs not just at the margins but also, because the hot lava core transmits the pressure throughout the flow, all over the area covered by the flow unit – indeed, the thickening may be greater in the middle than at the edges. This process is called **inflation**, and can be a very important alternate to the formation of breakout flows in the growth of a lava flow field.

When an old flow unit ceases to move and forms a new flow unit as a result of a breakout somewhere along the edge of the old flow, or when a new flow unit is initiated at the vent, we have the beginning of the formation of a **compound lava flow field** (Fig. 9.10). Each old flow unit continues to cool once magma has ceased to flow through it, and its stationary upper surface becomes rigidly frozen to its levées. However, as long as the core of the flow unit has experienced only minimal cooling it can still act as a pathway from the original vent to the currently active sites of emergence of lava onto the surface, and it is now what is described as a **lava tube** (Fig. 9.11). The fact that only a little heat is lost by conduction through the walls of a lava tube means that a network of tubes in a compound flow field can carry lava very much further from the original vent than it could possibly flow in a normal channelized flow.

Fig. 9.10 Compound pahoehoe lava flow forming on January 18, 2005, on the south flank coastal plain of Kilauea volcano, Hawai'I. (Photograph by Jon Castro, courtesy of the photographer and U.S. Geological Survey, Hawaiian Volcano Observatory.)

Fig. 9.11 Two scientists from the Hawaiian Volcano Observatory prepare to take a sample of lava from a lava tube seen through the skylight shown in the lower center of the image. The skylight formed by the collapse of the tube roof. Note the extensive compound flow field around the skylight. Image taken on the south flank coastal plain of Kilauea volcano, Hawai'I during the early 1990s. (Photograph by Elisabeth Parfitt.)

9.4 Lava flow rheology

The term **rheology** was introduced in Chapter 2 in connection with the way the rocks in the mantle flow when stressed. We need to consider a similar set of issues in connection with lava flows, but the situation is more complex because in the course of its emplacement the lava in a flow unit changes, from being almost completely molten when it is erupted, to being partly solid and partly molten when its front comes to rest, to being completely solid some time later after its core has cooled below its solidus temperature.

Most lava flows contain significant numbers of gas bubbles, and as the lava cools from the liquidus to the solidus increasing numbers of mineral crystals nucleate and grow within it. The presence of both of these components alters the way the purely liquid bulk of the material is able to shear and deform when stresses are applied to it. Surface tension forces act to keep the gas bubbles as nearly as possible spherical in shape, especially when they

are small, and so mostly they behave in the same way as crystals in acting as obstacles to the liquid flow. As long as the volume fraction of the lava that consists of these crystals and small bubbles is less than about 20%, they have little effect other than increasing the **viscosity** of the lava. The bulk behavior of the lava is still the same as that of a simple fluid such as water, in which the rate of deformation, the **strain rate**, is directly proportional to the applied stress. Fluids with this property are called **Newtonian** fluids, and an example is shown as the line labeled N in Fig. 9.12. The ratio of the stress to the strain rate is called the **Newtonian viscosity** of the fluid. However, when the volume fraction occupied by the bubbles and crystals becomes greater than about 30%, other effects appear. The lava now begins to acquire a threshold resistance to being sheared. This means that the stress applied to it must exceed a certain level called the **yield strength** before the lava will even begin to deform and flow. Once this initial resistance to deformation has been overcome, the way in which the bulk of the liquid responds may take one of a number of forms. The simplest of these is a response in which the strain rate is directly proportional to the applied stress in excess of the yield strength. This behavior is shown as the line labeled B in Fig. 9.12 and fluids with this property are called **Bingham plastics**. The slope of the line labeled B in Fig. 9.12 is the **plastic viscosity** of the fluid, and is a constant that, together with the yield strength, completely characterizes the properties of the fluid. Also shown on Fig. 9.12 are the curves labeled T and D: these show the flow properties of a **thixotropic** and a **dilatant** fluid, respectively. These fluids clearly have a more complex response to stresses trying to make them flow.

An important property of all of the **nonNewtonian** (i.e., the Bingham, thixotropic and dilatant) fluids that have a yield strength is illustrated by the lines labeled F1 and F2 in Fig. 9.12. These represent the same fluid moving under two different sets of conditions. Line F1 represents the flow moving down a shallow slope; the stress acting on it is the component of its weight acting along the slope, and because the slope is small this stress is also small. In fact this stress is only just greater than the yield strength of the fluid, and so it moves only slowly, indicated by the small strain rate (effectively the

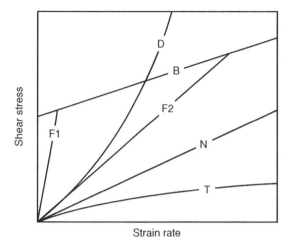

Fig. 9.12 The relationship between shear stress and strain rate for various types of fluid. The straight line labeled N represents a Newtonian fluid like water, for which the strain rate is linearly proportional to the shear stress and the ratio (stress/strain rate) is a constant, the Newtonian viscosity. Thixotropic (T) and dilatant (D) fluids show decreasing or increasing viscosity as the shear stress increases. Bingham plastics (e.g., line B) have a finite yield strength that must be overcome by the stress before shearing starts. The slopes of the lines F1 and F2 give the apparent viscosity of the Bingham plastic when it is subjected to two different stresses, one just larger than the yield strength and the other quite a lot larger. As the applied stress increases, the apparent viscosity decreases, from very large values when the yield strength is just exceeded to much smaller values when the stress is large.

speed). To an observer who knows nothing about the rheology of this fluid, it seems that its viscosity is the slope of the line F1 that connects the origin to the point representing the current stress and strain rate, and this viscosity is rather large. Now imagine that the fluid moves onto a slope that is twice as steep. The stress acting on the flowing material doubles, as shown by the line labeled F2. But this means that the stress is now much greater than the yield strength, and the resulting strain rate, and hence the speed, is also much greater. In fact the viscosity now appears to be given by the slope of line F2, which is several times smaller than the slope of F1. This property, that the effective viscosity of a nonNewtonian fluid is not a constant but instead depends on the physical environment, the ground slope in this case, is just one of the things

that makes predicting the behavior of lava flows very difficult.

9.5 Rheological control of lava flow geometry

It has generally been assumed that there is no need to consider rheological properties any more complex than those of Bingham fluid to describe lava flows. Indeed, early work on the properties of flows recognized that if it was assumed that lava was a Bingham plastic material, there was a natural reason for the observation that flows spread sideways to reach a well-defined width and then form levées that do not change their shape. This is illustrated in Fig. 9.13, which shows the components of the stress controlling the shape of a levée at one margin of a flow. In the downslope direction

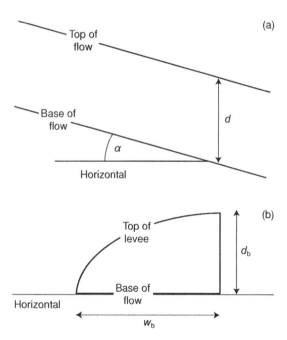

Fig. 9.13 The stresses acting on a lava flow levée. (a) In the downslope direction the stress at the base of the levée is proportional to the thickness d and the sine of the slope α. (b) In the across-slope direction it is the changing thickness of the levée that exerts the stress on its base and controls the relationship between its width w_b and its maximum thickness d_b. The requirement that the basal stress just equals the yield strength leads to the parabolic shape shown.

(a)

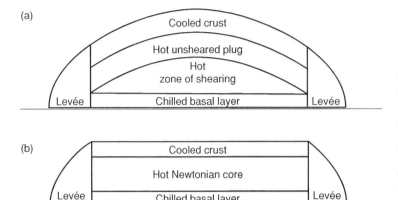

(b)

Fig. 9.14 Models of the internal structures of lava flow lobes. (a) The structure expected if all of the lava behaves as a Bingham plastic: the parabolic levée profile extends all the way to the centre of the central channel. (b) The structure expected when the hot lava in the central channel is nearly Newtonian, and so has a very different rheology from the levées.

(Fig. 9.13a), the lava can move only if its depth d is great enough that the shearing stress at the base of the flow exceeds the yield strength of the lava, S_y. The shearing stress S is given by

$$S = \rho g d \sin \alpha \qquad (9.2)$$

where α is the slope of the ground over which the flow is moving, ρ is the bulk density of the lava flow and, as usual, g is the acceleration due to gravity. So the minimum lava depth that can move, which defines the maximum thickness of the levée at its boundary with the central channel, is found from eqn 9.2 to be

$$d_b = S_y/(\rho g \sin \alpha) \qquad (9.3)$$

At right angles to the downslope direction, it is the change in stress at the base of the levée due to its curved upper surface that controls the levée shape, and it can be shown that the shape is a parabolic curve (see Fig. 9.13b). The width w_b of each levée is given by

$$w_b = S_y/(2 \rho g \sin^2 \alpha) \qquad (9.4)$$

Combining the last two equations it can be seen that

$$d_b/w_b = 2 \sin \alpha \qquad (9.5)$$

which shows that on shallower slopes we expect levées to be relatively thin and wide.

The earliest treatments of lava flows as Bingham plastics assumed that each flow consisted entirely of material with the same rheological properties. One consequence was that the parabolic shapes of the levées were expected to extend right into the middle of the central channel (see Fig. 9.14a), with an upper layer of lava that, like the levées, does not suffer enough shearing stress to make it deform. So although it is still hot enough to be molten, this upper layer rides as an apparently rigid "plug" on top of the lower layer of lava that is stressed enough to make it flow. These conditions make it possible to derive expressions for the width and center-line depth of the channel in terms of the yield strength of the lava. A second consequence of assuming a Bingham rheology was that the plastic viscosity of the lava could be linked to the yield strength (because lavas with high crystal or bubble contents would be expected to have large values for both of these properties) and both properties could be linked to the composition of the lava (on the grounds, again reasonable, that lavas with a given composition tend to erupt at similar temperatures and with similar crystal and bubble contents). Some flows of high-silica content lavas – rhyolites and dacites – do sometimes have cross-sectional shapes similar to that shown in Fig. 9.14a. Indeed, many such flows move only a little further downslope from the vent than they spread sideways, and there is so much resistance to flow from the high viscosity that some lava is forced uphill from the vent, so that a bulbous feature called a **lava dome**, only slightly elongated in the downslope direction, is

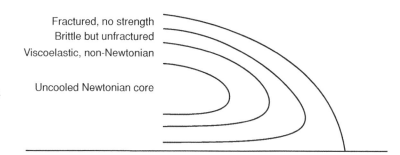

Fractured, no strength
Brittle but unfractured
Viscoelastic, non-Newtonian

Uncooled Newtonian core

Fig. 9.15 Illustration of the changing properties of the lava in a lava flow lobe as a result of the variation of temperature with distance from the flow's cooling boundaries.

formed. It is possible that the single Bingham rheology model is an adequate approximation for these features.

However, the realization that the majority of lava flows contain a very hot central core, at least until the time when they stop flowing, implies that there is no good reason to expect the lava deep in the channel, which is responsible for the advance of the flow, to have the same properties as the cool lava forming the levées. The cross-sectional profile of a channelized flow is more likely to be as shown in Fig. 9.14b, with the channel containing a layer of lava chilled against the original ground surface, above which a raft of cooled lava, distorted and sheared at its edges where it experiences friction with the stationary levées, is carried along on top of a Newtonian fluid core.

The fact that there are very great differences between the physical properties of the different parts of a lava flow is central to understanding flow shape and movement. Figure 9.15 illustrates the key facts. The material at the front and sides of a flow consists, in general, of at least three layers. The very cool outer layer contains many cracks that have formed due to the stresses of cooling, and the presence of the cracks means that this layer has no strength at all. Inside this is a layer where the lava is below its solidus temperature, but has not developed a network of interconnecting cracks, so that it has some brittle strength. Further still from the surface is a layer that has a temperature above the solidus. This material is viscoelastic and will deform in a plastic fashion if stressed slowly but, like the brittle layer, will develop cracks if it is stressed suddenly. Finally the lava furthest from the surface is well above its solidus temperature and has no significant strength, so that like any Newtonian

fluid it will flow if given the slightest chance to do so. Under any given set of stresses, it may be the brittle-elastic layer, the viscoelastic layer or the Newtonian layer that controls the apparent rheology of the flow as a whole. Working out the distribution of stresses inside a lava flow unit to decide how, and if, it will move is a problem that has not yet been completely solved.

9.6 Lava flow motion

It is reasonable to assume that, at least in basaltic lava flows, the hot lava in the central channel has a negligible yield strength. In that case we can assume that the lava in the central channel can be described by just one parameter, its constant Newtonian viscosity, η. It can then be shown that if the lava in the channel has depth d_c, which we assume (as is generally true) to be much less than the width of the channel, its average flow speed U is given by

$$U = (\rho\, g\, d_c^2 \sin \alpha)/(3\, \eta) \tag{9.6}$$

as long as the motion of the lava is smooth and laminar. If the flow speed is large enough, the motion of the liquid in the channel may become disordered and turbulent, in which case the speed is given by

$$U = [(2\, g\, d_c \sin \alpha)/f]^{1/2} \tag{9.7}$$

where f is a friction factor equal to about 0.01. It should be stressed that most of the types of lava flow yet observed in eruptions on Earth have been laminar, although carbonatite flows are commonly turbulent, and some low-viscosity basaltic flows, especially when flowing down very steep slopes or over cliffs, are locally turbulent. Also, there is every reason to think that many of the extremely low-viscosity **komatiite** flows that were much more common in the Earth's early history flowed in a very turbulent fashion, and we see evidence for turbulent lava motion on the Moon and Mars, as described in Chapter 13.

Now that we can relate the speed of a flow to its depth, we are in a position to establish the final relationship that determines the geometry of a lava flow. This is the requirement that, as long as the mass of lava being erupted from the vent every second, M_f, does not change, the same flux of lava must be flowing at every point in the flow. The flux is the product of the cross-sectional area of the flow and the flow speed, and the area is the depth of lava in the channel multiplied by the channel width, w_c. Thus we have

$$M_f = \rho \, w_c \, d_c \, U \qquad (9.8)$$

and although all three of w_c, d_c, and U may be changing (and even ρ may change a little if some of the gas bubbles in the flow burst and collapse), they must do so in such a way that the product of all four quantities remains constant.

The implication of these equations can be summarized as follows. Lava is erupted from a vent with a given viscosity and a negligible (or at least very small) yield strength. It flows downslope, and immediately begins to cool at its upper and lower edges. The cool material on the top surface of the flow is carried to the front, falls onto the ground and is pushed sideways to form a levée on either side of the central channel. The partly solidified levée material behaves as a Bingham plastic with a yield strength that determines the levée width and thickness. The core of hot material in the central channel behaves as a Newtonian fluid and pushes the levées aside until the combination of the width of the channel, the flow speed of the lava, and the depth of the lava can just accommodate the flux coming from the vent. The simplest situation is one in which the depth of lava in the channel is just equal to the depth of the inner edge of the levées. As the front of the flow advances, the material

reaching it must have flowed even further from the vent and so the wave of cooling will have advanced further into it causing a larger fraction of it to have cooled. This means that the yield strength of the levée material will increase with distance from the vent and the flow will get thicker. Meanwhile the continuing penetration of the cooling effects means that, even though the core of the flow is still very hot, less of its thickness is hot enough to deform easily, and so its advance speed slows down. The width of the channel is thus forced to change, usually in the direction of getting wider, so that the flux of lava from the vent is still accommodated by the flow. Finally all of this process must go on even though the flow may be advancing down the flank of a volcano, on which the ground slope varies with distance from the vent. And furthermore, it is quite possible that there are changing conditions in the magma reservoir supplying the vent which mean that the erupted mass flux also changes with time.

9.7 Lengths of lava flows

It will be clear from the above description that analyzing the motion of lava flows, and predicting how they will evolve as they advance down the slopes of a volcano, is not easy. One important property of flows does seem to be fairly predictable, however, and this is the maximum distance to which a single cooling-limited flow unit can travel. Recall eqn 9.1 that described the depth to which cooling can penetrate a flow. It might be expected that when most of the thickness of a flow was cooled it would stop moving. To investigate this we take the ratio between the cooling depth and some characteristic thickness scale of the flow. The number that is used is called the **equivalent diameter** of the flow, d_e, and it is defined as four times the cross-sectional area of the flowing lava divided by its wetted perimeter. The reason for this definition can be explained as follows.

Imagine lava flowing in a lava tube that has a circular cross-section with radius r. Then its cross-sectional area is πr^2 and, if the lava completely fills it, all of the perimeter is wetted by the lava, so the wetted perimeter is $2\pi r$. Then d_e is equal to four times πr^2 divided by $2\pi r$, which is $2r$. But $2r$ is, of course, the actual diameter of the tube, so in this case the equivalent diameter is equal to the actual diameter. Now consider lava flowing with a depth d_c in an open rectangular channel of width w_c. The lava wets the sides and the base of the channel but the upper surface is not in contact with anything fixed to the ground, and so the wetted perimeter is $(w_c + 2d_c)$. The cross-sectional area is equal to $(w_c d_c)$, and so d_e is equal to $[(4w_c d_c)/(w_c + 2d_c)]$. Now assume the commonest case for flows, that the channel width w_c is much greater than its depth d_c. Then the $2d_c$ part of the denominator can be neglected so that d_e is approximately equal to $4d_c$. Some exact values are $d_e = 3.33d_c$ when the channel width is 10 times the depth, $d_e = 3.64d_c$ when the width is 20 times the depth, and $d_e = 3.85d_c$ when the ratio is 50, a typical value for basaltic flows in Hawai'I.

With this definition of d_e, it is possible to define a dimensionless number called the Grätz number, Gz, as

$$Gz = d_e^2/(\kappa\, t) \qquad (9.9)$$

Clearly when a flow is first formed, no cooling has occurred, t is very small and the Grätz number is extremely large. As the flow advances with time, Gz decreases. After compiling information on cooling-limited lava flows with a range of different compositions, Pinkerton & Wilson (1994) found that individual flows stop moving when Gz has decreased to a critical value Gz_c close to 320. In the case of volume-limited flows it may happen that, although the vent has ceased to release lava, the central channel of the flow drains out to extend the front of the flow, and in these cases too the process is found to cease when the Grätz number decreases to a value equal to about 320.

This finding about the Grätz number underlines the basic limitation on lava flow advance. Equation 9.1 implies that in a time t conductive cooling will have penetrated a distance $\sim 2.3(\kappa t)^{1/2}$ into a flow. Cooling occurs at the base as well as the top, so the total thickness of cooled lava is $\sim 4.6(\kappa t)^{1/2}$. But eqn 9.9 implies that a flow stops when $(\kappa t) = d_e^2/320$, which, with $d_e = \sim 3.85d_c$, typical of Hawaiian flows, means $(\kappa t) = d_c^2/21.59$. Taking

square roots $(\kappa t)^{1/2} = d_c/4.65$, i.e., $d_c = 4.65(\kappa t)^{1/2}$, essentially identical to the above expression for the total depth of cooling. Thus the idea that an open channel flow stops moving when its Grätz number reaches 320 appears to be perfectly consistent with the idea that it has lost heat by conduction through both its upper and lower boundaries.

In fact, however, there are a number of complications not considered in the above analysis. Figure 9.1 shows that open channel flows tend to form a "raft" of relatively cool lava in the middle of the channel, with zones of shearing on either side where fresh, incandescent lava is exposed. This lava will be radiating away heat much faster than it can be conducted through the rigid cooled central raft, and so the above calculation based only on conduction of heat will underestimate the heat loss, and we might therefore expect flows to stop moving sooner than anticipated on the basis of conductive cooling alone. Also, the factor 2.3 which appears in eqn 9.1, although the one commonly used in treatments of heat flow, is nevertheless not unique. It represents the depth below the cooling surface where 90% of whatever total amount of cooling will eventually take place has already happened. One might equally well choose, say, 85%, in which case the factor in eqn 9.1 would be 2.07 instead of 2.3 and we would be comparing the value 4.65 obtained from the Grätz number with 4.14 instead of 4.6 – not such an impressive match. The fact that numerous open channel lava flows observed in the field do stop when their Grätz number reaches about 320 suggests that many of the details neglected in the simple model must cancel out. Understanding exactly why this is requires a more detailed model of lava flow motion and cooling than has yet been developed. This is even more true for flows in which the cooled raft in the central channel becomes connected to the levées creating a fully insulated lava tube.

9.8 Surface textures of lava flows

So far the large-scale structures of lava flows have been considered. If one is faced with having to walk across a lava flow, then other, small-scale issues become important. The first is that, even if the flow is no longer moving, its interior may still be molten. Then what matters is the strength of the cooled crust, largely a function of the crust thickness given by eqn 9.1. If the age of a flow and the total thickness of the flow unit are known we can usually make a reliable estimate of the advisability of walking on it; however, this estimate may be influenced by circumstances if one finds oneself unexpectedly surrounded by flows in the middle of a developing compound flow field!

If one is walking on recent lava flows, another practical issue is the texture and roughness of the surface. Lavas of basaltic and similar composition can develop two main textures, usually described by the Hawaiian words **'a'a** and **pahoehoe**. The pahoehoe texture consists of a series of folds in an otherwise smooth surface as shown in Fig. 9.16. The folded surface may itself become folded, leading to a very complex texture, but apart from the fact that the very outermost skin of the lava tends to shatter, on cooling, into tiny glassy platelets, any normal footware will be enough to protect one's feet on a pahoehoe lava flow.

In stark contrast to this, exactly the same lava composition may form an 'a'a surface (Fig. 9.17), consisting of irregular blocks a few to several tens of centimeters in size, with the individual blocks having extremely rough, sharp facets that can tear

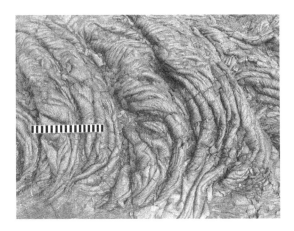

Fig. 9.16 The classic ropy texture of a pahoehoe lava flow surface. (Photograph by Lionel Wilson.)

to shreds the soles of the strongest boots after just a few minutes' walking. Worse still, the blocks are not well packed on the surface of the flow and tend to move or rotate under one's feet, so that a broken leg or ankle is a real possibility.

On flows of lava of intermediate composition, more viscous than basalt, the distinction between pahoehoe and 'a'a texture becomes lost and instead the flow surface consists of loose, irregular blocks, larger than those on 'a'a flows and lacking the extreme roughness. These are called **block lavas**.

Fig. 9.17 The rough, clinkery texture of an 'a'a lava flow surface. (Photograph by Pete Mouginis-Mark, University of Hawai'I.)

The origin of these various lava textures again involves the rheological properties of the cooling lava surface. As a flow moves, its surface is constantly being stretched. In general, the outer part of the flow consists of a brittle outermost layer beneath which the lava is viscoelastic. If the rate of deformation is small, as in all of the surface of a small or slow-moving flow unit, the lava behaves as a plastic and stretches. The lava near the surface flows less easily than that beneath it and as a result it buckles to form the characteristic folds; but nowhere do major fractures occur in the lava. If the deformation rate is greater than some critical value, however, the response of the lava changes to that of a brittle solid and fractures form, first on a small scale, giving rise to angular facets, and then, as cooling penetrates deeper, on a larger scale, producing loose, irregular blocks. The exact sizes of the blocks, and the details of their small-scale roughness, depend in a complex way on the ways in which the strain rate varies with time and the rheological properties change with depth below the surface.

This dependence of texture on deformation rate explains why, for example, both pahoehoe and 'a'a textures may be found on the same channelized basaltic lava flow. The edges (the levées) of the flow are at rest and the lava in the middle of the central channel is flowing fastest. The shearing stresses, and hence the rates of deformation, are very large at the front of the flow where channel material is being diverted into the levees, which explains why levees commonly consist of 'a'a, especially on large flows where the speeds, and hence deformation rates, are largest. Near the middles of the central channels of large flows, and everywhere on small, slow-moving flow units, however, the deformation rates are small enough that the pahoehoe texture persists.

This correlation of texture with deformation rate leads to a common type of compound lava flow field, especially on basaltic volcanoes. When a lava flow field begins to develop on a region with only a shallow slope, the speed of the lava in any one flow unit will be small (see eqns 9.6 and 9.7). Also, because the total lava supply will be divided up through the network of tubes within the growing flow field, each flow unit will have only a small mass flux and so the size (both the width and thickness –

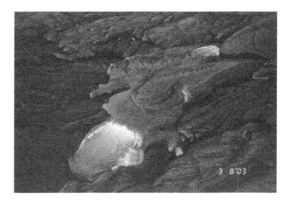

Fig. 9.18 A series of pahoehoe toes forming a compound lava field. (Photograph by Pete Mouginis-Mark, University of Hawai'I.)

see eqn 9.8) of the flow unit will be small. These facts combine to ensure that each flow unit does not advance very far, and consists entirely of pahoehoe lava. These flow units are called **pahoehoe toes** (Fig. 9.18). Exactly the same process can occur in compound lava flow fields on the ocean floor, but there the presence of the surrounding water provides buoyancy that offsets some of the weight of the lava, making the little flow units thicker, for a given width, than they would be on land. The resulting bulbous structures are then called **pillows** (Fig. 9.19) and we speak of fields of **pillow lava**.

There is a final lava flow texture that deserves mention. This is really a variation on 'a'a lava, but it occurs when lava emerges from a tube to form a new flow unit. If the tube is completely full of lava, then there is contact between the roof of the tube and the lava that emerges to form the top layer of the flow. This lava, having been in good contact with the tube roof, has suffered some cooling and so is viscoelastic. If conditions are just right, the upper skin of the lava is pulled up by the frictional drag into a series of tilted spikes that form a characteristic pattern on the flow surface (Fig. 9.20). The regularity of this pattern is reminiscent of the grooves that sometimes form on toothpaste as it emerges from a tube, and gave rise to the name **toothpaste lava**; however, there is no other similarity between toothpaste lava and toothpaste, and this type of lava, although only likely to be present in small amounts on any given lava flow field, is more destructive of boots (and skin if you fall on it)

Fig. 9.19 A sequence of pillows formed by undersea eruptions, now exposed in an uplifted cliff at Acicastello, Sicily. (Photograph by Lionel Wilson.)

Fig. 9.20 The very rough toothpaste-texture lava surface forming on the south flank coastal plain of Kilauea volcano, Hawai'I. (Photograph by Christina Heliker, courtesy U.S. Geological Survey, Hawaiian Volcano Observatory.)

than any other type. Equally, because the spikes are distributed over a relatively flat surface, this lava type is much easier to walk over than 'a'a provided one does have stout footware.

These lava surface textures have a significance that goes beyond the damage that they do to volcanologists and their boots. Surfaces with different textures scatter sunlight and radar waves in distinctive ways. The rougher the surface the more uniformly the radiation falling on it is scattered, whereas the smoother the surface the greater the chance that the reflection will be more like that from a mirror. Radar works by detecting energy scattered back to a detector mounted on the same platform as the transmitter, and so with energy scattered in all directions a significant amount returns to the receiver from an 'a'a surface whereas much less is likely to do so from a pahoehoe surface. Thus 'a'a lava always looks bright to radar whereas pahoehoe, unless the surface of the flow has been

disturbed by, for example, inflation, looks dull. We have seen that the surface texture is a function of the rheology and the flow conditions, and so radar observations of flows from aircraft or satellites, especially for flows that are very inaccessible on the ground, can be a useful tool.

9.9 Effects of ground slope and lava viscosity

The formation of channelized lava flows and compound lava flow fields is particularly characteristic of basaltic lavas. If we consider the viscosities (Table 2.1) of more evolved lavas such as andesites, dacites, and rhyolites, eqn 9.6 shows that, on a given slope, these lavas must be "very" much thicker if they are to move with a given speed. When account is taken of the fact that the speed and thickness are also related via the erupted mass flux (eqn 9.8) it is found that viscous lava flows are thicker, wider, and move much more slowly than basaltic flows on similar slopes. Indeed, the most viscous lavas, rhyolites, have so much difficulty in moving away from the vent that, unless the surface on which they erupt is very steep, they essentially accumulate in all directions around the vent to form a steep-sided **dome** (Fig. 1.3), and it is more the steep slope of the dome surface than the slope of the underlying ground that drives the lava to move. This great thickness of viscous flows and domes reduces the

Fig. 9.21 Lava cascading into the Alae pit crater during the eruption of Mauna Ulu, Hawai'I, in 1969, showing how the widths of lava flows do not change greatly as they move over ground of variable slope. (Photograph by J. Judd, Hawaiian Volcano Observatory, U.S. Geological Survey, courtesy of the National Oceanic and Atmospheric Administration, National Geophysical Data Center.)

rate at which the lava in the middle cools. Nevertheless, the critical Grätz number criterion (eqn 9.9) is still found to control the maximum growth of these features.

At the other extreme we need to consider what happens to a fluid basaltic lava flow when the slope of the ground on which it is traveling changes. It is found that if the slope gets very much steeper (e.g., Fig. 9.21) the width of the flow does not change much, and the combination of eqns 9.6 and 9.8 shows that the depth of the flow decreases (by the cube root of the slope increase) and the speed increases by the same factor. Conversely, if the slope gets much shallower it is found that the flow spreads sideways a great deal as well as advancing at the front of the flow. Both the depth and the speed of the lava decrease the further it spreads, and the cooling criterion soon causes the advance to cease. Continued arrival of lava from the steeper part of the flow upslope results in a series of "waves" of lava spreading out over the same ground, forming a characteristic **perched lava pond** (Fig. 9.22).

9.10 Summary

- Lava flows can form either directly, by overflow of liquid magma from a vent, or indirectly, from the accumulation of clots of fragmented magma falling from a fire fountain. The latter are called rootless flows, and when very short rootless flows form from very viscous magma they are called rheomorphic flows.

- Many flows organize themselves into a central channel, within which lava is moving, and a stationary bank or levée on each side of the flow, which forms when cool material from the flow front is pushed aside as the flow advances. Other flows, called sheet flows, have minimal levées, are wider, and advance over a broad front.

- Some lava flow units stop advancing because the supply of magma from the vent stops. These are volume-limited flows. Other flows stop because the lava near the front has become too cool to deform even though the vent is still delivering magma. These are cooling-limited flows. If the supply of magma continues, either a new flow unit forms at the vent, or a breakout occurs from an existing flow, in which part of the levée collapses and magma emerges to form a new flow unit. In this case the magma flowing within the original flow unit is thermally insulated by the cooled upper surface and the earlier flow becomes a lava tube. A third possibility is that all of the levées of an existing flow unit form fractures into which small amounts of lava move, thus allowing the upper surface to rise so that the whole flow inflates.

- New lava flow units form in other ways. The supply rate from the vent may suddenly increase,

Fig. 9.22 A perched lava pond forming near the prehistoric cinder cone Pu'u Kamoamoa on Kilauea volcano, Hawai'I, in 1983. The pond forms as lava moves onto nearly flat ground so that the forward speed becomes very slow and the flow starts to spread nearly uniformly in all directions. (Photograph by J.D. Griggs, courtesy U.S. Geological Survey, Hawaiian Volcano Observatory.)

causing lava to surge over an existing levée. Alternately, the inner part of a levée may break loose and move down-channel for some distance before blocking the channel and causing an overflow. In a long-lived eruption, all of these various ways of making new flow units may operate, to create a compound flow field.

• Because channelized lava flows form levées that prevent indefinite lateral spreading, they behave in many ways like nonNewtonian fluids, especially Bingham plastics that have a yield strength as well as a viscosity. Some cool, viscous lavas, especially those containing large numbers of crystals or gas bubbles, may be truly nonNewtonian liquids. For other, more fluid, basaltic lavas it is better to model the levées as one liquid, a Bingham plastic, and the central channel lava as a separate, more nearly Newtonian liquid.

• Although there are many potential complications, excepically for tube-fed lava flows, it appears that cooling-limited flows stop moving when the value of a particular dimensionless number, the Grätz number, decreases from an initially large value near the vent to a critical value of about 320. This condition corresponds to waves of cooling having penetrated to the center of the flow from the upper and lower boundaries. The same Grätz number limitation on flow advance applies to flows of evolved composition such as dacites and rhyolites, despite their tendency to

form thick, slow-moving lava flows or to build up around the vent as domes.

• The fact that the outermost surface of any lava flow is so cool that it behaves as a nonNewtonian fluid with strain-rate-dependent properties leads to a range of characteristic textures of lava flow surface, from smooth, folded pahoehoe surfaces to extremely rough 'a'a rubble, and to block lava.

9.11 Further reading

Calvari, S. & Pinkerton, H. (1999) Lava tube morphology on Etna and evidence for lava flow emplacement mechanisms. *J. Volcanol. Geotherm. Res.* **90**, 263–80.

Fink, J.H. & Anderson, S.W. (2000) Lava domes and coulees. In *Encyclopedia of Volcanoes* (Ed. H. Sigurdsson), pp. 307–19. Academic Press, San Diego, CA.

Griffiths, R.W. & Fink, J.H. (1993) Effect of surface cooling on the spreading of lava flows and domes. *J. Fluid Mech.* **252**, 667–702.

Hon, K., Kauahikaua, J., Denlinger, R. & Mackay, K. (1994) Emplacement and inflation of pahoehoe sheet flows: observations and measurements of active lava flows on Kilauea Volcano, Hawai'I. *Geol. Soc. Amer. Bull.* **106**, 351–70.

Hulme, G. (1974) The interpretation of lava flow morphology. *J. Roy. Astron. Soc.* **39**, 361–83.

Kilburn, C.R.J. (2000) Lava flows and flow fields. In *Encyclopedia of Volcanoes* (Ed. H. Sigurdsson), pp. 291–306. Academic Press, San Diego, CA.

Lejeune, A. & Richet, P. (1995) Rheology of crystal-bearing silicate melts: an experimental study at high viscosities. *J. Geophys. Res.* **100**, 4215–29.

Pinkerton, H. & Norton, G. (1995) Rheological properties of basaltic lavas at sub-liquidus temperatures: laboratory and field measurements on lavas from Mount Etna. *J. Volcanol. Geotherm. Res.* **68**, 307–23.

Pinkerton, H. & Wilson, L. (1994) Factors controlling the lengths of channel-fed lava flows, *Bull. Volcanol.* **56**, 108–20.

Self, S., Thordarson, T., Keszthelyi, L., Walker, G.P.L., Hon, K., Murphy, M.T., Long, P. & Finnemore, S.A. (1996) A new model for the emplacement of Columbia River basalts as large, inflated pahoehoe lava flow-fields. *Geophys. Res. Lett.* **23**, 2689–92.

Tallarico, A. & Dragoni, M. (2000) A three-dimensional Bingham model for channeled lava flows. *J. Geophys. Res.* **105**(B11), 25969–80.

9.12 Questions to think about

1 What is likely to be the difference between a lava flow formed by direct overflow from the vent and a rootless lava flow formed by the eruption, at the same volume flux, of a more gas-rich batch of the same magma?

2 Two compound lava fields form on a certain volcano. One forms by a large number of lava flow units being erupted side by side, whereas the other forms by earlier flow units becoming lava tubes and feeding later flows. If the total volume of lava erupted in the two cases is the same, which one is likely to have the greater area?

3 Two lava flows form from the same type of magma at the same temperature erupted at the same rate on the same topographic slope, but one is much more vesicular than the other. What effect is this likely to have on the morphology of the flow?

10 Eruption styles, scales, and frequencies

10.1 Introduction

Volcanic eruptions show a tremendous variety of styles, products, scales, and frequencies. Each eruption is, of course, unique and the sheer variety of activity can give the impression that each eruption is completely different from any other. Our intention in this book is, however, to emphasize the similarities in the underlying physics of eruptions rather than the uniqueness of individual events: to show that eruptions which may superficially appear very different are, in fact, very similar in terms of their basic mechanisms. For this reason the preceding chapters have emphasized the physical processes which control the basic character of eruptions (such as whether an eruption is effusive or explosive, sustained or transient) and of their products (pyroclastic deposits or lava flows). These are the most fundamental characteristics of volcanic eruptions and can be understood without reference to the specific details of an eruption. However, the more detailed character of any given eruption is determined by a number of factors, the most important of which is the composition of the magma involved. Chemical composition influences an eruption by affecting the physical properties of the erupting magma such as the viscosity and gas content. The first half of this chapter concerns how chemical composition affects eruption character and products.

Volcanic eruptions vary widely not only in character but in the scale and frequency of activity. An observer watching activity at Stromboli, a volcano in the Aeolian Islands north of Sicily, often has the opportunity to witness eruptions every few minutes to tens of minutes (Figs 1.12 & 1.13). By contrast hundreds of thousands of years may elapse between volcanic eruptions of other volcanoes. For example, the last three eruptions from the magma chamber underlying Yellowstone caldera in the western USA have occurred at time intervals of about 600,000 years. There is a fundamental link between the frequency and scale of eruptions, however, so that the frequent eruptions at Stromboli produce very small quantities of volcanic material (typically a few cubic meters to a few tens of cubic meters), whereas the infrequent eruptions of volcanic centers such as Yellowstone generally produce vast quantities of volcanic material. The last major eruption at Yellowstone occurred about 600,000 years ago and produced ~1000 km^3 of pyroclastic material, that is about 100 **billion** (10^{11}) times as much volcanic material as a typical eruption at Stromboli! The second half of this chapter looks at what is currently understood about the physical processes which control the frequency and scale of volcanic eruptions.

10.2 Chemical composition and styles of volcanic activity

As we have already seen, broad distinctions can be drawn between volcanic eruptions on the basis of whether they are effusive or explosive, steady or transient in character (Chapters 5–7). In practice, however, volcanologists have developed a system of naming which refers more specifically to the character of any individual eruption. So, for instance, the sustained lava fountaining eruptions commonly observed on Hawaiian volcanoes (Fig. 1.1), which generate long, fluid lava flows, came to be called **Hawaiian** eruptions and this term is used to denote any eruption of this character wherever it occurs in the world. The term **Plinian** is used to denote any eruption which, like a Hawaiian eruption, is explosive and sustained in character but which generates a tall eruption plume and pyroclastic fall deposits (Fig. 1.2). Other terms commonly used to describe different volcanic eruptions include Strombolian, Vulcanian, subPlinian, ultra-Plinian, and hydromagmatic. These terms have already been described in Chapter 1 and used throughout this book.

When volcanologists examined and named different types of eruption they found that there are general links between the composition of the magma erupted and the types of eruptive activity seen. So, for instance, Hawaiian and Strombolian eruptions usually involve basaltic magmas; Vulcanian eruptions usually involve compositions ranging from basaltic andesite to dacite; and Plinian eruptions occur with high-silica andesites to rhyolites (Fig. 10.1). By contrast, phreatomagmatic activity and effusive activity can occur in association with magma of any composition. The links between magma composition and eruption style are only general patterns; it is

still possible, for instance, to have a basaltic Plinian eruption if the erupted magma is unusually volatile-rich or if it incorporates water from an extensive groundwater system as it nears the surface.

During an eruption of any given type of magma the eruption style may change, so composition is not the only control on eruption style. During a basaltic eruption, for instance, there may be a change in character from Hawaiian to Strombolian or in a rhyolitic eruption from Plinian to ignimbrite-forming to effusive. Thus for any magma composition a certain range of behaviors is likely. It is important to recognize also that, despite the different eruption styles observed for different magma compositions, there are still similarities between the types of behavior seen for each composition. The change from Hawaiian to Strombolian activity in basaltic eruptions, for example, mirrors the change from Plinian to Vulcanian activity commonly observed in intermediate eruptions because both reflect a change from sustained to transient explosive activity. Both changes may occur for similar reasons (a reduction in the magma volume flux) even though the style of the eruptions themselves differ.

Nevertheless, the association of certain eruption styles with certain magma types shows that

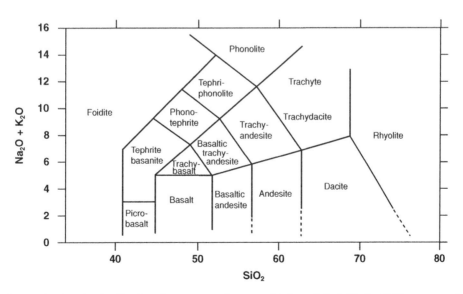

Fig. 10.1 Diagram showing the classification of magma types in terms of their total alkali ($Na_2O + K_2O$) content and silica content in weight percent. (Based on fig. 1 in Rogers, N. and Hawksworth, C. (2000) Composition of magmas. *Encyclopedia of Volcanoes*. Academic Press, pp. 115–131, copyright Elsevier (2002).)

chemical composition has a strong influence on the detailed nature of eruptions. In what way, then, does magma chemistry influence the physics of eruptions (and therefore the character and products of eruptions)? Two properties, the viscosity of the magma and the magma gas content, are strongly linked to magma composition and play an important role in controlling the physics of specific eruptions. The gas content ultimately determines whether an eruption is explosive or effusive, but it also plays a role in determining the magma viscosity, and it is the viscosity which determines how fast, and hence with what volume flux, a magma rises through a given sized fissure system toward the surface.

The viscosity of magma is in fact controlled by various interconnected factors in addition to the volatile content: these are the silica content of the magma, its temperature, its crystal content and the gas bubble content. High silica content and low temperature make a magma more viscous. As it happens, high silica content magmas melt at lower temperatures than those with lower silica contents such as basalts and so are inherently cooler, so that these two factors enhance one another. The presence of crystals within a magma tends to increase the magma viscosity. But crystals begin to form as a magma cools (e.g., when it nears the surface or is stored for long periods in a magma chamber) and so again this factor tends to combine with the previous two.

Chapter 5 looked at the solubility of various volatile species in different magmas. The two most important volatiles in magma are usually water (H_2O) and carbon dioxide (CO_2). The solubility of CO_2 is not strongly affected by magma composition, but the solubility of H_2O is (Fig. 5.2). This compositional influence on H_2O solubility means that, in general, the amount of H_2O dissolved in magmas is greatest in the most evolved (i.e., the most silica-rich) magmas (Table 10.1).

At first sight one might think that this effect would make high-silica melts less viscous than primitive melts, but in fact all it does is to partially offset the high viscosity caused by the lower temperature and higher silica content. And as a high-silica melt nears the surface, encounters lower pressures and exsolves water into gas bubbles, the reduction in the water content of the melt dramatically increases the liquid viscosity even further.

Table 10.1 Typical total water contents (wt%) in various magmas.

Basalts*	0.1–1.5
Andesites	1– 4
Dacites and rhyolites	3–7

*Excluding those in subduction settings where water contents can be as large as 4 – 6 wt%.

The presence of the gas bubbles themselves also has an effect on the viscosity, one which depends on the rate at which the magma is sheared. The rate of shear of a fluid is measured by the change in its flow speed over a given distance within the fluid. At low shear rates (e.g., a slow-moving fluid in a wide channel or conduit) the presence of bubbles will have the effect of increasing magma viscosity, but at higher shear rates (a fast-flowing fluid in a narrow channel or conduit) the bubbles readily deform and have the effect of reducing the viscosity. High-silica magmas, with inherently high viscosities due to all the other factors mentioned, tend to fall into the low shear-rate category, and so this effect only makes their viscosities even greater still. It is the inherently low-viscosity primitive magmas that benefit from this "shear-thinning" process and may have their viscosities reduced somewhat in this way.

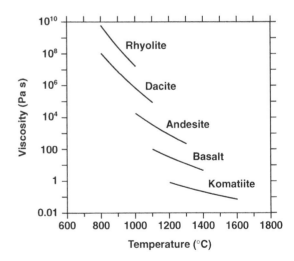

Fig. 10.2 Viscosity as a function of temperature for a range of common magma compositions. (Based on fig. 4 in Spera, F.J. (2000) Physical properties of magmas. *Encyclopedia of Volcanoes*. Academic Press, pp. 171–190, copyright Elsevier (2002).)

Figure 10.2 shows the general trend resulting from all of the above factors, in which the viscosity of primitive magmas is smallest and viscosity progressively increases for more evolved or silica-rich magmas. For comparison, the viscosity of water is 10^{-3} Pa s; this means that the least viscous magmas have viscosities 100 to 1000 times as great as water whereas rhyolites are more than 10 billion (10^{10}) times as viscous as water!

10.3 Chemical composition and effusive eruptions

10.3.1 Conditions of effusive eruption

One fundamental distinction between eruption styles is whether an eruption is effusive or explosive. The explosive character of eruptions results from the fact that rising magma contains gas dissolved within it (Chapter 5). As a result of this the majority of subaerial, and even some submarine, eruptions on Earth are explosive. There are four main circumstances in which effusion may occur rather than explosive activity.

• If the gas content of the rising magma is very small then exsolution will occur but the resulting gas bubble volume fraction will be insufficient to cause fragmentation (Chapter 6). Computer modeling suggests that the gas content must be less than about 0.02 wt% in order for this circumstance to occur. As shown above, even primitive magmas have typical gas contents of at least 0.1 wt%, and so effusive eruptions are very unlikely to be caused solely by a small initial magma gas content.

• If the magma loses enough of its gas at some stage during its ascent to the surface the gas content of the erupting magma may fall below the level at which fragmentation can occur. Gas loss may occur during storage in a magma chamber or through permeable conduit walls during ascent of the magma to the surface. There may be some indirect chemical controls on this gas loss. For instance, the gas content of a magma controls the depth at which supersaturation and exsolution can occur (section 5.3) and thus whether a gas phase is present in the magma chamber at all. Significant gas loss through the conduit walls can occur only if the magma is

Table 10.2 Calculations of the minimum water and carbon dioxide contents in erupting magma required to allow explosive volcanic eruption to occur at a range of water depths.

Water depth (m)	Total water pressure (MPa)	Minimum H_2O content	Minimum CO_2 content
30	0.3	0.12	0.18
100	1.0	0.36	0.62
300	3.0	0.99	1.87
1000	10.0	3.10	6.50
3000	30.0	9.27	22.42

rising relatively slowly and thus is less likely in fast rising, low-viscosity magmas. Effusive eruptions can also result when slowly rising magma that has been progressively degassed by a series of transient explosions is finally erupted onto the surface.

• Submarine eruptions are commonly effusive in character (section 9.2). Here effusion occurs not because the gas content of the magma is low but because exsolution of gas from the magma is suppressed due to the pressure of the water overlying the vent. Table 10.2 gives the pressure at various depths in the ocean and the total (exsolved plus remaining dissolved) amounts of either water or carbon dioxide that a magma would need to contain to guarantee an explosive eruption. Comparison of Tables 10.1 and 10.2 demonstrates that only ~500 m of water depth is needed to suppress explosive activity completely in most basalts, and that even the most gas-rich rhyolite could not erupt explosively under more than ~2300 m of water.

• Effusion may also occur if the viscosity of the erupting magma is sufficiently great to inhibit fragmentation. This can occur in eruptions of very viscous magmas in the dacite to rhyolite compositional range (Figs 10.1 & 10.2). Effusion of such magma produces steep-sided lava domes (sometimes higher than they are wide; Fig. 1.3) containing gas bubbles in which the pressure is significantly greater than atmospheric pressure. Chilling of the outer surface of the dome increases the magma viscosity even further and prevents bubbles bursting at the surface. However, if part of such a dome collapses due to mechanical weakness, the release of pressure may trigger fragmentation giving rise to a pyroclastic surge or density current, and thus converting what began as a relatively harmless effusive eruption in which lava advances slowly into a very much more dangerous explosive one.

Thus effusive eruptions can occur due to a range of different circumstances. Magma composition does play a role in controlling whether an eruption is effusive or explosive but factors such as the storage history of the magma and the environmental conditions under which the eruption occurs are often equally or more important. This is particularly true in extraterrestrial eruptions: the atmospheric pressure at the surface of Mars is about 200 times less than on Earth, and on the Moon and Io there is essentially no atmosphere, so that eruptions on these bodies (the Moon in the very distant past, Mars in more recent geological times and Io now) must almost always have an explosive component, and indeed Io has the largest volcanic eruption plumes ever seen in the Solar System (Chapter 13). In contrast the pressure at the surface of Venus ranges from 4 to 9 MPa, similar to that under 500–1000 m of ocean on Earth (see Table 10.2), making explosive eruptions very much less likely there than on Earth.

10.3.2 Chemical composition and lava flows

The primary products of effusive eruptions are lava flows. The behavior of lava flows is strongly influenced by the composition of the erupting lava because lava viscosity is a major control on flow dynamics. These links have been discussed in sections 9.4–9.6.

10.4 Chemical composition and explosive eruptions

10.4.1 Transient and sustained explosive activity

The majority of subaerial eruptions on Earth are explosive in character. A fundamental distinction which can be drawn between such eruptions is whether they are sustained or transient in nature. Strombolian and Vulcanian eruptions differ considerably in violence and in the deposits they generate

but they are both transient in nature, Strombolian eruptions being associated with basaltic magmas and Vulcanian eruptions generally with more evolved magmas. Similarly Hawaiian and Plinian eruptions are both types of sustained explosive eruption although they also differ in character (compare Figs 1.1 & 1.2), in products, and in the composition of the magma involved.

We have already seen that whether an explosive eruption is transient or sustained is strongly influenced by the rise speed of the magma. This is because composition influences viscosity, which in turn controls the rise speed of gas bubbles within the magma and hence their ability to overtake one another and coalesce (Chapters 5 and 7). Three main issues are important here.

1 Magma viscosity affects the ability of bubbles to move upward through the rising magma. The greater the magma viscosity the slower the rise of the bubbles relative to the magma, which makes bubble coalescence less likely.

2 The magma viscosity also affects the rise speed of the magma itself, greater viscosities tending to reduce the rise speed. This increases the time available for bubbles to rise relative to the magma and thus for bubble coalescence to occur.

3 The gas content of the magma also affects the likelihood of bubble coalescence occurring. Greater gas contents lead to a greater number density of bubbles in the magma and more bubble growth during rise, both of which also increase the opportunities for bubble coalescence to occur. Thus, as gas content tends to be greater for more silica-rich magmas this tends to increase the opportunity for coalescence.

The net effect of these factors is not simple to deduce because they can counteract each other. For example, a basaltic magma tends to have both low viscosity and low gas content, so that factor 1 tends to increase the opportunity for bubble coalescence (and transient eruption) whereas factors 2 and 3 tend to decrease bubble coalescence. Figure 10.3 shows the results of computer modeling carried out to assess the influence of the above factors on basaltic activity. Figure 10.3a shows the effect described in point 1 with increasing viscosity making coalescence less likely for a given rise speed. Figure 10.3b shows the effect described in point 3

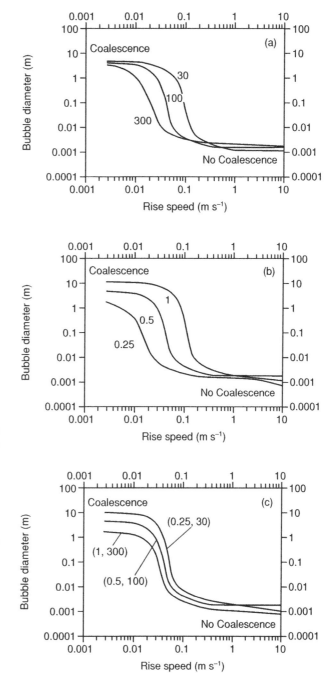

Fig. 10.3 Variation of the size of the largest gas bubble that can form in a rising magma as a function of the rise speed for various combinations of magma viscosity and volatile content. (a) The volatile content is kept constant and the curves are labeled with the magma viscosity in Pa s; increasing the viscosity makes coalescence more difficult. (b) The viscosity is held constant and the curves are labeled with the exsolved volatile content of the magma; decreasing the gas content makes coalescence more difficult. (c) The viscosity and volatile content vary together, and the influence of the viscosity is seen to outweigh the influence of the volatile content. (The figure is based on figs 2 & 3 in Parfitt, E.A. & Wilson, L. (1995) Explosive volcanic eruptions – IX: the transition between Hawaiian-style lava fountaining and Strombolian explosive activity. *Geophys. J. Internat.*, **121**, 226–232. Copyright Wiley-Blackwell Publishing Ltd.)

in which coalescence increases with increasing gas content for a given rise speed. Figure 10.3c shows one example of the combined effects of composition on the likelihood of bubble coalescence. Here viscosity and gas content increase together, with

the combined effect that coalescence becomes progressively less likely for a given rise speed, i.e., the effect of increasing viscosity is dominant.

Magma viscosity varies so greatly between different magmas (Fig. 10.2), compared with the relative variation in magma gas content or in magma rise speeds, that viscosity is probably the dominant factor controlling whether significant bubble coalescence can occur in different magmas. This can be demonstrated by some simple calculations such as those performed for basaltic magmas in section 5.5.3.

Let us consider the case of bubble rise in an intermediate or very evolved high-silica magma. Say that in both cases magma exsolution begins at a depth of 5 km beneath the surface. Typical rise speeds for intermediate and evolved magmas are thought to be in the range 0.001 to 0.015 m s^{-1}. This means that it takes 3.3×10^5 to 5×10^6 seconds for the magma to rise from 5 km depth to the surface. Whether bubble coalescence can occur during this ascent depends on the rise speed of bubbles through the rising magma (section 5.5.3). If we consider the intermediate magma first and assume that it has a magma viscosity of 10^3 Pa s then a bubble 1 mm in radius will rise at a speed of ~6.5 $\times$ 10^{-6} m s^{-1} and thus can rise a distance of between 2.1 and 32.5 m through the overlying magma during the time it takes the magma to rise to the surface. If the bubbles are larger, say 1 cm in radius, their rise speed is ~6.5 $\times 10^{-4}$ m s^{-1} (eqn 5.12) and the distance traveled by the bubbles relative to the magma is correspondingly greater at 214 to 3250 m. By contrast, in a more evolved magma with a viscosity of 10^6 Pa s the rise speed of a bubble 1 mm in radius decreases to 6.5×10^{-9} m s^{-1} and the distance traveled through the magma during ascent to the surface is only 2.1×10^{-3} to 0.03 m. Even at a bubble radius of 1 cm the bubble rise speed is only 6.5 $\times$ 10^{-7} m s^{-1} and the distance risen by the bubbles only 0.2 to 3.3 m. Thus in this simple example bubble coalescence is likely to occur in the intermediate magma case but is very unlikely in the case of the high-silica magma even at very low magma rise speeds. The low rise speeds of gas bubbles in these magmas explains why transient explosions are commonly associated only with basaltic and intermediate magmas but not with evolved magmas.

10.4.2 Chemical composition and transient explosive activity

We have seen that transient explosions occur when magma rise speeds are slow. The two main types of transient eruption – Strombolian and Vulcanian – differ from each other in a number of ways. Individual Strombolian explosions produce very small volumes of material, eject this material from the vent at relatively low speeds, and usually occur more closely spaced in time than Vulcanian eruptions. Strombolian eruptions usually involve basaltic magmas whereas Vulcanian eruptions are typically associated with the eruption of intermediate magmas. Thus, there seems to be a strong connection between the composition of the erupting magma and the character of the transient explosion which results. In what way is the chemical composition controlling the eruption dynamics?

Field observations and mathematical modeling of Strombolian and Vulcanian eruptions suggest that the differing "violence" of the eruptions is related to the strength of the "cap" on the magma column prior to eruption (see section 7.2.1 and Fig. 7.1). In Strombolian eruptions, the gap between explosions is too short to allow much cooling of the magma at the top of the magma column, and the cooled "skin" that does develop tears easily as gas bubbles accumulate beneath it causing weak explosions (Figs 1.14 & 7.1). In Vulcanian explosions the magma at the top of the magma column cools much more between explosions forming a solid "cap" (Figs 7.1 & 7.2) and so the pressure beneath it must build to considerably greater levels prior to explosion. This is probably because of the different viscosities of the magmas involved. In basaltic magmas the low viscosity allows relatively rapid rise of gas bubbles through the magma and this means that the surface of the magma column is disrupted and removed frequently so that there is not time for a thick crust to develop, i.e., the rise speed of the bubbles controls the time available for the skin on the magma column to develop and hence the violence of the resulting explosion. Thus basaltic magmas commonly give rise to Strombolian eruptions. The rise speeds of bubbles in intermediate magmas will be lower due to the higher viscosity of the magma (sections 5.5.3 and 10.4.1). The slower rise and accumulation rate of gas at the top of the magma column then means that the top of the magma column has more time to cool than in the Strombolian case. This greater cooling means that the pressure necessary to cause failure of the cap is greater. The brittle failure of this cap gives rise to Vulcanian explosions (Fig. 1.16).

In reality Vulcanian explosions are not always caused solely by magmatic gases; interaction with groundwater also plays a role (section 7.3). The strong link between composition and volcanic activity suggests, though, that the viscosity of the magma and its influence on bubble rise speed play an important role even when groundwater is involved in an eruption.

10.4.3 Chemical composition and sustained explosive eruptions

We have seen that sustained explosive eruptions occur when the rise speed of magma is sufficiently great to prevent significant segregation of magmatic gas bubbles from the magma in which they originated (section 5.5). A range of sustained explosive eruptions can occur which vary considerably in character. Hawaiian eruptions involve the eruption of relatively coarse clasts at relatively low exit velocities, produce low eruption plumes and dominantly generate lava flows. These eruptions are associated with basaltic magmas. Other sustained explosive eruptions ranging from subPlinian, through Plinian to ultra-Plinian are more normally associated with intermediate to evolved magma. They generate greater plume heights, produce finer clast-size distributions, and are dominated by fall deposits which are much more widely dispersed than those produced in Hawaiian eruptions. It has been commonly assumed that the difference in viscosity of the erupting magma is what controls whether an eruption is Hawaiian or Plinian. However, both the viscosity and the magma gas content are lower in basaltic magmas and both are likely to play some role in determining the details of the eruption dynamics.

ROLE OF VISCOSITY

A key difference between Hawaiian eruptions and the spectrum of Plinian eruptions is the degree of fragmentation of the erupting clasts (section 6.6). Hawaiian eruptions produce coarse clasts which can be carried typically only a few hundred meters above the vent in incandescent lava fountains (Fig. 1.1) and which then fall back around the vent forming cones and lava flows (see section 9.2). In Plinian eruptions the degree of fragmentation of

the magma is much greater and the clasts which are erupted are considerably smaller. This allows transport of much of the erupted mass upwards in a convecting eruption plume and dispersal over a wide area (section 8.2). So a key issue is the factors which control the degree of fragmentation and hence the clast-size distribution of the erupting material.

The fragmentation process is still not well understood, but considerable advances in understanding were made during the 1990s. It is thought that fragmentation can occur through two primary mechanisms: rapid acceleration or rapid decompression of magma. Rapid decompression is a likely trigger for fragmentation and explosive activity in situations where there is a rapid reduction of confining stress on the magma – such as when a lava dome collapses (section 8.4.2.2; Fig. 8.10) or during slope failure such as that which triggered the initial lateral blast during the May 1980 eruption at Mount St Helens. Rapid acceleration is considered to be the more likely cause of fragmentation in Hawaiian and many Plinian eruptions. In this case, vesiculation due to exsolution of gas from rising magma is delayed until a high degree of supersaturation is reached. Rapid exsolution then causes the development and rapid acceleration of a magmatic foam (section 5.6). Rapid acceleration results in high strain rates which induce stresses across the bubble walls sufficient to cause brittle failure and fragmentation. The viscosity of the magma involved is considered to be crucial to the nature of this process. The viscosity of magmas can change dramatically as water exsolves from them. This effect is small in basaltic magmas and so basaltic fragmentation occurs progressively by the thinning and tearing of bubble walls and produces relatively large and fluid lava clots. Exsolution of water from more evolved magmas, though, causes a dramatic increase in the magma viscosity (section 10.2 and Fig. 10.4). The high viscosity of these magmas means that rapid acceleration leads to strain rates which are high enough to cause the magma to fail in a brittle fashion. This causes more complete fragmentation and the generation of smaller clasts. The highest strain rates are likely to cause the highest degree of fragmentation. The difference between elastic and brittle fragmentation can be likened to the behavior of "silly putty". When "silly putty" is stretched slowly (i.e., at low strain rates) it continues to stretch until

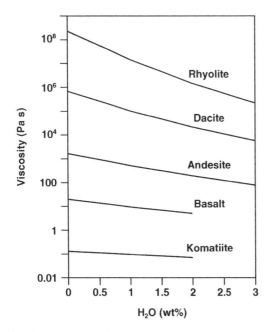

Fig. 10.4 Variation of the viscosities of various magmas with their dissolved water content at a constant temperature. Decreasing the water content increases the viscosity, especially when the magma is silica-rich. (Based on fig. 5 in Spera, F.J. (2000) Physical properties of magmas. *Encyclopedia of Volcanoes*. Academic Press, pp. 171–190, copyright Elsevier (2002).)

it eventually tears apart. If it is stretched rapidly (i.e., at high strain rates) the putty breaks after only a small amount of stretching producing a sharp edge, i.e., it has "broken" in a brittle fashion. For basaltic magmas strain rates and magma viscosity are too small to allow brittle failure to occur and the magma stretches and tears producing large, fluid lava clots. In more evolved magmas the increase in viscosity caused by the exsolution of water means that strain rates may be high enough to cause brittle fracturing of the magma resulting in the formation of small clasts. This difference is fundamental to controlling the style of the resulting eruption.

ROLE OF GAS CONTENT

The magma gas content may also play a role in controlling the style of sustained explosive eruptions. For instance, basaltic magmas differ from more evolved ones in having low gas contents as well as low viscosities. It is possible that the lower gas content of basaltic magmas also affects the fragmentation process. The amount of gas within the magma affects the energy released during ascent and thus is likely to affect the amount of acceleration occurring prior to fragmentation (section 6.4.1) and hence to affect the strain rates experienced by the magma, with higher gas contents causing higher strain rates and greater fragmentation.

The gas content of the magma also influences the exit velocity of material in sustained eruptions by affecting the depth at which fragmentation occurs and the total energy release (section 6.4.1), with lower gas contents leading to lower exit velocities. Typical exit velocities in Hawaiian eruptions, where gas contents are low, are ~100 m s^{-1} whereas in Plinian eruptions they are more typically ~300 m s^{-1}. This difference affects the ability of the erupting jet to entrain air and develop a stable eruption plume. The low gas content of basaltic magmas makes the development of a stable eruption plume unlikely, and the lava fountain characterizing a Hawaiian eruption is, in effect, the basaltic equivalent of a collapsed eruption column (section 6.7.1).

10.5 Summary of compositional controls on eruption character

The previous sections have shown that chemical composition can play a crucial role in determining the detailed character of volcanic eruptions. The link between chemical composition and eruption style is to a large degree due to the links between chemical composition and two key physical properties of the erupting magma – the viscosity and magma gas content. Primitive magmas generally have low viscosities and low magma gas contents while progressively more evolved magmas have progressively higher viscosities and gas contents. This section summarizes the links between composition and eruption style described in the previous sections.

• Chemical composition has relatively little influence in determining whether an eruption is effusive or explosive in character. This is more strongly influenced by the storage history of the magma and the environmental conditions at the eruptive vent. Composition does, however, play a critical role in determining the nature of effusive eruptions because magma viscosity (which is strongly linked to composition – Fig. 10.2) strongly influences the dynamics of lava flows (see sections 9.4–9.6).

• Whether an explosive eruption is transient or sustained in character is largely determined by the rise speed of the magma and the ability of gas bubbles to rise through, and segregate from, the magma. Transient explosions occur when gas can segregate and rise through the magma in the form of a few large bubbles whereas sustained explosive activity occurs when little segregation of gas from the magma occurs and a large number of small bubbles is present. Composition plays a significant role in determining whether gas rise and segregation can occur. In low and intermediate magmas viscosity is sufficiently small to allow gas bubbles to rise through the magma and transient explosions can occur. The high viscosity of high-silica magmas such as rhyolites prevents gas bubble segregation and thus suppresses transient explosions caused by magmatic gases (they can, of course, still be associated with transient explosions caused by interaction with meteoric water).

• Two main types of transient explosion are known – Strombolian and Vulcanian – and they are strongly associated with certain magma types. Strombolian eruptions occur with basaltic magmas whereas Vulcanian explosions are associated with intermediate magmas. The differences in style of these two types of transient explosion are thought to be caused by differences in the amount of cooling experienced by the top of the magma column between explosions. In Strombolian explosions the low viscosity of the magma allows rapid rise of gas bubbles through the magma and hence the surface of the magma column has little time to cool before more gas bubbles rise and disrupt it. The bursting of these bubbles gives rise to the characteristically weak explosions of Strombolian events. For more viscous magmas, however, the rise speed of gas bubbles is slower and accumulation of gas beneath the "plug" at the top of the magma column is slower. This means that the plug cools more

prior to removal and removal only occurs when sufficient gas pressure builds to cause brittle failure of this plug. This gives rise to the more violent explosions characteristic of Vulcanian events.

• For sustained explosive activity a key difference exists between the size distribution of the material ejected in Hawaiian and Plinian eruptions. Hawaiian eruptions produce large, fluid lava clots whereas Plinian eruptions produce smaller clasts. This difference is crucial in determining the very different styles of these two types of eruption. The links between clast-size distribution and fragmentation are not well understood but current ideas suggest that the difference in degree of fragmentation between Hawaiian and Plinian eruptions is mainly due to the differences in viscosity between the magmas involved. In Hawaiian eruptions the low viscosity of the magma causes relatively low strain rates during ascent and progressive fragmentation occurs. For more viscous magmas the strain rates are higher and fragmentation is a brittle and more complete process resulting in smaller clasts. Magma gas content may also influence the degree of fragmentation by affecting the acceleration and strain rates experienced by magma during ascent.

10.6 Magnitudes and frequencies of volcanic eruptions

At any given moment in time typically about 20 volcanoes will be erupting on Earth. Usually we are aware of this fact only if we live close to an active volcano or the eruption is dramatic enough to make the news. Through news coverage and television documentaries most people have some idea of recent damaging eruptions such as the 1980 eruption of Mount St Helens in the western USA (Fig. 1.2), the 1991 eruption of Mount Pinatubo in the Philippines, the eruptions on Montserrat which started in 1995, or the 2002 eruption of Nyiragongo in the Democratic Republic of Congo. People usually also have some idea of the destructive effects of "large" eruptions which have occurred further in the past. For instance, the 1883 eruption of Krakatau (the eruption was so loud that it woke people sleeping in southern Australia more than

Fig. 10.5 View of the ruins of the Roman city of Herculaneum, destroyed in the AD 79 eruption of Vesuvius. The present-day city of Ercolano, built on top of the deposits from the eruption, and the volcano Vesuvius itself are visible in the background. (Photograph by Lucia Gurioli, University of Hawai'I.)

3200 km away!) or the AD 79 eruption of Vesuvius which destroyed Pompeii and Herculaneum (Fig. 10.5).

10.6.1 The magnitude of historic volcanic eruptions

Volcanologists keep detailed records of the Earth's volcanic activity. Such records are compiled using a range of information including eyewitness accounts, geophysical data collected by volcano observatories, geological mapping and remote sensing studies. Details of past activity have been painstakingly reconstructed using information such as historical records and geological mapping. One such record is that of the Global Volcanism Program of the Smithsonian Institution in Washington, DC (http://www.volcano.si.edu/gvp/index.htm). This lists known eruptive activity which has occurred during the past 10,000 years. The scale of eruptions in this catalog is classified using an index called the **Volcanic Explosivity Index** or VEI. This is a widely used method of classifying the magnitude and intensity of volcanic eruptions.

The magnitude of an eruption is defined as the total volume or mass of material erupted. The intensity is a measure of the volumetric or mass eruption rate.

The VEI produces a single number between 0 and 8 which gives a combined measure of magnitude and intensity. It therefore assumes a link between these two properties of an eruption. Table 10.3 illustrates the VEI classification system. Note that in this system the intensity of the eruption is denoted by the height of the eruption column (because column height is directly related to mass eruption rate – eqn 6.7). Table 10.4 shows the classification of a number of historical eruptions using this system.

Although the VEI system is widely used it has a number of drawbacks. One is that it is concerned with explosive eruptions, so that any eruption which is primarily lava-producing would be classified as having a low VEI value even if the eruption had a high eruption rate, produced a large volume of lava, and was very destructive. Furthermore, the VEI assumes a connection between the magnitude and intensity of an eruption which is not borne out by observations. A more reliable system is to use independent measures of magnitude and intensity. One such system, for example, defines the magnitude of an eruption as

$$\text{magnitude} = \log_{10}(M_e) - 7 \qquad (10.1)$$

where M_e is the total erupted mass (in kg), and the intensity of an eruption as

$$\text{intensity} = \log_{10}(M_f) + 3 \qquad (10.2)$$

where M_f is the mass flux (in kg s^{-1}). The integers in these equations are chosen simply for convenience: few eruptions involve masses less than 10^7 kg or mass fluxes less than 10^3 kg s^{-1}. This system has the advantage that it allows direct comparison of explosive and lava-producing eruptions. Table 10.4 illustrates the use of this system for classifying a number of historical eruptions. For primarily explosive eruptions the VEI value is similar to the magnitude value, but notice the difference in these values for lava-producing eruptions such as the 1950 Mauna Loa eruption and the 1991–93 Etna eruption.

Table 10.4 includes examples of both the typical small magnitude and low intensity activity which is going on all the time on the Earth and of the largest eruptions to have occurred in the recent past. In

Table 10.3 The Volcanic Explosivity Index (VEI) classification scheme.

Volcanic Explosivity Index	General eruption description	Qualitative eruption description	Erupted volume (m^3)	Eruption column height (km)
0	Nonexplosive	"Gentle, effusive"	$< 10^4$	< 0.1
1	Small	"Gentle, effusive"	10^6	0.1–1
2	Moderate	"Explosive"	10^7	1–5
3	Moderate/large	"Explosive"	10^8	3–15
4	Large	"Explosive"	10^9	10–25
5	Very large	"Cataclysmic,	10^{10}	> 25
6	Very large	paroxysmal,	10^{11}	up
7	Very large	or	10^{12}	to
8	Very large	colossal"	$> 10^{12}$	~55

Table 10.4 The magnitude, intensity and Volcanic Explosivity Index (VEI) of a number of historical eruptions: M_e is the total erupted mass and M_f the peak mass flux.

Eruption	Erupted volume (km^3)	M_e (kg)	M_f (kg s^{-1})	Column height (km)	VEI	Magnitude	Intensity
Etna (Italy) 1991–1993	0.25	5×10^{11}	6×10^4	~1	2	4.7	7.8
Pinatubo (Philippines) 1991	7.5	1.1×10^{13}	4×10^8	35	6	6	11.6
Mount St Helens (USA) 1980	1.2	1.3×10^{12}	2×10^7	19	5	4.8	10.3
Soufriere (St Vincent) 1979	0.05	10^{11}	3×10^7	18	3	4	10.5
Kilauea (USA) 1959	0.03	6.6×10^{10}	1.5×10^5	> 0.6	2	3.8	8.2
Bezymianny (Russia) 1956	~2	10^{12}	2.2×10^8	36	5	5.3	11.3
Mauna Loa (USA) 1950	0.39	10^{12}	7×10^6	?	0	5	9.8
Novarupta (Alaska) 1912	15	3×10^{13}	1×10^8	25	6	6.5	11.0
Santa Maria (Guatemala) 1902	12	2×10^{13}	1.7×10^8	34	6	6.3	11.2
Tarawera (New Zealand) 1886	2	2×10^{12}	2.2×10^8	34	5	5.3	11.3
Krakatau (Indonesia) 1883	> 10	3×10^{13}	~5×10^7	25	6	6.5	10.7
Askja (Iceland) 1875	0.17	~10^{11}	8×10^7	26	4	4.0	10.9
Tambora (Indonesia) 1815	> 50	2×10^{14}	2.8×10^8	43	7	7.3	11.4
Laki (Iceland) 1783	15	3×10^{13}	2.4×10^7	12	4	6.5	10.4
Taupo (New Zealand) ~ AD 180	35	8×10^{13}	1.1×10^9	51	6+	6.9	12.0
Vesuvius (Italy) AD 79	6	6×10^{12}	1.5×10^8	32	6	5.8	11.2

Data taken from Richter, D.H., Eaton, J.P., Murata, K.J., Ault, W.U. and Krivoy, H.L. (1970) Chronological narrative of the 1959–60 eruption of Kilauea Volcano, Hawai'I. *U.S. Geol. Surv. Prof. Pap.*, **537-E**, 73 pp.; Sparks, R.S.J., Wilson, L. and Sigurdsson, H. (1981) The pyroclastic deposits of the 1875 eruption of Askja, Iceland. *Philos. Trans. Roy. Soc. London Ser. A*, **299**, 241–273; Wadge, G. (1981) The variation of magma discharge during basaltic eruptions. *J. Volcanol. Geotherm. Res.*, **11**, 139–168; Newhall & Self (1982); Sparks, R.S.J. and Wilson, L. (1982) Explosive volcanic eruptions – V. Observations of plume dynamics during the 1979 Soufriere eruption, St Vincent. *Geophys. J. Roy. Astron. Soc.*, **69**, 551–570; Walker, G.P.L., Self, S. and Wilson, L. (1984) Tarawera 1886, New Zealand – a basaltic plinian fissure eruption. *J. Volcanol. Geotherm. Res.*, **21**, 61–78; Cas, R.A.F. and Wright, J.V. (1987) *Volcanic Successions – Modern and Ancient*. Chapman and Hall, London, 528 pp.; Wilson, L. and Walker, G.P.L. (1987) Explosive volcanic eruptions – VI. Ejecta dispersal in Plinian eruptions – the control of eruption conditions and atmospheric properties. *Geophys. J. Roy. Astron. Soc.*, **89**, 657–679; Pyle, D.M. (1989) The thickness, volume and grainsize of tephra fall deposits. *Bull. Volcanol.*, **51**, 1–15; Simkin, T. and Seibert, L. (1994) *Volcanoes of the World. Geoscience Press*, 349 pp.; Parfitt, E.A. (1998) A study of clast size distribution, ash deposition and fragmentation in a Hawaiian-style volcanic eruption. *J. Volcanol. Geotherm. Res.*, **84**, 197–208; and Pyle (2000).

Fig. 10.6 A thin lava flow surrounding a house in the town on Kalapana on the sotheast flank of Kilauea volcano, Hawai'I. (Photograph by Jim Griggs, courtesy U.S. Geological Survey, Hawaiian Volcano Observatory.)

the former category are eruptions such as the ones listed for Kilauea and Mauna Loa volcanoes in Hawai'I. The listed Kilauean eruption is just one of 30 which have occurred there since 1950 alone. These eruptions are primarily lava-producing and are of relatively small volume. This is reflected in the low VEI value for the eruptions and their low magnitude and intensity values. Although these eruptions are relatively small and frequent they can still be locally destructive. For instance, the ongoing eruption which started in January 1983 had by January 2001 destroyed ~180 buildings and 13 km of highway. This includes the almost complete destruction of the village of Kalapana between 1986 and 1990 (Fig. 10.6). Similarly Table 10.4 lists one of many frequent eruptions of Etna in Sicily. Eruptions there are also primarily lava-producing. Eruption styles vary between Strombolian, Hawaiian, and Vulcanian and relatively violent explosions there do, on occasion, kill people standing too close to the eruptive vents. For instance, in September 1979, nine tourists were killed and many more injured by an explosion at Bocca Nuova, one of Etna's summit craters. On many occasions lava flows produced during eruptions on Etna have threatened to destroy or have destroyed villages located on the flanks of the volcano and a number of methods have been tried to divert or stop the flow of lava there (see section 11.5). In 1992, for example, a lava flow came within 700 m of destroying the village of Zafferana Etnea. The pattern of frequent eruptions of small magnitude and intensity of these three centers is repeated at many other active centers around the world such as Hekla in Iceland and Piton de la Fournaise on Réunion Island in the Indian Ocean.

The largest eruptions listed in Table 10.4 are, by contrast, rare and highly destructive events. The 1815 eruption of Tambora on Sumbawa Island in Indonesia, for example, was responsible for the deaths of ~92,000 people, 90% of them from starvation and disease in the aftermath of the eruption. The eruption in April 1815 produced an eruption column which reached a height of between 40 and 50 km. The spreading eruption cloud was so dense that it plunged a region up to 600 km downwind of the volcano into darkness for periods of up to 2 days. The dispersal of particles from the cloud through the stratosphere is thought to have been responsible for short-term climate change, the year following the eruption, 1816, being known as the "year without a summer" (see section 12.1).

10.6.2 The magnitude of volcanic eruptions in the geological record

Our perception of volcanic eruptions is that those such as the 1980 Mount St Helens eruption and the AD 79 eruption of Vesuvius are large, and indeed they are when only the time scale of human history is considered. Eruptions such as the 1815 Tambora eruption and the AD 180 Taupo eruption are even larger and represent the upper limit of the magnitude and scale of recent eruptions. However, volcanologists have studied deposits from eruptions which date much further back in the geological past, and find that the largest eruptions in the geological records dwarf the eruptions which humans have witnessed first hand.

Table 10.5 lists a number of the largest volcanic eruptions identified in the geological record. It should be noted that the enormous scale of the deposits from these kinds of eruptions means that they are much less well studied than those from smaller, historical eruptions. Even so, comparison with the historical eruptions in Table 10.4 shows the large scale of these eruptions compared with eruptions observed in the recent past. For example, the May 1980 eruption of Mount St Helens seems a

Table 10.5 Examples of some of the largest volcanic eruptions occurring in the geological record.

Deposit/age	Volcanic center/caldera	Erupted volume (km³)	Type of deposit
Fish Canyon, ~28 Ma	La Garita, Colorado	5000	Ignimbrite
Huckleberry Ridge, ~2 Ma	Yellowstone, Wyoming	2450	Ignimbrite
Individual Grande Ronde flow, ~16 Ma	Columbia River Flood Basalts, USA	> 2000	Lava flow
Cerro Galan, ~2.2 Ma	Cerro Galan, Argentina	2000	Ignimbrite
Toba, ~75 ka	Toba, Sumatra	1500	Ignimbrite
Carpenter Ridge, ~27 Ma	Bachelor, Colorado	1350	Ignimbrite
Roza Flow, ~15 Ma	Columbia River Flood Basalts, USA	700	Lava flow

Data taken from Lipman, P.W. (2000) Calderas. In *Encyclopedia of Volcanoes* (Ed. H. Sigurdsson), pp. 643–662. Academic Press, San Diego, CA; Reidel, S.P. and Tolan, T.L. (1992) Eruption and emplacement of flood basalt: an example from the large-volume Teepee Butte member, Columbia River Basalt Group. *Geol. Soc. Am. Bull.*, **104**, 1650–1671; Self, S., Thordarson, T. & Keszthelyi, L. (1997) Emplacement of continental flood basalt lava flows. In *Large Igneous Provinces: Continental, Oceanic, and Planetary Flood Volcanism* (Eds J.J. Mahoney & M.F. Coffin), pp. 381–410. Geophysical Monograph 100, American Geophysical Union, Washington, DC; Simkin, T. & Seibert, L. (2000) Earth's volcanoes and eruptions: an overview. In *Encyclopedia of Volcanoes* (Ed. H. Sigurdsson), pp. 249–261. Academic Press, San Diego, CA.

relatively large eruption but the 2 Ma eruption of Yellowstone is equivalent to 2000 Mount St Helens eruptions!

The largest eruptions recorded in the geological record are of two very different types. The large eruptions of volcanic centers such as Yellowstone in the western USA produce pyroclastic deposits, both fall deposits and ignimbrites, with the latter being dominant. The other class of large-volume eruptions are **flood basalt eruptions**. These produce very large-scale lava flows: flows often extend hundreds of kilometers from their vent systems and individual flows are typically 20 to 100 m thick. Individual flood basalt flows may have volumes in excess of 2000 km³ (Table 10.5) and are found in sequences of flows, one on top of the other (Fig. 10.7), such that the volume of a single flood basalt province can be as much as 10^6 km³.

Fig. 10.7 Part of the Deccan Volcanic Province, western India, an enormous stack of ~65 Ma-old flood-basalt lava flows. The flows, the cores of which are the dark layers in the image, are typically ~30–40 m thick. The Deccan is one of the world's large igneous provinces. (Image courtesy of Stephen Self.)

10.6.3 The frequency of volcanic eruptions

Each individual volcano tends to have its own pattern of activity, with some volcanoes erupting much more frequently than others. Intervals between eruptions are commonly tens of minutes at Stromboli but thousands to hundreds of thousands of years for large volcanoes erupting rhyolites. However, there is a tendency, for individual volcanoes and for volcanoes as a whole, for eruptions of small magnitude to occur frequently whereas larger eruptions are rarer. This is why the magnitude of eruptions experienced during human history is considerably smaller than the scale of eruptions which are found in the geological record. Eruptions of very large magnitude are (fortunately) very rare and so on the small time frame of human history we have generally experienced small magnitude eruptions. Figure 4.19 showed the number of eruptions of a given magnitude which are expected to occur per thousand years. About 100 eruptions of the scale of the 1980 Mount St Helens eruption could be expected per 1000 years, or one per decade. On the scale of the 1883 Krakatau eruption about 10 eruptions could be expected per 1000 years or one per century. Events on the scale of the 1815 eruption of Tambora occur about once in a millennium. The figure shows a distinct change in slope at VEI values greater than 7. This is probably a reflection of the incomplete nature of our records of large eruptions rather than a real feature of volcanic behavior.

10.6.4 Magma chambers and eruption magnitude and frequency

As seen in Chapter 4, magma is generally stored at some level within the crust prior to eruption. It is plausible to imagine then that a link would exist between the magnitude of an eruption and the size of the magma chamber which feeds it. For instance, small but frequent basaltic eruptions from volcanoes in Hawai'I and Iceland occur from magma chambers with typical volumes of only 35–150 km^3, determined by the methods described in section 4.2.3. Such chambers could, therefore, only produce this volume of magma during an eruption assuming the chamber could be fully evacuated. Typical eruption volumes are, in fact, only a small fraction of the chamber volume. For example, the magma chamber beneath Hekla in Iceland is estimated to have a volume of up to 145 km^3. The 1991 eruption at Hekla produced 0.15 km^3 of lava or ~0.1% of the magma chamber volume. The chamber volume at Kilauea is ~50 km^3. Typical erupted volumes at Kilauea are between 10^{-3} and 0.1 km^3, or 0.002 and 0.2% of the chamber volume. Large pyroclastic density current eruptions (Table 10.5) producing thousands of cubic kilometers of material require storage in magma chambers which are much larger than the small basaltic chambers even if all the magma could be evacuated from them. If, as in the basaltic case, only ~0.1% of the total chamber volume could be erupted, then an eruption producing 1000 km^3 of material would require storage in a chamber with a volume of 10^6 km^3 (equivalent to a sphere of radius ~60 km).

Further evidence for a general link between magma chamber size and eruption magnitude comes from comparing caldera size with erupted volumes. Calderas form by collapse of the magma chamber roof as magma is withdrawn during eruptions which generate more than ~10–50 km^3 of material. It is thought that the caldera diameter is a reflection of the diameter of the underlying magma chamber. When the area of a caldera produced in a given eruption is compared with the volume of material erupted a good correlation is observed (Fig. 10.8). For example, the caldera formed by the 2.2 Ma Cerro Galan eruption which produced ~2000 km^3 of material (Table 10.5) was 25 by 35 km in diameter, an area of ~690 km^2, and the caldera formed in the 5000 km^3 La Garita eruption (Table 10.5) measured 35 by 75 km, an area of ~2000 km^2. The ratios of the erupted volume to caldera area for these two eruptions are ~2.9 km and 2.4 km, respectively, suggesting that these were approximately the vertical extents of the parts of the chambers evacuated.

10.7 Elastic and inelastic eruptions

Simple models such as the one described in section 4.4 can explain the general behavior of volcanic systems and predict the pattern of behavior expected at a single volcano. For a magmatic system with a

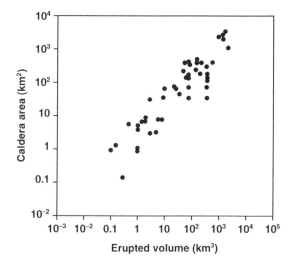

Fig. 10.8 Illustration of the strong correlation between the area of the caldera formed in a large-volume eruption and the volume of magma erupted. Modified from fig. 2 in Smith, R.L. (1979) Ash-flow magmatism. *Geol. Soc. Am. Spec. Pap.*, **180**, 5–27.

chamber of a given size the model predicts a pattern of behavior in which the chamber inflates as magma is added to it, erupts a certain volume of magma when the failure point is reached, and then reinflates once eruption ceases, initiating a new cycle of activity. Variations in the exact failure conditions (due to irregularities in the chamber wall, for example) and of the magma supply rate to the chamber will lead to variations in the repose time

between events and the volume erupted. This simple picture mirrors the observed behavior of many small basaltic systems. Figure 10.9 shows, for example, part of the tilt record of Kilauea volcano for part of 1983 and 1984. A tiltmeter at the summit of the volcano continuously measures the slope or "tilt" of the ground surface. When the magma chamber beneath the summit is inflating the ground above it tilts upwards and outwards in response. When an intrusion or eruption removes magma from the magma chamber deflation occurs and the ground tilt direction reverses. Figure 10.9 shows a series of inflation and deflation events which occurred at the summit of the volcano during 1983 and 1984 in response to a series of eruptions occurring on the volcano's flank. Forty-seven such eruptions occurred between 1983 and 1986, and Fig. 10.9 shows the tilt variations associated with eruptions 3 to 19 in this series. The tilt record shows the cyclic pattern of magma chamber inflation and deflation predicted by the simple elastic magma chamber models just described. Prior to each eruption the magma chamber fills with fresh magma and the ground gradually tilts outwards. Each eruption is associated with a rapid inward tilting as the magma chamber deflates. The tilt record in Fig. 10.9 also shows the variability in repose times between each eruption and in the amount of inflation and deflation associated with each eruption. Compare, for instance, the different amounts of deflation associated with eruptions 18 and 19.

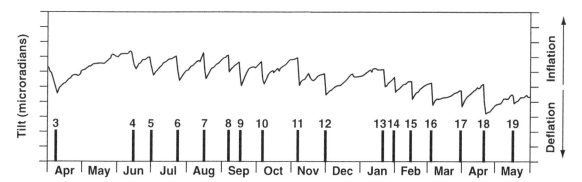

Fig. 10.9 Summit tilt changes reflecting a series of inflation and deflation events in the summit magma chamber of Kilauea volcano, Hawai'I, corresponding to a series of eruptive episodes in 1983 and 1984. Modified from fig. 1.2 in Wolfe, E.W., Neal, C.A., Banks, N.G. & Duggan, T.J. (1988) Geological observations and chronology of eruptive events. pp. 1–97 in The Puu Oo eruption of Kilauea volcano, Hawai'I: Episodes 1 through 20, January 3, 1983, through June 8, 1984. U.S.G.S. Prof. Pap. 1463.

The greater amount of deflation during eruption 18 is reflected in the greater volume of lava produced in this eruption – twelve times as much lava as during eruption 19! The similarity between the actual behavior shown in Fig. 10.9 and the behavior predicted by simple elastic models such as that described above is remarkable.

The type of eruptive activity just described is "elastic", i.e., the inflation and deflation of the magma chamber is cyclic and not associated with any significant permanent deformation; eruption ceases when the overpressure generated prior to eruption has been relieved. Not all volcanic systems, however, behave in such a simple way. In many eruptions this simple pattern of behavior is significantly modified because an eruption causes inelastic, i.e., irreversible, deformation of the volcanic edifice. These are eruptions in which activity continues even after the initial overpressure has been relieved and in which continued eruption may eventually reduce the pressure inside the chamber to the point where the roof collapses. Such "inelastic" eruptions seem to occur when erupted volumes exceed at least ~10 to 50 km^3. These inelastic eruptions include some of the largest eruptions known – the ignimbrite-forming eruptions in Table 10.5 are all examples of inelastic activity and are all associated with caldera collapse. Inelastic events can also occur, though, in small basaltic systems – the current caldera at Kilauea volcano, for example, formed during an eruption in 1790. As inelastic eruptions are ones in which the largest volume of material can be produced by any given volcanic center, it is important to understand the circumstances in which an eruption may continue beyond its elastic limit.

There are two main conditions in which an eruption may be able to continue after its initial overpressure has been relieved. The first occurs on shield volcanoes. If the vent at which eruption occurs is at a lower elevation than the top of the magma chamber then magma can continue to drain from the magma chamber through the feeder dike system even after the overpressure has been relieved. This is the mechanism envisaged at Kilauea volcano to explain caldera formation in 1790. Kilauea possesses two long rift zones – areas of dike intrusion which extend laterally from the summit. Kilauea's east rift zone extends to a total length of 125 km, 75 km of which is offshore and has built a submarine ridge (Fig. 10.10). Eruptions on this ridge can occur at elevations which are lower than the elevation of the top of the summit magma chamber, and it is inferred that large submarine eruptions periodically cause excessive drain-down of the magma chamber and hence caldera formation.

The condition that allows inelastic activity and caldera formation in association with large explosive eruptions is very different. It depends on the idea that a gas phase exists in the upper part of the magma chamber at the time when the initial overpressure has been relieved. If no such gas phase is present in the chamber once the overpressure is relieved then the eruption must cease because there is no pressure available to push the magma upwards through the dike/conduit system. If, however, the magma is initially supersaturated in volatiles, or becomes so as the pressure in the chamber falls (during the elastic phase of the eruption), then the formation and growth of the gas bubbles will drive the gas–magma mixture out of the chamber and allow eruption to continue even as the chamber pressure continues to decline. In fact, this becomes a self-perpetuating behavior because the further reduction in the chamber pressure triggers further exsolution and so on. If this behavior continues for long enough then eventually the reduction in pressure causes failure of the chamber roof and caldera formation. The amount of magma which can be erupted will depend on the detailed eruptive behavior. If exsolution of gas from the magma ceases before caldera collapse occurs then the limiting factor will be the amount of gas initially dissolved in the magma and the chamber depth. If caldera collapse occurs, the force of the unsupported roof rocks pushing down on the remaining magma in the chamber could force a high proportion of that magma out.

10.8 Eruptions of exceptional magnitude

10.8.1 Introduction

The largest eruptions seen in the geological record are ignimbrite-forming eruptions and flood basalt

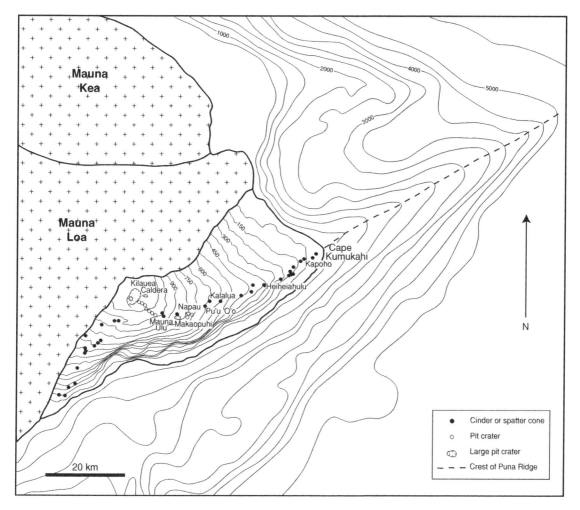

Fig. 10.10 Map of a portion of the island of Hawai'I showing the subaerial and submarine topography of Kilauea volcano. The submarine ridge which extends offshore is known as the Puna Ridge and is the submarine extension of Kilauea's East Rift Zone. (Adapted from fig. 1 published in *Journal of Volcanology and Geothermal Research*, Vol 113, Parfitt, E.A., Gregg, T.K.P. and Smith, D.K., A comparison between subaerial and submarine eruptions at Kilauea volcano, Hawai'I: implications for the thermal viability of lateral feeder dikes, 213–242, copyright Elsevier (2002).)

eruptions (Table 10.5). The fact that the largest eruptions are confined to only these two eruption types suggests that there is something special about them which produces such exceptionally large volumes of material. What is it about these eruptions which allow them to produce such large volumes and what controls the upper limit on the scale of these eruptions? The two types of eruption differ greatly from each other and so in answering these questions we will consider each type separately.

10.8.2 Large ignimbrite-forming eruptions

The largest ignimbrite-forming eruptions (Table 10.5) are all associated with caldera formation. They are "inelastic" eruptions of the type described in section 10.7 and are able to produce an excessive volume of material because they are not limited by elastic processes. The very largest eruptions probably form part of a continuum of eruption scales from a minimum size of around 10 km^3. The scale of the eruption is probably controlled primarily by the size of the underlying magma chamber, with larger chambers simply having more magma to erupt. Why, though, are these eruptions always ignimbrite-forming? Commonly, the eruptions start with Plinian activity but evolve into ignimbrite-forming events, the majority of the erupted volume being generated during the ignimbrite-forming phase. We saw in section 6.7 that Plinian eruptions can evolve into ignimbrite-forming eruptions if the gas content of the magma declines or if the mass flux increases significantly during the eruption. It seems likely that the key issue here is that the magma is unusually gas-rich. This not only ensures that the initial phase of the eruption is Plinian, but also guarantees that gas exsolution from the magma continues beyond the critical point where the initial excess pressure in the magma chamber is relieved. This ensures that fractures begin to form in the roof rocks and incipient caldera collapse starts. The fractures grow in a very inelastic way and, depending on their exact locations and orientations, can become much wider than the dikes which fed the initial Plinian phase. The consequent increase in mass eruption rate as the fractures widen, coupled with the eventual decline in volatile content of the erupting magma as deeper levels in the chamber are tapped, leads to the change to ignimbrite formation and ensures that this happens at a very high mass flux. The term "super-volcano" has recently been coined to describe volcanoes with large-volume magma reservoirs that can erupt in this way, but there is nothing fundamentally different between their eruption mechanisms and those of less devastating events.

10.8.3 Flood basalt eruptions

Flood basalts differ from the eruptions we have discussed thus far in that they appear to erupt magma directly from the base of the lithosphere without any significant crustal storage occurring. It is known that flood basalt eruptions are associated with mantle plumes; they appear to be generated when a mantle plume first impinges on the lithosphere. Current ideas suggest that large volumes of magma generated in the plume head are accumulated at the base of the lithosphere and then erupted directly

to the surface through feeder dike systems. There is currently disagreement, however, about the exact eruption mechanisms. One view is that, once the dike system reaches the surface, magma is able to erupt from it at very high rates (of the order of 10^9 kg s^{-1}) in eruptions which last a matter of days. This style of eruption can be seen as analogous to the "elastic" eruptions described above, and could be possible because the great width of a dike extending completely through the crust would minimize the frictional energy losses of magma rising through it. Eruption would cease when the overpressure driving the eruption is relieved. The large volumes of individual eruptions would then reflect the large volumes of magma produced in the plume head and stored at the base of the lithosphere.

Another explanation is also possible, however. It has been suggested that flood basalt eruptions may represent much longer duration eruptions in which lava is erupted much more slowly than previously thought. This type of activity can be likened to that seen during certain Hawaiian eruptions. Most eruptions at Kilauea volcano, for example, produce small volumes of lava in eruptions of short duration (a few hours or days). Sometimes, though, an eruption occurs in which magma is erupted continuously for years or decades at a fairly constant, slow rate. These eruptions differ from the small cyclic eruptions described in section 4.4. In these long-duration eruptions a dike system produces a continuous link from the mantle through the magma chamber to the surface, and magma can be erupted steadily through it without periods of storage and inactivity. During such eruptions the eruption rate is determined by the rate at which magma is transferred out of the mantle, and the erupted volume is not limited by the usual constraints based on chamber size. The eruption can continue as long as the dike system remains open. It is possible that during flood basalt eruptions a similar circumstance prevails: once emplaced, a dike system is able to supply magma steadily to the surface at a rate which is dictated by the rate at which magma is produced in, or is able to segregate from, the mantle plume, rather than simply the rate at which it can flow up the dike. The emplacement of the flood basalt lava flow field is then seen as an extremely large-scale version of what happens during the formation of a pahoehoe compound lava flow field. Large lava flows are emplaced, then inflated, and finally used as lava tubes to feed more large flows. In such a case the limiting factor on the total erupted volume would be how much magma could be generated in the plume head and how continuously it could be supplied to the dike system. Any significant break in supply could cause cooling and solidification of the dike system and interrupt activity until stresses built up within the lithosphere to the point that a new dike propagated.

10.9 Summary

Observational data have demonstrated three fundamental features of the behavior of volcanic systems:

1 that there is a link between the magnitude and frequency of activity such that small eruptions occur frequently and larger eruptions occur less frequently;
2 that the volume of magma erupted from a given volcano is commonly linked to the size of the magma chamber feeding it such that large chambers feed large eruptions;
3 that the largest eruptions in the geological record are of two distinct types: ignimbrite-forming eruptions and flood basalt eruptions.

These points can be explained as follows:

• Simple models of magma chamber failure explain the first two points. For a chamber to erupt it first inflates as magma is added to it. Once the pressure exceeds a critical point determined by the strength of the chamber walls, the walls fail and an eruption or intrusion will occur. The larger the chamber the more magma must be added to it before failure occurs. Assuming that the rate at which magma is supplied to a chamber does not vary very greatly between different volcanic systems, this means that the repose time between events will be greater for larger magma chambers, i.e., the frequency of eruptions from large magma chambers is smaller than from smaller magma chambers. Furthermore, as the volume of

magma added to the chamber prior to eruption will be approximately the same as the volume which is erupted (as long as the behavior is elastic) then such models also explain why larger chambers generate larger eruptions.
• The largest volcanic eruptions are of only two types because these eruptions represent special cases in which conditions are such that excessively large amounts of magmatic material can be erupted.
• Very large volume ignimbrite-forming eruptions are "inelastic" events in which eruption causes caldera formation to occur. Such eruptions can produce volumes which greatly exceed the volumes generated in "elastic" eruptions. They can occur if there is a gas phase present in the magma chamber when the initial overpressure has been relieved. The expansion of the gas phase drives the magma–gas mixture out of the chamber. This can continue as long as gas continues to exsolve from the magma. The chamber pressure will continue to decline as more magma is removed and if it declines sufficiently the roof of the chamber may fail causing caldera collapse. This roof collapse can then cause huge quantities of magma to be driven from the eruption at very high rates causing very large volume ignimbrite-forming eruptions.
• Flood basalt eruptions are associated with the initial impingement of a mantle plume on the lithosphere and represent events in which magma is erupted directly from the base of the lithosphere with no significant shallow storage occurring. Opinion differs about the mechanism of these eruptions. One view is that the eruptions occur at extremely high eruption rates but last no more than a matter of days. In the other view eruptions may occur at much slower rates over a period of years to decades. Both possible styles of eruption are analogous to behavior observed in small-scale basaltic eruptions. The former would be equivalent to the elastic eruptions described in section 10.7 and the volume erupted would be limited by the size of the magma storage area at the base of the lithosphere. The latter would be limited by how long magma could be continuously supplied through the feeder dike system from the mantle source zone. A break in supply would

allow the magma in the dike to cool and solidify, causing the end of the eruption. A new eruption would only occur when a new dike system could be formed.

10.10 Further reading

Blake, S. (1981) Volcanism and the dynamics of open magma chambers. *Nature* **289**, 783–5.

Cashman, K.V., Sturtevant, B., Papale, P. & Navon, O. (2000) Magmatic fragmentation. In *Encyclopedia of Volcanoes* (Ed. H. Sigurdsson), pp. 421–30. Academic Press, San Diego, CA.

Druitt, T.H. & Sparks, R.S.J. (1984) On the formation of calderas during ignimbrite eruptions. *Nature* **310**, 679–81.

Newhall, C.G. & Self, S. (1982) The Volcanic Explosivity Index (VEI): an estimate of explosive magnitude for historical volcanism. *J. Geophys. Res.* **87**, 1231–8.

Parfitt, E.A. & Wilson, L. (1995) Explosive volcanic eruptions – IX. The transition between Hawaiian-style lava fountaining and Strombolian explosive activity. *Geophys. J. Int.* **121**, 226–32.

Pyle, D.M. (2000) Sizes of volcanic eruptions. In *Encyclopedia of Volcanoes* (Ed. H. Sigurdsson), pp. 263–9. Academic Press, San Diego, CA.

Simkin, T. & Seibert, L. (2000) Earth's volcanoes and eruptions: an overview. In *Encyclopedia of Volcanoes* (Ed. H. Sigurdsson), pp. 249–61. Academic Press, San Diego, CA.

Smith, R.L. (1979) Ash-flow magmatism. *Geol. Soc. Am. Spec. Pap.* **180**, 3–27.

Spera, F.J. (2000) Physical properties of magmas. In *Encyclopedia of Volcanoes* (Ed. H. Sigurdsson), pp. 171–90. Academic Press, San Diego, CA.

Spera, F.J. & Crisp, J.A. (1981) Eruption volume, periodicity, and caldera area: relationships and inferences on development of compositional zonation in silicic magma chambers. *J. Volcanol. Geotherm. Res.* **11**, 169–87.

Wallace, P. & Anderson, A.T. (2000) Volatiles in magma. In *Encyclopedia of Volcanoes* (Ed. H. Sigurdsson), pp. 149–70. Academic Press, San Diego, CA.

Wilson, L. & Head, J.W. (1981) Ascent and eruption of basaltic magma on the Earth and Moon. *J. Geophys. Res.* **86**, 2971–3001.

10.11 Questions to think about

1 Why are eruptions on the ocean floor more likely to be effusive than explosive?

2 What is the main reason, overall, that evolved magmas are more likely to have very explosive eruptions than more basaltic magmas?

3 If a long-lived eruption is explosive, what factors control whether it is intermittently explosive or continuously explosive?

4 Why do we use more than one scheme for categorizing eruptions?

5 What are the general trends of the relationships between magma reservoir size, volume of magma erupted, and frequency of eruption?

11 Volcanic hazards and volcano monitoring

11.1 Introduction

The historic and archeological record is littered with examples of disasters caused by volcanic eruptions: the burial of Pompeii and Herculaneum by the AD 79 eruption of Vesuvius, the collapse of the Minoan civilization as a result of an eruption at Santorini around 1650 BC, and the devastation produced by the tsunami waves generated during the eruption of Krakatoa in 1883. This chapter looks at the range of hazards that can be caused by volcanic eruptions, at how scientists try to assess the hazards presented by any one volcano, at methods used to monitor active volcanoes, and at successes and failures in predicting volcanic activity.

11.2 Types of volcanic hazard

Volcanic eruptions can present a wide range of hazards to human life and property and to the wider environment. Some hazards are direct, such as the destruction of property by lava flows or death caused by being overrun by a pyroclastic density current. Other hazards are indirect, such as starvation due to destruction of crops or changes in climate caused by volcanic activity (see Chapter 12).

11.2.1 Lava flows

Lava flows do not generally cause the deaths of people directly. This is because, except in the cases of fluid basaltic flows very near the vent, the vast majority of lava flows do not move as fast as moderately fit people can walk. Thus, unless an eruption

starts very close to a habitation where people are sleeping, or the people involved are ill or infirm, they are likely to be able to escape, as long as they walk away from the downslope path that the flow is following. Vastly more people have been killed by the products of explosive eruptions than by being overtaken by lava flows.

Lava flows can, however, cause the total destruction of property. In some cases flows that are thick enough do this by simply burying buildings, generally crushing them in the process (Fig. 11.1). In other cases a building may be strong enough to withstand the pressure exerted on its walls as a flow piles up against it, and thus the flow may eventually just surround the building, but the intense heat radiated from the flow will ignite virtually any flammable material and so the building will burn (Fig. 11.2). Given the slow advance rate of flows, it is common

Fig. 11.1 Buildings in the village of Kalapana on the southeast flank of Kilauea volcano, Hawai'I, extensively damaged by lava flows. (Photograph by Jim Griggs, courtesy of U.S. Geological Survey, Hawaiian Volcano Observatory.)

Fig. 11.2 Building in the village of Kalapana on the southeast flank of Kilauea volcano, Hawai'I, set on fire by encroaching lava flow. (Photograph by Jim Griggs, courtesy of U.S. Geological Survey, Hawaiian Volcano Observatory.)

for not only people but also all of their easily movable possessions to be evacuated from threatened buildings, and in a few instances some of the building itself may be removed in time. A famous case is that of a small historic church, the Star-of-the-Sea painted church in the village of Kalapana, Hawai'I. This building was braced and lifted bodily in time to be saved from the lava flows from Kilauea's Pu'u 'O'o-Kupaianaha eruption in 1991 (Fig. 11.3). Virtually all the other buildings in this village were destroyed, however.

Fig. 11.3 The St Mary's Star of the Sea Catholic Church, Kalapana, Hawai'I, being moved to a safe location to avoid advancing lava flows. (Photograph by D. Weisel, courtesy of U.S. Geological Survey, Hawaiian Volcano Observatory.)

11.2.2 Pyroclastic falls

In the most energetic explosive events producing eruption columns generating fall deposits, pumice clasts up to more than a meter in average size can be erupted. However, as seen in Chapters 6 and 8, these will be deposited extremely close to the vent, and so will pose only a very localized threat to people and buildings. Also, because subsidence to form calderas, or at least significant depressions, is common around the vents of large-volume energetic eruptions, the vent regions often contain lakes, and so are not usually the sites of large population centers. A major exception to this is the Campi Flegrei Caldera, which contains part of the suburbs of the city of Naples in Italy.

The major danger from eruptions generating fall deposits comes from the accumulation of large volumes of fine-grained pyroclasts onto the roofs of buildings over a wide area downwind of the vent. The clasts will generally have been transported high into the eruption cloud and will have taken tens of minutes to hours to reach their final destination on the ground. It is not the temperature of the material that matters, therefore, but just the stress that its weight exerts on the building's structure, causing it to collapse (Fig. 11.4). This is especially true if rain falls on the material before it can be removed from the roof. A loosely packed layer of vesicular pumice may have a bulk density of as little as 600 kg m^{-3}; filling only half the vacant space with rainwater raises the density to 1350 kg m^{-3}, more than doubling the load on the building beneath. With a more conservative density of 1000 kg m^{-3}, a layer of damp pyroclasts 1 m thick would exert a pressure of 10 kPa, which is very close to the maximum recommended design load for a modern building such as a reinforced concrete warehouse.

It is not just buildings that are affected by fall deposits; cars and larger vehicles are also at risk in various ways. Accumulation of ash around vehicles near Mount St Helens during the 1980 eruption melted their tires. And during the 1990 eruption of Pinatubo volcano in the Philippines, preferential accumulation of ash on upper surfaces of the tail structures of aircraft parked in the open at a USA air base caused them to tip over, so that their noses rose into the air and their tails hit the

Fig. 11.4 An accumulating pyroclastic fall deposit from the 1991 eruption of Mount Pinatubo volcano caused the collapse of part of the Officers' Club at Clark Air Base, Philippines. (USGS Photograph by T.J. Casadevall, courtesy of U.S. Geological Survey.)

Fig. 11.5 A World Airways DC-10 aircraft tilted onto its tail by the weight of air fall pyroclasts during the June 15, 1991 eruption of Mount Pinatubo volcano, Philippines. (Photograph by R.L. Rieger., courtesy of U.S. Geological Survey.)

ground (Fig. 11.5), causing a great deal of structural damage.

11.2.3 Ash in the atmosphere

It is not just the accumulation of pyroclasts on the ground that matters. The region from tens to hundreds of kilometers downwind of an eruption column and eruption cloud will contain small falling silicate particles. The number density of these particles may not be great enough to make them easily visible from the cockpit of any aircraft flying below

the level of the top of the cloud, and so unless the pilot is warned, the aircraft may fly into the cloud of particles. Aircraft jet engines work by forcing air into the turbine system and using it to burn fuel. If that air contains silicate particles, they not only reduce the air flow but they also melt in the burning fuel and some of the silicate liquid is smeared onto the inside of the engine casing, where it solidifies into a glassy coating (Fig. 11.6). As this layer builds up, it further restricts air flow and reduces the efficiency of the engine. The immediate reaction of the pilot, noticing the loss of power, may be to throttle up the engine to recover power, but this will raise its temperature and make the problem worse. In the extreme case, the engine may stop. Indeed, several events occurred in the 1990s in which commercial aircraft had all of their engines shut down in flight in this way. Fortunately, all of them managed to restart at least one engine before crashing. However, in some cases, a final problem for the pilots landing these planes was to discover, on nearing the ground, that the cockpit windows had been "sand-blasted" nearly opaque by the impacts of small pyroclasts. The effects of small particles in the atmosphere are felt on the ground and at low levels too: surface vehicles and helicopters being used to evacuate people during eruptions find that their engines cannot function.

There is a second hazard from ash-size particles settling to the ground. They can be inhaled by people

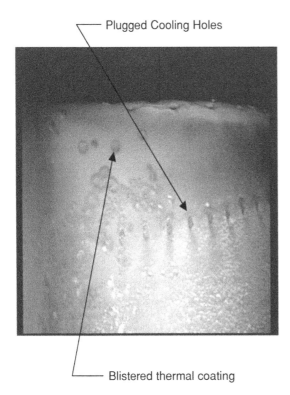

Plugged Cooling Holes

Blistered thermal coating

Fig. 11.6 Damage to the interior of a jet engine from a DC-8 aircraft that encountered ash from an eruption of Hekla in February 2000. (Photograph courtesy of Thomas Grindle (NASA DFRC), Frank Burcham (ASM, Inc.) and David Pieri (JPL).)

in the fallout area. Humans and animals require an input of oxygen just as much as aircraft engines do. In the cases of living creatures, the air must be able to reach the alveoli in their lungs. A layer of moist ash on the lung surface stops this happening – it behaves like a layer of wet cement – and can cause death as easily as asphyxiation or drowning. Even if death does not occur, there can be long-term damage to the lungs comparable to that associated with exposure to asbestos fibers or coal dust.

11.2.4 Pyroclastic density currents and surges

These are by far the most dangerous of volcanic hazards in terms of their immediate effect on people, animals, and buildings. The obvious feature is the high temperature of the gas and clasts, at least 600 K and as much as 1000 K in pyroclastic density

currents. No living thing can survive being engulfed by pyroclasts at these temperatures. Surges may have lower temperatures, given that they are likely to have involved more interaction of the juvenile materials with the atmosphere, but they are still often hotter than 300 K.

The speeds of density currents and surges, commonly at least 100 m s^{-1}, mean that it is essentially impossible to imagine running – or driving in any surface vehicle – away from one. A jet-engined aircraft (but not a helicopter) can fly faster than a pyroclastic density current can travel, but without access to a vertical-take-off type of aircraft with its engines already running it would be hard for anyone to make use of this fact. These high speeds of pyroclastic density currents also mean that they exert high pressures on obstacles such as buildings that they encounter. The pressure exerted on a wall at right angles to a pyroclastic density current with a bulk density of $\rho = 1000$ kg m^{-3} and a speed of $u = 100$ m s^{-1} will be $0.5\,\rho\,u^2 = 5$ MPa, hundreds of times greater than typical load-bearing strengths of the strongest building materials and so it is not surprising that even much more dilute pyroclastic surges cause great structural damage (Fig. 11.7).

Perhaps the only possible protection against a pyroclastic density current is topography. We saw in Chapter 8 that high-speed currents can surmount ridges several hundred meters high if there is no alternate. However, given the opportunity, a density current will travel down a valley rather than climb a ridge. So, given a fast vehicle and enough advance warning, one can imagine driving to the top of a hill or ridge in the hope that the density current will travel down a valley adjacent to the ridge, or flow around the hill. Unfortunately, as will be discussed later in this chapter, it is not likely that the general population in a volcanic area will have the required early warning. And even if the pyroclastic density current proper does not reach the top of the hill or ridge, it is quite likely that the associated surge cloud, almost equally lethal to people, will do so (Fig. 11.8).

11.2.5 Lahars

Whenever rain falls onto pyroclastic materials, whether in a fall or a density current deposit, the potential exists to fluidize the volcanic material.

Fig. 11.7 A Roman column knocked down by the first pyroclastic density current that reached the city of Herculaneum during the AD 79 eruption of Vesuvius. The column is embedded in the deposit and is exposed in a tunnel. (Photograph by Lucia Gurioli, University of Hawai'I.)

Fig. 11.8 A dilute pyroclastic surge cloud crossing a ridge. The top of the ridge is at the top of the slope covered with bright deposits from earlier surges. The topography of the ridge has deflected and channeled the main pyroclastic density current from which the surge cloud is derived so that it is traveling down the valley behind the ridge and cannot be seen in this view. (© NERC, 1997. Montserrat Volcano Observatory photograph.)

The mixture of water and solid clasts is called a **lahar** (Fig. 11.9). If most of the solid material is fine-grained, the lahar is sometimes called a mud-flow, but of course mudflows can also be formed from nonvolcanic materials. Lahars generated from pyroclastic density current deposits can be quite hot: mixing equal volumes of pyroclasts at, say, 650 K with water at 300 K yields a mixture in which the water is just boiling. However, lahars involving fall deposits are generally cold. An altern-ate mechanism to produce lahars is for pyroclasts to be emplaced on top of a layer of ice or snow; if the clasts become waterlogged and sink through the liquid water as it is being produced by melting, this maximizes the rate of heat transfer.

Lahars have rheological properties somewhat similar to those of lava flows, being nonNewtonian fluids, but they generally have much lower visco-sities and can move at speeds up to a few tens of meters per second. They can travel considerable distances – at least tens of kilometers. A terrible example was the lahar from the summit region of the Nevada Del Ruiz volcano in Colombia in 1985 which swept ~70 km down a river valley on the flank of the volcano at up to 15 m s^{-1} killing more than 23,000 people in the town of Armero. Although lahars can flow almost as fast as water, their densities are typically 1.5 to 2 times greater than the density of water, so that they cause more destruction than similar-sized water floods. Also, after it is emplaced, the water–ash mixture sets rather like concrete, and so both rescue and clean-up operations are extremely difficult. An added problem is that formation of lahars is often likely to continue to be a threat long after an eruption is

Fig. 11.9 A lahar deposit from an eruption of Mount Pinatubo, Philippines. The deposit forms the floor of the channel, and the lahar material forms a veneer on the channel walls. (Photograph by Pete Mouginis-Mark, University of Hawai'I.)

over: if the ash emplaced by the eruption is unconsolidated, then every time that there is a major rainstorm, it is likely that a new lahar will occur until all of the deposit is eroded away. The many lahars that were generated over a several year period following the 1991 eruption of Pinatubo in the Philippines provide a good example: they led to the evacuation of hundreds of thousands of people long after the eruption proper was over.

11.2.6 Jökulhlaups

When an eruption occurs under a glacier, a characteristic sequence of events can occur. The weight of the overlying ice is commonly great enough that the magma does not fragment in a volcanic explosion but is extruded along the contact between the base of the glacier and the ground, essentially as a sill. However, melting of the overlying ice produces water, and this interacts with the magma, cooling its surface and leading to some of the processes, such as fragmentation of the magma surface into glassy particles, associated with fuel–coolant interactions (see section 7.3.2). In the subglacial case the process is very much less violent, but it does maximize heat transfer from the magma to the ice.

Although the magma may melt completely through the ice and produce a hydromagmatic explosive eruption, the main hazard in these events is the accumulation of a large volume of water that can eventually escape from under the edge of the glacier in a catastrophic flood (Fig. 11.10). These floods are quite common in parts of Iceland, and the Icelandic word jökulhlaup is used to describe them. A recent well-monitored example was the 1996 Gjálp fissure eruption under the Vatnajökull ice-cap in southern Iceland. Here ~3.5 km^3 of water accumulated over a 5 week period and then drained from under the ice in just 2 days, the flood destroying roads and bridges in its path (Fig. 11.10).

11.2.7 Volcanic gases

We have seen that the commonest volcanic gas is water vapor which, except when it is very hot, is harmless. Actually this statement is not quite true, because although water has no direct adverse effects on plants and animals on the ground, it may have very adverse effects when high eruption columns carry it into the upper atmosphere – the stratosphere. This is because water molecules become involved in a complex chain of chemical reactions that involves sunlight. One of the products of these interactions is the layer of ozone that absorbs many wavelengths of ultraviolet light that, if they reach the ground, can cause skin cancers. Excessive amounts of water vapor, like other gases present in trace amounts, can modify the reactions that maintain the ozone in ways that are still not fully understood.

Fig. 11.10 The jökulhlaup from the 1996 Gjálp eruption, Iceland. The flood has destroyed part of the bridge over the river Gygjukvisl on Skeidararsandur, South Iceland. (Photograph taken on November 5, 1996, by Magnus Tumi Gudmundsson, Institute of Earth Sciences, University of Iceland.)

Unfortunately, every other common volcanic volatile is directly dangerous to most living creatures. The next commonest volcanic gas is carbon dioxide. As a cold gas this is dangerous mainly because we cannot breath it – indeed, carbon dioxide is the waste gas that animals breathe out. After it has cooled to the ambient temperature, carbon dioxide is ~50% denser than air and so it collects in topographic hollows, especially if there is no wind to stir up the atmosphere. Any person or animal walking into a depression filled with carbon dioxide will rapidly become unconscious as no new oxygen goes into their bloodstream and the existing supply is used up. An additional problem is that, for a small fraction of the human population, carbon dioxide inhibits part of the central nervous system, so that even if one realizes what is wrong, one cannot climb out of the depression to safety.

Carbon dioxide is a corrosive, as well as poisonous, acid gas. In certain places, this volatile is released from shallow magma bodies and seeps upward to collect as a dissolved gas in the water in lakes. The added gas coming from below makes the water at the bottom of the lake a little denser than that above, so that it stays at the bottom. If some event such as a landslide, or even just a very heavy downpour of rain on one part of the lake, disturbs this density stratification, the water in the lake may overturn. Water from the bottom, which is saturated in gas, rises to the top where the pressure is much less. Here it becomes supersaturated, and explosively exsolves a dense cloud of carbon dioxide. The cloud of gas is denser than air and will hug the ground like a pyroclastic density current and travel downhill. Many people died of a mixture of chemical burns and asphyxiation in four villages around Lake Nyos in western Cameroon in 1986 when an event of this kind took place. Since then, a system of pipes has been installed in the lake to syphon water continuously from the bottom to the surface to allow it to lose gas slowly and steadily, instead of in catastrophic overturn events.

Other gases can be equally lethal in various ways. The volcano Hekla, in Iceland, often erupts magmas that are rich in the halogen element fluorine. This gas is even denser than carbon dioxide, and so even more likely to collect in hollows in the topography. On a number of occasions, sheep (and the grass that they eat) have been poisoned in large numbers by accumulations of this gas. Almst equally poisonous are the high concentrations of sulfur dioxide and hydrogen sulfide that basaltic volcanoes sometimes release. Sulfur dioxide can dissolve in atmospheric water drops and react with oxygen to form sulfuric acid. Tiny droplets of this kind are called **aerosols**, and when present in large amounts can alter the way the atmosphere reflects and absorbs sunlight, thus changing the climate. We discuss the effects of the sulfur dioxide haze from the 1783 eruption of Laki volcano in Iceland in the next chapter.

11.3 Hazard assessment

There are several aspects to assessing volcanic hazards. The first step is to decide what constitutes a hazard, and there are both physical and statistical aspects to this. Many centers of population are located close to volcanoes known to have been active in historic time – the city of Naples near Mount Vesuvius in Italy is a good example. For people in areas such as these, being overrun by a lava flow or ash flow is an obvious potential threat; but there is a less obvious hazard: being killed in the crash of an aircraft with ash-choked engines, which could happen to those same citizens while on vacation in many other parts of the world.

This introduces the idea of local, regional, and global volcanic hazards. Global hazards involve volcanoes injecting large amounts of gas, aerosols, or small ash particles into the atmosphere and changing the climate, as considered in Chapter 12. Regional hazards include the effects on aircraft operations of ash falling from eruption clouds, and such phenomena as the collapse of volcanoes erupting in the ocean, which can generate tsunami waves traveling great distances. Local hazards include all of the products of eruptions that are emplaced close to the volcanic source – lava flows, pyroclastic density currents, and fall deposits.

Historically, it is local hazards that have received most attention because they are generally the easiest to identify. Even so, prior to the 20th century the best indicators of volcanic activity were quite likely to be the memories of local inhabitants, and this underlines one of the potential problems in dealing with hazards. If a volcanic eruption (or any other kind of catastrophe for that matter) has not happened within the memory of one's grandparents, it is simply not perceived as a likely threat. Yet many kinds of natural disaster, including eruptions, have return periods (the most likely intervals between occurrences) of at least tens of thousands of years. It is only since the emergence of systematic geological mapping, and the development of a basic theoretical understanding of what kind of eruption produces a given type of deposit, that the true potential hazard of many volcanic systems has been perceived. And where the likelihood of an

eruption taking place within one generation is extremely small, it could be argued from a pragmatic sociological viewpoint that assessing a volcanic hazard is not the most urgent issue if other problems, such as flood, earthquake, tsunami, disease and war, are more likely to affect the local population.

If a volcano erupts sufficiently frequently, there are likely to be well-preserved deposits from at least the most recent eruptions. Geologists can identify and map out these deposits, so that the common styles and scales of activity are readily apparent. The most likely ranges of distances from the volcano at which people will be at risk from the various kinds of activity can be defined and drawn on a map of the area. Furthermore, eruption products can be dated in various ways, by radiocarbon or tree-ring dating of dead vegetation trapped in the deposits, or other isotopic dating methods, and the typical repose periods between events of a given scale can be established. Conversely, for volcanoes that erupt extremely infrequently, the state of preservation of even the most recent deposit may be very poor, to the point where they are unrecognizable. In the cases of many of the mountains of the Andes and central America, for example, it was not realized until the advent of modern geology that they "were" volcanoes. For these types of volcanoes we have to fall back on our knowledge of how other similar volcanic centers in similar tectonic settings have behaved to assess the likely extent of the hazards. Once the range of likely future behavior patterns of a particular volcano has been determined, the next step is to estimate when the next eruption is likely to occur. On short time scales this is best done by geophysical monitoring, which is considered in the next section. On longer time scales, it is done mainly on the basis of the statistics of previous eruptions. The best that can be produced is the probability of an eruption of a given magnitude occurring within a given period of time.

One way of making a statistical appraisal of possible future activity is to classify previous eruptions by the Volcanic Explosivity Index (VEI) as defined in Table 10.3 and to compile a list of how many eruptions of a given VEI rating have occurred in a given interval of time. Table 11.1 shows an example

Table 11.1 Data on eruption frequency and probability for 92 eruptions of Mount Vesuvius.

Volcanic Explosivity Index value	Eruption type	Number of eruptions	Probability value
1	Lava flow	21	0.228
1.5	Strombolian crisis	15	0.391
2	Strong Strombolian activity	17	0.576
2.5	Strong Strombolian activity + lava fountains	10	0.685
3	Small eruption column	15	0.848
4	Subplinian eruption	5	0.902
5	Plinian eruption	8	0.989
6	Ultra-Plinian eruption	1	1.000

Taken from section 7.4.1 in Dobran, F. (2001) *Volcanic Processes – Mechanisms in Material Transport*. Kluwer Academic/ Plenum Publishers, New York, 590 pp. With kind permission of Springer Science and Business Media.

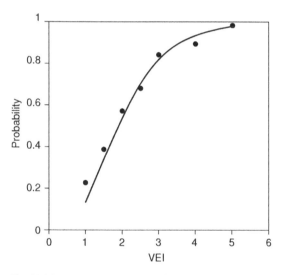

Fig. 11.11 Probability of a given **Volcanic Explosivity Index** (VEI). (Based on fig. 7.19 in *Volcanic Processes: Mechanisms in Material Transport*, by F. Dobran, Kluwer Academic/Plenum Publishers, New York, 2001, 590 pp. With kind permission of Springer Science and Business Media.)

from Dobran (2001) treating a data base of 92 eruptions at Mount Vesuvius. The probability value, P, for each VEI in Table 11.1 is the cumulative fraction of eruptions for which the intensity is less than or equal to that VEI. For example, the P value for VEI ≤ 2.5 is $(21 + 15 + 17 + 10)/92 = 0.685$. Plotting the P values against their corresponding VEI values, a graph such as Fig. 11.11 is found, which can be fitted by the double exponential function

$$P(\text{intensity} \geq \text{VEI}) = 1 - \exp[-6.634 \, \Delta t$$
$$\exp(-1.18 \, \text{VEI})] \qquad (11.1)$$

where Δt is the time interval between eruptions in years. Some examples of the implications of this are that, in any 10 year period, the probability of a Plinian eruption with VEI = 5 occurring is only 0.16, or about 1 in 6, whereas the chance of a Strombolian eruption with VEI = 2.5 is close to 0.97, i.e., extremely likely.

11.4 Monitoring volcanoes and short-term eruption prediction

Volcanoes that erupt frequently are commonly monitored in various ways for two reasons: to learn more about the internal structure of the volcano and to look for signs of activity. The two processes are linked, because the more we understand about a particular volcano, the more sense we are likely to make of indicators of impending eruptions.

Traditionally, monitoring volcanoes has implied making measurements at ground level on the volcano. Unfortunately, all continuous monitoring exercises involving instrumentation on a volcano are expensive and vulnerable. Usually at least several instruments have to be left in place on the volcano and the measurements are either recorded on site in some simple computer-controlled data storage device or returned to a distant base via a radio or satellite link. All parts of the system have to be provided with power in the form of batteries or solar panels or both. In the event of an eruption, the equipment may well be destroyed; and because the instruments are left unattended they can be damaged by animals or tourists (or simply stolen!). For all these reasons, fewer volcanoes are permanently instrumented than many volcanologists would like. Nevertheless, a growing number of volcanoes have permanent scientific stations – volcano observatories – located on them, notable examples being Kilauea in Hawai'I, Sakurajima in Japan, and Piton de la Fournaise in Réunion. We now review the various ground-based monitoring methods in common use.

We have seen that magma may be stored at shallow levels in volcanoes or may ascend directly from great depths. As any dike propagates it causes fracturing of the rocks at its tip. On the microscopic scale this is a brittle, erratic process and produces quite a lot of acoustic noise as rock surfaces fail and split, and this sound propagates as seismic waves in the surrounding rocks. These waves decay in amplitude as they spread out, and so a dike propagating at great depth is harder to detect than one near the surface. Thus magma rising into a magma reservoir from the mantle may well not generate any seismic noise noticeable at the surface, especially if it is rising through rocks that have been heated and softened by earlier dikes. However, the magma reservoir will have to inflate to accommodate the new magma, and this may cause new cracks to open, or old ones to move, around its boundary. This process also produces seismic signals. Finally, once magma is moving through an open dike, its motion is not quite steady, and vibrations of the walls of the dike occur generating a particular kind of seismic noise called **harmonic tremor**. Thus keeping several seismometers (e.g., Fig. 11.12) permanently in place around a volcanic center is a standard way of monitoring both the potential for activity and its onset.

The fact that magma must accumulate in a shallow reservoir, or rise into a shallow dike or conduits, before an eruption can start provides another technique for anticipating eruptions: monitoring the inflation of the volcano. The way that changes can occur in the tilt of the ground around an inflating magma reservoir was discussed in Chapter 4, and Figs 4.7–4.9 illustrate the kinds of tilt

Fig. 11.12 A compact, three-component broadband seismometer system. The three detectors sense the three spatial components of seismic vibrations over a wide range of frequencies and are stacked vertically in a ~9 cm outside-diameter stainless steel case. This type of instrument can be set up in the surface vault of a volcano observatory, in a subsurface vault or in a vertically drilled hole. (Image courtesy of Dr Cansun Guralp, Guralp Systems Ltd.)

and uplift that can occur. A complication with interpreting these kinds of data is that movement of magma within a volcano may cause one part to deflate as another part is inflating. All that is detected at the surface is the net effect of the movements, and so a unique interpretation may not be possible.

The arrival of new magma in a shallow reservoir, especially if it causes some deformation leading to the opening of old cracks or the production of new ones, may lead to the enhanced release of volcanic gases. Routinely monitoring gas release is a sensible precaution at volcanoes that are located near inhabited centers or receive a lot of tourists, because even if eruptive activity is not imminent, changes in gas flow can occur. The commonest gases to be monitored are carbon dioxide and sulfur dioxide. This is because juvenile water vapor, although usually released in larger amounts than the other two gases, is easily confused with atmospheric water vapor. Various methods are used. Gas samples can be sucked directly into containers from the soil or from open fractures or fumaroles for return to a laboratory for analysis. The disadvantage of this method is that the gases may be hot, making it hazardous to approach the sampling site and introducing the risk of chemical reactions between the

Fig. 11.13 A Fourier transform infrared spectrometer taking data during an eruption of Mount Etna in 2001. The instrument is collecting light emitted from the eruption column. Volcanic gases in the atmosphere absorb some of the light at certain characteristic wavelengths, allowing their presence to be detected and their concentrations to be measured. (Image courtesy of Clive Oppenheimer, Cambridge University.)

gases, if more than one species is present, as the sample cools. Real-time analysis of gases can now be carried out with miniaturized infrared LICOR devices. These can be used on the ground or flown in aircraft passing above volcanic centers. Other instruments that can operate from aircraft or the ground (Fig. 11.13) measure the amounts of carbon dioxide and sulfur dioxide in the atmosphere by detecting the amount of light from the Sun (or a distant bright artificial source) absorbed by these gases. Instruments that work at ultraviolet wavelengths are called correlation spectrometers (COSPECs) and those that work in the infrared are called Fourier transform infrared spectrometers (FTIRs).

With the advent of Earth-orbiting satellites, various methods have been explored for monitoring volcanoes remotely. The characteristics of the more recent satellite platforms are near-polar orbit at heights of 700–800 km above the surface, which gives an interval of about 16 days between revisits of the same point on the Earth's surface. A wide variety of multispectral detectors are used, with resolutions typically a few tens of meters on the ground. These instruments can be used to look for changes in the temperature of the surface, an obvious indicator of hot volcanic materials being present at shallow depth, and there have been a number of successful first recognitions of eruption in the process of breaking out, especially in unpopulated areas. Satellites can also monitor the appearance of the surface: changes in the color, texture, or reflectivity might indicate, for example, the poisoning of vegetation by enhanced release of volcanic gases. The main limitation of these techniques at the moment is the long interval between obtaining images of the same point on the surface with the same detector, so that a reliable comparison can be made to detect changes. However, once the outbreak of an eruption has been detected, every available sensor on every available satellite can in principle be turned onto the eruption site, and in this way the activity can be followed in enough detail to allow reliable short-term forecasts to be made. This is particularly useful for warning aircraft of developments in eruption cloud dispersal, for example. One modern development where the 2–3 week interval between observing the same point on the ground is less of a problem is the use of radar images to detect inflation or deflation of volcanoes. Radar can be used to detect distance changes by the time taken for the electromagnetic waves to travel from the transmitter on a satellite to the ground and back to the receiver on the same satellite. But the transmitter and receiver can also keep track of the phase of the waves, i.e., where in the course of one cycle of the waves the detection occurs. Changes in the phase can represent movements of the surface by as little as a few centimeters. Furthermore, the changes can be monitored over the entire area imaged, not just at a few specific locations as is typical of measurements made on the surface. Thus a synoptic picture of the changes can be obtained.

Continuous, or at least frequent, monitoring of the various properties of a volcano in the ways described above helps to establish its normal state and so gives an indication of when it is behaving abnormally, i.e., when it is in a state of unrest. Depending on the extent of the recorded history of a given volcano, the changes in its behavior sometimes give a good indication of when, or at least if, it is going to erupt. But there is almost never a

Table 11.2 Volcano alert systems.

Color coded alert levels		Numerical alert levels		
Color	Implication		Indicative phenomena	Volcano status
Green	No eruption anticipated. Volcano is in quiet, "dormant" state	0	Typical background surface activity; seismicity deformation and heat flow at low levels	Usual dormant or quiescent state
Yellow	An eruption is possible in the next few weeks and may occur with little or no additional warning. Small earthquakes and/or increased levels of volcanic gas emissions have been detected locally	1	Apparent seismic, geodetic, thermal, or other unrest indicators	Initial signs of possible unrest. No eruption threat
Orange	An explosive eruption is possible within a few days and may occur with little or no warning. Ash plume(s) are not expected to reach > 9 km above sea level. Increased numbers of local earthquakes are evident. Nonexplosive extrusion of a lava dome or lava flows may be occurring	2	Increase in number or intensity of unrest indicators (seismicity, deformation, heat flow, etc.)	Confirmation of volcano unrest. Eruption threat
Red	A major explosive eruption is expected within 24 hours. Large ash plume(s) are expected to reach at least 9 km above sea level. Strong earthquake activity is detected even at distant monitoring stations. An explosive eruption may be in progress	3	Minor steam eruptions. High or increasing trends in indicators of unrest, significant effects on volcano, possibly beyond	Minor eruptions started. Real possibility of hazardous eruptions
		4	Eruption of new magma. Sustained high levels of unrest indicators, significant effects beyond volcano	Hazardous local eruption in progress. Large-scale eruption now possible
		5	Destruction with major damage beyond active volcano. Significant risk over wider areas	Large hazardous volcanic eruption in progress

sufficiently repeatable pattern of activity that anything other than a statistical forecast can be given. For many volcanoes this gives rise to levels of alert, either color coded or numerical, as shown in Table 11.2.

11.5 Hazard mitigation

Reducing the severity of the outcome of an eruption can be approached in two ways. Certain kinds of precautions can be taken before an eruption starts, and various other actions can be taken during the eruption. Of course, monitoring the volcano, in the various ways described in the previous section, is itself one form of pre-eruption precaution. If sufficiently early and sufficiently reliable warnings of an impending eruption can be given, the local population can be evacuated safely. It has to be said, however, that the word "reliable" is important here: there are a number of examples of evacuations after which the expected eruption did not occur, and the social and economic damage done to the evacuated people was very great.

In some cases, especially in areas of high rainfall and hence high surface erosion, the topography of a volcano is such that there are obvious paths that lava flows, pyroclastic density currents, lahars and mudflows are likely to take. In these cases it is possible to erect barriers in an attempt to stop the flows, or at least divert them from populated areas. As seen earlier the forces that lavas flows and, especially, pyroclastic density currents can exert on obstacles in their paths are very large, and there have been few attempts to erect barriers intended to stop them, although a few at least partially successful attempts at lava flow diversions have been made on Mount Etna. Barriers against mudflows and lahars have had more success, for example in the case of lahars from Mount Pinatubo.

If there is no opportunity to evacuate beforehand, then the actions that can be taken to reduce the impact of an eruption are limited. If the eruption is explosive and involves the formation of a fall deposit, sweeping roofs clean of accumulating ash will minimize the chances of structural collapse. But if the density of ash in the atmosphere is too great, the health hazard will outweigh the economic benefit of staying to protect property. In that case the best choice is to walk (assuming no working vehicles are available) away from the axis of deposition at right-angles to the direction of the wind, bearing in mind that if the activity evolves to produce pyroclastic density currents or surges, high ground may offer a small advantage.

11.6 Summary

- The hazards from volcanoes can be direct, in the form of the immediate effects of lava or ash on people and buildings, or indirect, for example in the form of direct destruction of crops or agricultural land causing starvation, or effects on the climate (see Chapter 12).
- Lava flows rarely kill people or animals but commonly completely destroy property and ruin agricultural land.
- Pyroclastic fall deposits are mainly a threat to buildings unless the density of fine ash particles in the air becomes a health hazard to people

and animals, although flying into the distal parts of atmospheric eruption clouds poses a major threat to aircraft.
- Pyroclastic density currents and surges are fatal to people and animals, and flows exert enough force on obstacles to destroy most buildings. The speeds of flows and surges are so large that realistically they cannot be outrun and pre-eruption evacuation is the only defense.
- Volcanic mudflows and lahars produced by mobilization of pyroclastic deposits by water have a similar mechanical destructive power to pyroclastic density currents.
- All volcanic gases except water vapor are chemically poisonous and all of them are unbreathable. All except water vapor are heavier than air when at ambient temperature and so can collect in hollows with lethal consequences to animals and people.
- Hazard assessment for a volcano requires the recognition of the products of previous eruptions and the analysis of their emplacement to obtain an idea of the likely range of possibilities for future eruptions.
- Numerous volcanoes that erupt frequently, or are near inhabited areas, are continuously monitored by recording seismic activity, measuring the deformation of the volcanic edifice, or monitoring the gases being released from shallow magma.
- Even given detailed monitoring, it is still not easy to give accurate warnings of impending activity, and volcano alert systems give qualitative assessments of the probability of a given level of activity of a given type within a given time frame.

11.7 Further reading

Casadevall, T.J. (1992) Volcanic hazards and aviation safety: lessons of the past decade. *FAA Aviat. Saf. J.* **2**(3), 1–11.

Chester, D.K. (2005) Volcanoes, society and culture. In *Volcanoes and the Environment* (Eds J. Marti & G.J. Ernst), pp. 404–39. Cambridge University Press, Cambridge.

Chester, D.K., Degg, M., Duncan, A.M. & Guest, J.E. (2001) The increasing exposure of cities to the effects of volcanic eruptions: a global survey. *Environmental Hazards* **2**, 167–78.

Marti, J. & Ernst, G.G.J. (Eds) (2005) *Volcanoes and the Environment*. Cambridge University Press, Cambridge.

Murck, B.W., Skinner, B.J. & Porter, S.C. (1997) *Dangerous Earth, an Introduction to Geologic Hazards*. Wiley, Chichester.

Tilling, R.I. & Lipman, P.W. (1993) Lessons in reducing volcano risk. *Nature* **364**, 277–80.

11.8 Questions to think about

1 If you had to live 2 km away from a volcano and had a choice, would you choose a basaltic volcano or a rhyolitic one, and why?

2 Why is the danger from large silicic eruptions not limited to the immediate duration of the eruption?

3 Why is observation from Earth-orbiting satellites not the ideal answer to volcano monitoring?

12 Volcanoes and climate

12.1 Introduction

It has long been thought that volcanic eruptions can affect climate for some time after an eruption. For instance, contemporary accounts relate how the 44 BC eruption of Mount Etna caused a dimming of the Sun, which was blamed for crop failure and famine in Rome. In 1783 the American scientist/diplomat Benjamin Franklin described a "dry fog" which spread across much of Europe and reduced the amount of sunlight reaching the ground. He linked this dry fog and reduced sunlight to the very severe winter which occurred in 1783–4 and suggested that the 1783 Laki eruption in Iceland was responsible. Similarly the year 1816 is known as the "Year Without a Summer". The unusually cold weather of 1816 followed the April 1815 eruption of Tambora, a volcano on the island of Sumbawa in Indonesia. Accounts from 1816 talk about a "dry fog" or haze in the atmosphere which dimmed the Sun, just as Franklin described in 1783. As will be seen, this is a feature of the injection of ash and gas into the atmosphere during volcanic eruptions. Other similar accounts exist which suggest that there is a link between volcanic eruptions and periods of abnormal weather.

During the past 100 years a number of studies have been carried out to investigate the scientific validity of anecdotal accounts linking volcanic activity to climate change. These have shown that some volcanic eruptions do indeed cause a short-term change in climate, usually lasting ~2–3 years after the eruption. This chapter looks at some of the evidence which has been used to arrive at this conclusion. We will look at how volcanic eruptions affect climate, at the factors which control the impact of individual eruptions and, finally, at links between the largest volcanic eruptions and **mass extinction** events preserved in the geological record.

12.2 Evidence for the impact of volcanic eruptions on climate

Investigating climate change is notoriously difficult, as is evident from the current debate about whether human activity is causing global warming. Studies aimed at investigating links between volcanic activity and climate change have similar difficulties to overcome. Until recently, such studies relied on comparing records of volcanic activity with existing climate records. Since the late 1970s it has become possible to investigate the climate effects of volcanic eruptions directly using satellite monitoring. Detailed study of two eruptions – the 1982 El Chichón eruption in Mexico and the 1991 Mount Pinatubo eruption in the Philippines – have been the focus of very detailed study which has significantly improved our understanding of how volcanic eruptions affect climate. These findings will be discussed later. We will start, though, by looking at some of the evidence that scientists used to link volcanic eruptions to climate change before the advent of satellite monitoring.

A number of studies have used records of known volcanic eruptions which have been classified in such a way as to allow the most "significant" eruptions to be highlighted. The volcanic records are then compared with climate records to see if the "significant" eruptions coincide with periods of abnormal climate. For instance, some studies have used the Volcanic Explosivity Index (VEI) (see section 10.6.1) as a way of defining the magnitude of eruptions and then looked to see if the largest eruptions coincide with periods of unusual climate. Other studies have used Lamb's "Dust Veil Index" (DVI). This is an index derived specifically to look at the effects of volcanic eruptions on the transmission of sunlight through the atmosphere. It defines a DVI value for known eruptions, with larger values indicating a greater effect on light transmission. The DVI value for an eruption is derived from a range of information including historical accounts, optical phenomena, radiation measurements, and eruption volume. The "dry fog" which impeded the transmission of sunlight after the Laki and Tambora eruptions is now known to be caused by the injection of sulfurous gases into the atmosphere. Chemical reactions between the gas and water vapor in the atmosphere cause the conversion of the gas into droplets of sulfuric acid (H_2SO_4) which are eventually removed from the atmosphere as a natural form of "acid rain" (or, more correctly, "acid precipitation" as it includes snow as well as rain). Recently scientists have started to use data from ice cores drilled near the poles to develop records of which eruptions caused the greatest input of sulfurous gases into the atmosphere. When examined in detail, ice cores can be seen to be formed of a series of thin layers, each one representing snowfall during an individual year. Thus it is possible to count and date each layer in the core. The acidity of the ice in each layer can be measured. The deposition of acidic snow following volcanic eruptions leads to layers which are unusually acidic. Examination of ice cores from Greenland and Antarctica reveals the presence of **acidity spikes** which often can be correlated with known volcanic eruptions (Fig. 12.1).

Having categorized known volcanic eruptions using one of these methods, it is possible to then compare the most "significant" eruptions with climate records to see if there is correspondence between volcanic eruptions and periods of unusual climate. One such study examined the effects of the largest eruptions to have occurred in the 19th and 20th centuries. It was found that a few eruptions, such as the 1883 Krakatau eruption and the 1963 eruption of Agung, were followed by a period of several years of cooler weather. The amount of cooling was, however, only of the same scale as the nonvolcanically related temperature variations, and so it is not possible to prove that the cooling was volcanically induced. When the temperature deviations after a number of major eruptions were examined in detail, though, a consistent pattern emerged which suggested that the largest eruptions are related to a small, but consistent, decline in temperature for 2–3 years after an eruption (Fig. 12.2).

Table 12.1 shows another examination of volcanic and climate data. In this case data from ice-core records and DVI studies are compared with years in which trees in the western USA developed "frost rings". The study of tree rings (dendrochronology) is one of many techniques used by climatologists to build-up records of past climate. Trees produce annual growth rings. If the tree experiences a period of unusually cold weather during its growing season (i.e., not during winter) it may develop an unusually narrow ring known as a **frost ring**. By examining tree rings in living and dead trees, scientists have been able to develop dendrochronologies dating back ~10,000 years. Years in which frost rings occurred can be identified and dated precisely and can then be compared with records of volcanic activity to see if the two are linked. Table 12.1 shows a comparison between these two sets of data for the years 1600 to 1965. There are a number of points of note about this study.

• Considering just the ice-core and frost-ring data there are seven eruptions which are recorded in the ice core which correspond with frost rings. There are a further four frost rings which correspond with known major eruptions which do not have a signature in the ice-core record. So overall 11 of the 21 frost rings (52%) correspond to known large-scale eruptions.

• Not all of the volcanic eruptions which are recorded in the ice core correspond to frost rings. This could be for any number of reasons such as: (i) the eruption did not affect climate; (ii) the change in climate was one of warming not cooling; (iii) the effect on climate was too localized to affect trees in the western USA; (iv) the eruption is recorded in the ice core because it occurred near Greenland and was not actually a large enough

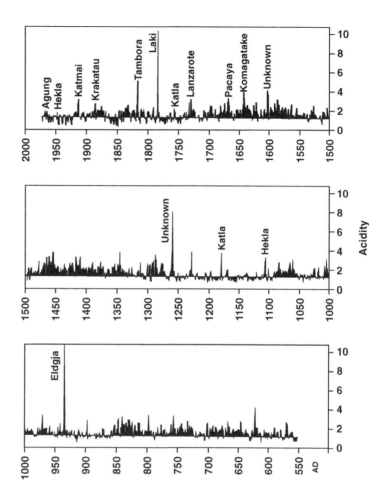

Fig. 12.1 Occurrences of sudden increases (spikes) of acidity in polar-ice cores marking known eruptions over the past 1500 years. (Adapted by permission from Macmillan Publishers Ltd: *Nature*, Hammer, C.U., Clausen, H.B. and Dansgaard, W., Greenland ice sheet evidence of post-glacial volcanism and its climatic impact, **288**, 230–235, copyright (1980).)

eruption to affect climate (e.g., Hekla, 1947 and Katla, 1755).

• Frost rings are formed in years in which no known eruption is evident in either the ice-core records or in historical records. It is unlikely therefore that the cold conditions which generated these frost rings were caused by a volcanic eruption.

• Overall the comparison suggests that, in some instances, volcanic eruptions may have been responsible for cooling which was severe enough to generate frost rings but that not all eruptions cause cooling and not all frost rings are the result of volcanic activity.

Though individual studies, such as the one just discussed, are not always conclusive, through a range of different studies it has become evident that volcanic eruptions can induce mild climate change, usually cooling, for a period of 2–3 years after an eruption. The difficulty of proving such a link comes in part from the limitations of available data sets, and partly because the effects we are trying to detect are subtle and often complex, e.g., cooling may occur in one region but warming in another. Our appreciation of the complexity of the effects of volcanic eruptions on climate has developed tremendously as the result of the new opportunities for detailed study offered with the advent of satellite monitoring.

12.3 Satellite monitoring of climate change after volcanic eruptions

Until relatively recently our understanding of whether volcanic eruptions affect climate was based on historical records and studies such as those just described. With the advent of the space

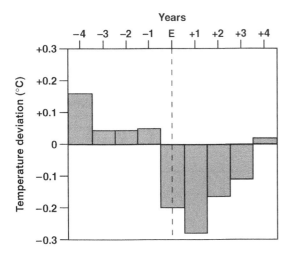

Fig. 12.2 Typical example of global temperature changes for a few years before and after a major explosive volcanic eruption, showing the small but significant amount of cooling caused by such events. (Based on fig. 2a in Self, S., Rampino, M.R. and Barbera, J.J. The possible effects of large 19th and 20th century volcanic eruptions on zonal and hemispheric surface temperatures. *J. Volcanol. Geotherm. Res.*, **11**, 41–60, copyright Elsevier (1981).)

age, however, it has become possible to detect the climate impact of volcanic eruptions directly. The 1982 El Chichón eruption in Mexico and the 1991 eruption of Mount Pinatubo in the Philippines (one of the largest eruptions of the 20th century) were the first eruptions for which detailed satellite monitoring of their effects on climate was possible. Study of these two eruptions has demonstrated unequivocally that volcanic eruptions can induce rapid, short-term climate change and has allowed detailed study of how eruptions cause climate change.

The 1991 Pinatubo eruption, for example, generated an eruption plume greater than 30 km high and injected large quantities of ash and gas into the atmosphere. Satellites were able to detect the conversion of sulfurous gases from the eruption into sulfuric acid aerosols (see section 11.2.7) and to detect the dispersal of the ash and aerosols within the stratosphere. This showed that the ash and aerosols had circled the globe within 22 days of the eruption. Satellites detected how the presence of the aerosols affected the Earth's albedo. The albedo is a measure of how much incoming sunlight is reflected back into space; a higher albedo means

more reflection of light and therefore less light reaching the surface. The satellites detected an increase in albedo immediately following the eruption indicating that the amount of sunlight reaching the surface had decreased. Furthermore, satellite instruments showed that the acidic aerosols were absorbing some of the incoming sunlight and that the stratosphere was warming as a result. Overall the eruption caused a detectable cooling of the Earth's surface with an average global decrease in temperature of ~0.5°C. The effects when examined in detail, however, proved to be considerably more complex, with some regions experiencing net cooling but others actually experiencing net warming. For instance, North America, Europe, and Siberia were warmer than normal during the winter of 1991–2 while Alaska, Greenland, the Middle East, and China were cooler. In fact, it was so cold in Jerusalem that it snowed there, which is extremely unusual. These effects resulted from changes in normal weather patterns induced by the stratospheric warming caused by the eruption.

12.4 The effects of volcanic eruptions on climate

In most cases the dominant effect of eruptions on climate is one of overall cooling, but the Mount Pinatubo eruption shows that the actual effects, when they can be recorded in detail, are more complex, and vary through time and with geographical location. This section discusses why volcanic eruptions affect climate and why different eruptions have different effects.

Accounts of many volcanic eruptions describe how a cloud of ash from the eruption blocks out the Sun and causes it to become as dark as night in the middle of the day. During the 1815 Tambora eruption, for example, areas as much as 600 km from the volcano experienced 2 days of darkness as a result of the spread of the ash downwind. The injection and progressive dispersal of ash in the atmosphere causes a less obvious but still significant blocking of some sunlight, with the reduction in the amount of sunlight reaching the surface thus causing surface cooling. While the effect of the release of ash is the most obvious immediate effect

Table 12.1 Comparison between volcanic activity and occurrence of frost rings in bristlecone pine trees in the western USA between 1600 and 1965. The first column indicates eruptions which produced a significant acidity spike in an ice core collected from Crete, Greenland. The second column indicates known historical eruptions with significant Dust Veil Index (DVI) values occurring during this time period. The third column indicates years between 1600 and 1965 in which frost-ring damage was recorded in bristlecone pine trees in the western USA. The final column indicates whether there is coincidence between eruptions and the occurrence of frost rings.

Greenland ice-core spikes	DVI	Frost-ring years	Match?
Agung, 1963	Agung, 1963	1965	Yes
Hekla, 1947	–	–	No
–	–	1941	No
Katmai, 1912	Katmai, 1912	1912	Yes
–	Santa Maria, 1902	1902	Yes
Krakatau, 1883	Krakatau, 1883	1884	Yes
–	–	1866	No
–	Merapi, 1837	1837	Yes
–	–	1831	No
–	–	1828	No
Tambora, 1815	Tambora, 1815	1817	Yes
–	–	1805	No
Laki, 1783	–	–	No
–	–	1761	No
Katla, 1755	–	–	No
Lanzarote, 1730–36	–	1732	Yes
–	Tongkoko, 1680	1680	Yes
Pacaya, 1671	–	–	No
–	Long Island, 1660	1660	Yes
Komagatake, 1640	Komagatake, 1640	1640	Yes
Unknown, 1601	–	1601	Yes

Data from Hammer, C.U., Clausen, H.B. and Dansgaard, W. (1980) Greenland ice-sheet evidence of post-glacial volcanism and its climatic impact. *Nature*, **288**, 230–235; LaMarche, V.C. and Hirschboeck, K.K. (1984) Frost rings in trees as records of major volcanic eruptions. *Nature*, **307**, 121–126.

of an eruption, it is not the most significant one in terms of climate change. Most ash particles injected into the atmosphere have only a short residence time (the length of time they spend in the atmosphere). They are typically removed from the atmosphere within days to weeks of the eruption. Smaller particles will stay in the atmosphere longer, but they are usually only a small fraction of the erupted mass and so their impact is minimal.

Volcanic eruptions release gas as well as ash into the atmosphere. Sulfurous gases, SO_2 and H_2S, released during an eruption will combine with water vapor in the atmosphere to form droplets or **aerosols** of sulfuric acid (H_2SO_4) which are typically ~1–2 μm in diameter. These acidic aerosols are far more important in affecting climate than are the ash particles. This is in part because they have a longer residence time than the ash. Aerosols form in the atmosphere over a period of a few weeks after the eruption and, due to their small size, have long fallout times (see eqn 8.3 and Fig. 8.4). Satellite studies show that the concentration of aerosols in the atmosphere after an eruption declines over a period of 2–3 years. Another reason that the aerosols are more important than the ash is because they are about ten times more effective at scattering incoming sunlight. So the presence of the aerosols causes much incoming sunlight to be scattered back out into space, reducing the amount of sunlight reaching the ground and causing surface cooling. Aerosol droplets not only scatter sunlight, they also absorb it. Absorption of sunlight (and longer wavelength radiation coming from the Earth's surface) by aerosols can cause significant warming of the stratosphere. Satellite observations after the 1991 Pinatubo eruption showed significant stratospheric warming. The size of the aerosols is an important factor in determining whether the overall effect is one of cooling or warming. If the radius of the aerosols is typically < 2 μm then cooling dominates; for larger aerosols warming will dominate. The size of the aerosols depends to some extent on the pre-existing state of the atmosphere, and this is why even if an eruption, such as the Pinatubo eruption, causes net global cooling the effects are more complex when examined in detail, and some regions may actually warm rather than cool.

The effect of any given eruption on climate depends on a number of factors such as the height of the eruption plume, the geographical location of the eruptive vent, the composition of the erupting magma, and the volume and duration of the eruption. These effects are discussed below.

12.4.1 The influence of plume height on climate change

The height to which ash and gas are carried in a volcanic eruption has a profound influence on the impact of the eruption because it determines how long the climate may be affected and also the size of the region affected.

The maximum height to which material is carried depends primarily on the mass flux of the eruption (see section 6.5.2). In relatively low mass-flux eruptions the ash and gas stay within the **troposphere** (the lowest layer of the atmosphere), but if the mass flux is high enough both ash and gas can be injected into the **stratosphere** (Fig. 6.7). The height of the boundary between the troposphere and stratosphere, called the **tropopause**, varies with geographical position and season but is typically 10–15 km. The ash and aerosols have a limited residence time which depends on their size and the height they reach within the atmosphere. As we have seen, ash and aerosols reaching the stratosphere have residence times of a few months and a few years, respectively. Ash and aerosols confined to the troposphere, however, have far shorter residence times because they are rapidly "rained-out" by the tropospheric weather systems. There is only likely to be a significant long-term effect if ash and aerosols are injected into the stratosphere and so have a relatively long residence time.

The height of the eruption plume has another important effect. In eruptions which are confined to the troposphere, the rapid removal of erupted ash and gas means that the material typically only spreads out in the atmosphere to affect areas on a regional scale. So eruptions which are confined to

the troposphere are only likely to affect climate on a regional scale. If, however, the plume reaches the stratosphere, the area affected can be very much greater. For example, Fig. 8.3 shows satellite data on the spread of the ash from the 1982 El Chichón eruption. Winds in the stratosphere disperse the ash and aerosols and areas far from the eruption site can be affected. Stratospheric injection allows ash and aerosols to affect climate within a whole hemisphere, or even on a global scale if the eruption site is located close to the Equator (see below).

In general, then, it is only eruptions which involve stratospheric injection which are likely to have widespread effects on climate. The consequences of eruptions confined to the troposphere may be significant, though, under certain circumstances, such as if the eruption is of long duration (see below), or if the effects in one region are severe enough to cause "knock-on" effects on the atmospheric circulation and weather systems in the rest of the affected hemisphere.

12.4.2 The effect of geographical location on climate impact

The geographical location of the eruption site also has an effect on the size of the region affected by an eruption. The atmospheric circulation of the northern and southern hemispheres is such that little mixing of air occurs between the two. This means that if an erupting volcano is located at middle to high latitudes, the ash and aerosols it releases into the stratosphere will be spread only within that hemisphere; there will be very little mixing of the material into the opposite hemisphere. If, however, the eruption occurs in an equatorial region, ash and gas can be released into both hemispheres and spread by the winds in each hemisphere so that the ash and aerosols can affect climate on a truly global scale. For instance, Fig. 12.3 shows the widespread dispersal of tephra during the 1883 Krakatau eruption.

12.4.3 The effects of eruption volume on climate impact

It is expected that the larger the eruption volume, the greater the effect on climate will be. Study of the largest historical eruptions has shown that they

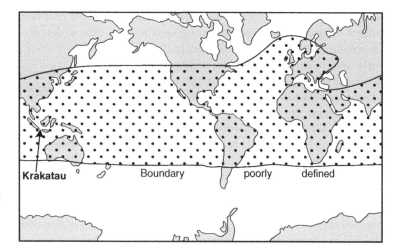

Fig. 12.3 The extent of atmospheric dispersal of tephra from the 1883 explosive eruption of Krakatau. (Based on fig. 17.4 in Francis P. (1993) *Volcanoes: a Planetary Perspective*. By permission of Oxford University Press.)

can cause global cooling of up to ~0.5°C for periods of 2–3 years after the eruption. The largest eruptions recorded in the geological record are much larger than these historical eruptions (see Chapter 10). It is natural to expect, therefore, that the effects of these larger eruptions would be proportionately greater. For instance, the Toba eruption 75,000 years ago, one of the biggest eruptions in the geological record, produced an erupted magma volume of 1500 km^3 (equivalent to ~4200 Mount St Helens 1980 eruptions!) and an estimated 3300 Tg of H_2SO_4 aerosols (Table 12.2). Simulations of the

likely effect of this eruption on climate have been made using computer models designed to look at the climate impact of a nuclear war, and suggest that the effects of a Toba-scale eruption are comparable with "**nuclear winter**" scenarios. Global cooling of 5–15°C is predicted; this scale of cooling is equivalent to the global temperature difference between now and the last Ice Age! The model probably overestimates the effect of such an eruption because the residence time of the ash in the atmosphere is short and because the sulfurous gases released could not all be converted to aerosols immediately because the stratosphere does not contain enough water vapor. Climate modeling is notoriously difficult and it is very hard, therefore, to make an accurate assessment of the effects of very large volcanic eruptions on climate. It is difficult to believe, though, that the effects of such a large eruption would not be profound.

12.4.4 The effects of magma composition on climate impact

We have seen that the aerosols formed as a result of volcanic activity have a far more profound effect on climate than does the ash. Only certain gases form aerosols and it is the release of the sulfurous gases, SO_2 and H_2S, which form the sulfuric acid aerosols that are most important. For this reason the composition of the erupting magma, particularly its sulfur content, is very important in determining the impact of an eruption. Table 12.2 shows

Table 12.2 The relationship between atmospheric aerosol loading and erupted volume during various eruptions. The erupted volume is given as the dense rock equivalent (DRE) value.

Eruption	Erupted volume (DRE) (km^3)	Estimated aerosol loading (Tg)
Toba, 75 ka	1500	3300
Laki, 1783	12.5	90–280
Tambora, 1815	50	180
Krakatau, 1883	10	50
Katmai, 1912	12	20
Agung, 1963	0.3–0.6	16–30
Fuego, 1974	0.1	3–6
Mount St Helens, 1980	0.3–0.4	0.3
El Chichón, 1982	0.38	12
Mount Pinatubo, 1991	4–5	30

estimates of the mass of sulfuric acid aerosols formed during various eruptions compared with the volume of the eruptions. The data in the table can be used to illustrate two important points. First, by comparing two eruptions of similar size, Mount St Helens and El Chichón, we can see the importance of composition: the Mount St Helens eruption produced only 1/40th of the mass of aerosols produced in the El Chichón eruption because of the lower sulfur content of the Mount St Helens magma. The difference in composition meant that the El Chichón eruption produced a measurable decline in global temperature whereas the Mount St Helens eruption did not. Furthermore, eruptions such as El Chichón and Agung produced an amount of aerosol loading similar to that of much larger eruptions such as Katmai and Pinatubo because of the higher sulfur content of the magma involved in the El Chichón and Agung eruptions. These observations show that eruption magnitude alone is not always a good guide to the impact of an eruption on climate. On the other hand, a very large eruption of sulfur-poor magma will still cause a massive aerosol loading just because of the sheer size of the eruption. The Toba eruption is a good example of this (Table 12.2). So in determining the impact of an eruption both composition and eruption size are important factors.

The sulfur content of magma varies widely but, in general, basaltic magmas have the greatest sulfur contents and can contain as much as 10 times as much sulfur as more evolved magmas. Thus, on the basis of composition alone, a basaltic eruption of a given volume would be expected to have a considerably greater effect on climate than an eruption of similar volume of a more evolved magma. Basaltic eruptions, however, do not generally produce high enough eruption plumes to cause significant stratospheric injection of ash and gas whereas eruptions of more evolved magma commonly do. So the eruptions with the greatest sulfur content (and thus the greatest potential to affect climate) are, on the face of it, also those with the least capacity to affect climate on a global scale. The situation is not quite this simple, however, for two reasons: (i) the largest basaltic eruptions, flood basalt eruptions, may cause stratospheric injection (see section 12.5.1) and (ii) the duration

of the eruption may also have a significant role to play in determining the effects of an eruption.

12.4.5 The effects of eruption duration on climate impact

Plinian eruptions which are able to inject ash and gas into the stratosphere typically have durations of less than a day. In such eruptions, then, the injection of ash and gas is short-lived and the atmosphere can begin to "recover" as soon as the eruption ends (as ash and aerosols begin to be removed by fallout and rain out). Basaltic eruptions can have much longer durations (it is not unusual for basaltic eruptions to last years or even decades) but proceed at much lower eruption rates than Plinian eruptions. Whereas the smaller eruption rates mean that plume heights are relatively low in basaltic eruptions, the continuous nature of the eruptions means that the atmosphere is being constantly loaded with acidic aerosols and thus that atmospheric recovery is delayed and the effects of the eruption can be prolonged well beyond the typical 2–3 year time span associated with recovery after Plinian eruptions. The effect of these milder but longer duration eruptions has yet to be adequately investigated. Furthermore, recent work suggests that larger basaltic eruptions might actually have more potential for stratospheric injection of gas than has previously been appreciated. If this is true it means that basaltic eruptions may be far more important in inducing climate change than Plinian eruptions because of the higher sulfur contents and the prolonged nature of the basaltic eruptions.

12.5 Volcanoes and mass extinctions

Historical records of volcanic activity and climate variation cover only a tiny fraction of Earth history. We have seen that eruptions during the historical past have affected climate. Compared with the scale of eruptions in the geological past these modern eruptions are of very small volume. Volcanologists are very interested, therefore, in how the larger eruptions in the geological past may have affected climate. In particular they are interested in possible links between volcanic activity and mass extinction events. Mass extinctions are events during which a large number of plant and animal species die out in a geologically short period of time. One idea for why these events occur is that they happen when rapid climate change causes environmental stress which in turn causes a collapse of the food chain. The most famous of these events is the extinction at the end of the Cretaceous period 65 million years ago in which the dinosaurs died out. This extinction has been linked by scientists to both flood basalt eruptions occurring in India at the time and to a meteorite impact in the Yucatan peninsula of Mexico. The meteorite theory is the currently accepted theory for what caused the extinction. This section examines ideas about the likely effects of the largest volcanic eruptions and looks at evidence linking them to mass extinction events.

As seen in Chapter 10, the largest eruptions in the geological record can be broadly divided into very large Plinian/ignimbrite-forming eruptions and flood basalt eruptions (Table 10.5). The former are single eruptions lasting at most a few days that involve evolved magmas and can have volumes as great as $\sim$2000–3000 km^3. The latter involve sequences of eruptions in which single events have volumes as great as 2000 km^3 and the province formed by a sequence can have an erupted volume as great as 2×10^6 km^3, being emplaced in a geologically short time period (1–2 Ma).

Initial interest in the climate impact of the largest eruptions centered on the large rhyolitic eruptions because they have the largest volumes and the highest eruption rates and hence the greatest plume heights. As seen in section 12.4.3, models of the climate effect of these eruptions suggest that they could produce the equivalent of a 'nuclear winter'. Basaltic eruptions do not usually cause stratospheric injection of ash and gas and so were not initially considered likely to cause significant climate impact. However, recent re-examination of the Laki eruption (the largest basaltic eruption to occur in historic time) suggests that it caused some stratospheric injection of aerosols. Mass fluxes during the Laki eruption are estimated to be a maximum of 2×10^7 kg s^{-1}. Mass fluxes during flood basalt eruptions are not known and are currently the subject of considerable debate by volcanologists. Some think

that the eruptions occur very rapidly (days) at high eruption rates ($\sim$10^9 kg s^{-1}), while others think that the eruptions occur over longer time periods (years, perhaps even decades) at slower rates ($\sim$10^7 kg s^{-1}). Eruption modeling, combined with the evidence that some stratospheric injection occurred during the Laki eruption, suggests that flood basalt eruptions are likely to cause some stratospheric injection regardless of which of these two views is correct. So it now appears that flood basalt eruptions are likely to be more important in causing climate change than the largest evolved magma eruptions as they would release far more sulfur into the atmosphere. Furthermore, as stated above, it may be that the longevity of basaltic eruptions proves to be very important in lengthening the duration for which climate is affected. For evolved magma eruptions and, indeed, asteroid impact scenarios, the rapid fallout of material means that the climate impact of the event is only likely to last 2–3 years. Is this long enough to trigger an environmental catastrophe sufficient to cause a major collapse of the food chain and hence a mass extinction? A longer, but initially less severe, basaltic eruption has the potential to affect climate for decades and thus seems far more likely to cause such a collapse. Furthermore, thus far ideas about mass extinctions and volcanism have tended to concentrate on climate change alone. The removal of acidic aerosols from the atmosphere will cause natural acid rain and the additional environmental stress caused by this in soil, rivers and oceans has yet to be assessed. Thus there are many more issues which need to be investigated before a complete picture can be developed of the likely environmental impact of flood basalt eruptions. Figure 12.4, however, gives us tantalizing evidence that flood basalt eruptions are a significant factor in triggering mass extinction events. It shows that when the geological record for the past 250 million years is examined there is a very strong correlation between the occurrence of flood basalt eruptions and mass extinction events.

12.6 Summary

- Evidence from the examination of historical records of volcanic activity and climate variation,

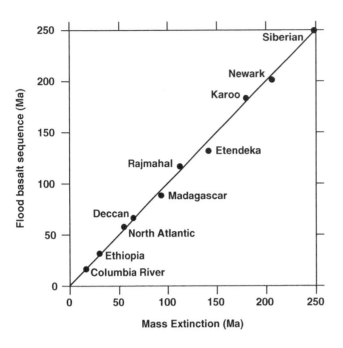

Fig. 12.4 The apparent correlation between the timing of named flood basalt eruptions and mass extinctions of species in the biological record. (Based on fig. 2 in Rampino, M.R. and Self, S. (2000) Volcanism and biotic extinctions. *Encyclopedia of Volcanoes*. Academic Press, pp. 1083–1091, copyright Elsevier (2002).)

and from satellite monitoring, shows that volcanic eruptions can cause short-term climate change. In general the effect of a volcanic eruption is to cause global cooling. The effect of historic eruptions has been to cause small but significant cooling (typically $\leq 0.5°C$) for 2–3 years after the eruption.

- A number of factors determine the effect of a given eruption on climate. These include the height reached by the eruption plume; the erupted volume; the geographical location of the volcano; the composition of the erupted magma; and the eruption duration.

- Although the general effect of an eruption is to cause overall surface cooling, when examined in detail the effects are more complex because cooling can disrupt normal weather systems. This means that some areas may actually experience net warming rather than cooling. Changes in rainfall patterns may also occur in the aftermath of a large eruption.

- Volcanic eruptions affect climate by injecting ash and volcanic gas into the atmosphere. Sulfurous gases react with water vapor in the atmosphere to form sulfuric acid aerosols. The ash and aerosols intercept some of the incoming solar radiation and scatter it back out into space. This causes a reduction in the amount of sunlight reaching the surface and hence surface cooling. The aerosols have the most significant effect because they are better scatterers of light and because they have a longer residence time in the atmosphere than the ash particles.

- The effects of the largest eruptions seen in the geological record are expected to be much more significant than those of historical eruptions. Current evidence suggests that flood basalt eruptions are likely to be more important than the largest rhyolitic eruptions in causing climate change because of the higher sulfur content of basaltic eruptions and their extended durations. There is a strong correlation between the occurrence of flood basalt eruptions and mass extinction events which suggests that environmental stress caused by the eruptions may act as a trigger for the extinctions.

12.7 Further reading

Courtillot, V., Jaeger, J.-J., Yang, Z., Feraud, G. & Hofmann, C. (1996) The influence of continental flood basalts on mass extinctions: where do we stand? *Geol. Soc. Am. Spec. Pap.* **307**, 513–25.

Marti, J. & Ernst, G.G.J. (Eds) (2005) *Volcanoes and the Environment*. Cambridge University Press, Cambridge.

McCormick, M.P., Thomason, L.W. & Trepte, C.R. (1995) Atmospheric effects of the Mt Pinatubo eruption. *Nature* 373, 399–404.

Rampino, M.R. & Self, S. (2000) Volcanism and biotic extinctions. In *Encyclopedia of Volcanoes* (Ed. H. Sigurdsson), pp. 1083–91. Academic Press, San Diego, CA.

Rampino, M.R., Self, S. & Stothers, R.B. (1988) Volcanic winters. *Ann. Rev. Earth Planet. Sci.* 16, 73–99.

Robock, A. (2000) Volcanic eruptions and climate. *Rev. Geophys.* 38, 191–219.

Self, S., Rampino, M.R. & Barbera, J.J. (1981) The possible effects of large 19th and 20th century volcanic eruptions on zonal and hemispheric surface temperatures. *J. Volcanol. Geotherm. Res.* 11, 41–60.

Sigurdsson, H. (1990) Assessment of the atmospheric impact of volcanic eruptions. *Geol. Soc. Am. Spec. Pap.* 247, 99–110.

12.8 Questions to think about

1 Which type of rhyolitic eruption, one that produces a high eruption cloud or one that produces a pyroclastic density current, is more likely to affect the climate?

2 What are the main ways in which volcanoes can influence the properties of the upper atmosphere?

3 Why do we now think that flood-basalt eruptions may be more important than rhyolitic Plinian eruptions in influencing climate?

13 Volcanism on other planets

13.1 Introduction

As seen in earlier chapters, the Earth exhibits volcanic activity because this is a major mechanism by which it can lose internal heat. On Earth heat loss through volcanism can be thought of as consisting of three parts: volcanism at divergent margins, volcanism at convergent margins and volcanism associated with hot spots. The first two are associated with the plate tectonic system while the third occurs well away from plate boundaries and is driven by deeper mantle convection not directly linked to surface plate motions. Thus volcanism associated with the plate tectonic system accounts for about 60% of volcanic heat loss and hot spots only 40%. As volcanic activity is such a fundamental mechanism by which the Earth loses internal heat

we might expect to see evidence for volcanism on any other bodies in the Solar System that are similar to the Earth. It is now known from investigations using manned and unmanned spacecraft since the early 1960s that volcanism has indeed occurred on many other bodies; those for which there is unambiguous evidence are Venus, Mars, our Moon, Jupiter's satellite Io and the asteroid 4 Vesta (asteroids are referred to by their numbers as well as names). We also strongly suspect that volcanism has occurred on the planet Mercury. Figure 13.1 shows all of these bodies to scale.

One fundamental observation that emerges is that the Earth appears to be the only one of these bodies that has unambiguous evidence of plate tectonics. We still do not entirely understand why this is so, but many arguments have been made

Fig. 13.1 Shown to scale, from left to right are the planets Mercury, Venus, Earth with its Moon, and Mars; Jupiter's satellite Io; and the asteroid 4 Vesta. These are the bodies in the Solar System known (suspected in the case of Mercury) to have been volcanically active. (NASA images.)

about the importance of the Earth having an extensive water layer on its surface. The recycling of the wet sediments carried down into the mantle on subducting oceanic plates helps to generate low-density continental rocks whose presence encourages subduction to continue, but this begs the question of what initiated subduction in the first place. The fundamental cause does not seem to be planetary size (Venus is nearly the same size as the Earth) or internal heat generation rate (Jupiter's satellite Io has a bigger heat source than the Earth and all of the other relevant bodies have smaller ones). This chapter concerns what we can learn by studying how volcanism on other planets differs from that on Earth. It will focus particularly on how the differing environmental conditions (such as atmospheric pressure, acceleration due to gravity, and surface temperature) on other bodies will affect the styles of volcanic activity that occur there.

13.2 Volcanically active bodies in the Solar System

First, let us be clear that volcanism as we normally think of it could be relevant only to Earth-like bodies consisting largely of silicate rocks: the planets Venus, Mars, and Mercury, our Moon, some of the satellites of the gas-giant planets, and some of the asteroids. The gas-giants (Jupiter and Saturn) and ice-giants (Uranus and Neptune) are dominated by massive, cloudy atmospheres consisting mainly of hydrogen and helium, and we have very little knowledge of the size or nature of the solid planet beneath the atmosphere. In the case of Jupiter, some theoretical models of the planet suggest that its atmosphere, dominated by hydrogen, increases in density under the weight of the overlying layers until it effectively becomes a dense solid near the middle. The mean densities of Uranus and Neptune do at least imply that there is a more normal mixture of rock, water, and ice deep beneath the clouds, but little is known beyond that.

However, the four large satellites of Jupiter – Io, Europa, Ganymede and Callisto – have densities consistent with the idea that silicate rocks, together with iron cores where relevant, form respectively 100%, 93.5%, 46%, and 45% of their masses. The other component, in the cases of Europa, Ganymede, and Callisto, appears to be completely dominated by water, frozen to ice at the surface. Saturn's largest satellite Titan probably consists of about 38% rock while Neptune's largest satellite Triton may contain about 50% rock. The nonrock component of these last two bodies is not just water – various other low molecular weight compounds such as methane, ammonia and nitrogen are present too. In all of the cases where these satellites contain a great deal of water and ice, the rock component has been concentrated towards the center of the body, so it is reasonable to think of them as silicate bodies covered by unusually deep, and partly frozen, oceans. In the case of some of the bodies where frozen water, i.e., ice, overlies liquid water that is rich in other volatiles, it appears that liquid water can break through the ice crust to form water flows that rapidly freeze. In some cases the volatiles appear to have caused explosive disruption of water into droplets thrown out at high speed like pyroclasts in a conventional eruption. This kind of activity is called **cryovolcanism** and has many analogies with silicate volcanism.

Every one of the silicate bodies, whether it is a "genuine" planet such as Venus or Mars, or happens to be a satellite of a gas-giant such as Io or Europa, is just like the Earth in having its share of radioactive elements producing heat. Io and Europa have an extra and very unusual heat source in the form of the flexing of their solid bodies by tides due to their parent planet Jupiter. It is therefore natural to expect that at least some of these bodies might have tried, at some stage in their history, to lose internal heat by volcanic action. Unfortunately, it is only in the case of Io that the silicate surface can be seen, this being hidden by the ice/water layers elsewhere. But on Io there is abundant evidence of vast amounts of volcanism, clearly driven mainly by the tidal flexing, in the form of lava flows and pyroclastic deposits. Furthermore, the lengths, widths, and thicknesses of the flows seen in images, together with temperature measurements from infrared sensors, show that the magmas being erupted are primary melts from the mantle, the equivalent of basaltic ocean-floor rocks on Earth.

Finally, we also need to consider the asteroids, the thousands of small (only hundreds of kilometers or less in diameter) bodies that escaped being incorporated into any of the larger planets and satellites. It is known, by examining the meteorites that reach Earth after being broken off these asteroids, that at least several tens of them formed quickly and very early in the history of the Solar System. This allowed them to incorporate some short-lived radioactive elements that acted as strong heat sources, and these asteroids were able to warm up to the point where they differentiated into iron cores and silicate mantles. Furthermore, there is also evidence, again from meteorites, that at least several of these asteroids formed basaltic crusts as a result of volcanic activity.

13.3 The effects of environmental conditions on volcanic processes

Previous chapters have discussed what is currently understood about the physical processes which control the character of volcanic activity. We took it for granted in doing so that a particular value for the acceleration due to gravity applies to the Earth, and that eruptions on land take place under a particular atmospheric pressure. However, when eruptions occur on the Earth's ocean floors, the weight of the overlying water exerts a sufficiently great pressure to greatly reduce the amount of gas that can exsolve from the magma, and this suppresses explosive eruptions except in gas-rich magmas or under circumstances where gas can be concentrated into a small part of the magma. So in thinking about volcanism on other bodies it is necessary to consider how the differing environmental conditions might affect the volcanic activity which occurs. This section considers in a general way how different environmental conditions are likely to influence volcanism.

The above comparison suggests that the first factor to consider is atmospheric pressure, or more exactly "the external pressure at the point where magma emerges at a planetary surface". The words are chosen carefully to take account of the fact that eruptions take place into both air and water on Earth. Later in this chapter it will be seen that they have almost certainly sometimes done both on Mars, and that if eruptions have occurred on Europa,

they have done so under very deep water. Even when only eruptions into a gaseous atmosphere are involved the pressure range we have to anticipate is enormous: from a maximum of 9 MPa on Venus down to a hard vacuum, i.e., essentially zero pressure, on the Moon, Mercury, and Jupiter's volcanically active satellite Io. The pressure under water on Earth is about 40 MPa in many deep parts of the ocean, and reaches about 110 MPa at the 11 km depth of the deepest oceanic subduction-zone trench. On Mars the maximum water depths may have been 3 to 4 km in the ocean that many scientists think filled Mars' northern lowlands in the early history of the planet, but the acceleration due to gravity is about 38% of that on Earth so the highest pressure would have been only ~15 MPa. The global ocean on Europa may be as much as 100 km deep and, although the acceleration due to gravity is only ~13% that of Earth, this makes the pressure typically a record 130 MPa.

The general effects of this wide pressure range on eruption conditions can be illustrated by considering the eruption of a basaltic magma containing 1 wt% water. Water is the commonest volatile in the Solar System, and 1 wt% is a reasonably common value on Earth. As shown in earlier chapters, the first thing than can happen as the magma rises toward the surface is that it becomes supersaturated in the volatile and starts to form gas bubbles. These expand as the pressure decreases (and new ones form). If the pressure becomes low enough that the gas bubble volume fraction exceeds about 75%, the magma will be disrupted into pyroclasts and an explosive eruption will happen. If not, a vesicular lava flow will be erupted. Table 13.1 shows what will happen to this typical basaltic magma on various bodies.

The variations are striking: this basaltic magma would not exsolve any gas at all under the Europa ocean and in the deeper parts of the Earth's oceans. It would produce vesicular lava flows in shallow oceans on Earth, under all likely oceans on Mars, and anywhere on the waterless surface of Venus; and it would only erupt explosively on land on Earth and Mars, and anywhere on the atmosphereless (and waterless) surfaces of Mercury, the Moon, and Io. The eruption speeds given in Table 13.1 are calculated using the methods described in Chapter

Table 13.1 The consequences of erupting a basaltic magma containing 1 wt% water on the bodies and under the conditions specified.

Location of eruption	Ambient pressure	Percentage of gas released	Nature of eruption
Europa (deep ocean always present)	130 MPa	0	Lava, no gas bubbles
Earth, under deep ocean	40 MPa	0	Lava, no gas bubbles
Earth, under shallow ocean	20 MPa	12.3	Lava, 9% vesicles
Mars, under deep ancient ocean	15 MPa	28.3	Lava, 24% vesicles
Venus (no ocean ever present)	9 MPa	49.8	Lava, 48% vesicles
Mars, under shallow ancient ocean	7.5 MPa	56.3	Lava, 56% vesicles
Earth, on land at sea level	0.1 MPa	97.9	Explosive eruption, ejecta speed 220 m s^{-1}
Mars (no oceans for most of history)	500 Pa	99.9	Explosive eruption, ejecta speed 347 m s^{-1}
Mercury, Moon, Io (no oceans ever)	~0	100.0	Explosive eruption, ejecta speed 484 m s^{-1}

6 and show that the lower the external pressure the more vigorous the explosion.

The second major factor controlling eruption conditions is gravity. Here the influence is subtle. The lithostatic pressure, P, at any given depth D below the surface of a planet where the acceleration due to gravity is g is given by $P = \rho g D$, where ρ is the mean density of the overlying rock mass. This statement neglects the fact that, especially near the surface of a planet, rocks can support stresses of several megapascals before fracturing or deforming slowly in a viscous fashion, but even so it gives a good approximation to the typical pressure at a given depth, and the approximation gets better as the depth increases. The implication of this relationship is that on a low-gravity planet (and the Earth has the largest acceleration due to gravity of all the bodies we have to consider) one must go to a greater depth below the surface to reach any given pressure.

The gravity influences volcanic structures in three ways. First, it controls the depths at which magma reservoirs are likely to be found. In Chapter 4 it was shown that the density variation with depth, and hence the level of neutral buoyancy, in a growing volcano was controlled by the progressive crushing of pore spaces as what was once a surface layer of vesicular lava or ash was buried ever deeper. The crushing process depends on pressure, and since a greater depth is needed to cause a given pressure increase on a low-gravity planet, reservoirs on all of the other planets are expected to be deeper inside volcanoes than on Earth. Second, the acceleration due to gravity influences the vertical sizes of magma reservoirs. It is currently thought that reservoirs grow until they reach a vertical height H such that the stress across the walls due to the difference, $\Delta\rho$, between the densities of the magma inside and the solid country rock outside is equal to the strength of the walls. This stress is proportional to $(\Delta\rho g H)$. However, the strengths of all volcanic rocks are similar, so the value of $(\Delta\rho g H)$ should be similar on all planets. Also the density differences between the solid and liquid states of all volcanic rocks are also rather similar, so the value of $\Delta\rho$ should be similar on all planets. The only way that both requirements can be satisfied is if H is inversely proportional to g: the lower the gravity, the greater the vertical extent of the magma reservoir. Third, the acceleration due to gravity influences the sizes of dikes and sills. This is really just an extension of the second effect, because the lengths of the long axes of dikes are determined mainly by the tensile strengths of rocks and magma density differences in the same way as the vertical extents of magma reservoirs, and to a first approximation the thicknesses of dikes and sills, are proportional to the lengths in a way governed by the elastic properties of the rocks (again the same for rocks on all planets) and not the gravity. So dikes and sills are expected to be bigger in all their dimensions on low-gravity planets. We shall see below that the sizes of volcanic features on the other planets amply support these conclusions.

13.4 The Moon

Relative to the other silicate planetary bodies, the Moon has had a very strange history. The rocks brought back from the Moon by the Apollo astronauts show that the composition of the Moon is similar to that of the Earth apart from the fact that all volatile compounds and elements have either partly or completely been lost. This is now interpreted to mean that, around the time the Earth had nearly finished forming, it collided with another planet, probably about the size of Mars. Most of the bulk of these two bodies coalesced to form the present Earth. Some material was thrown off at high speed, being partly melted in the process and so losing all of its volatile compounds, and cooled to form a ring of solid particles around the Earth. This ring accumulated to form the Moon, doing this so rapidly towards the end of the process that the outer few hundred kilometers of the Moon melted completely again to form a **magma ocean**. As this ocean cooled and crystallized a crust accumulated from light minerals forming a rock called anorthosite, while denser minerals such as olivine sank to join the unmelted interior. All of this took place within a time span of a few tens of millions of years.

From the Earth it is easy to see that much of the Moon's surface is covered by craters of a vast range of sizes, from more than 1000 km down to the smallest features visible, only a few hundred meters across. Before rock samples were obtained from the Moon, some people interpreted these craters, depressions with raised rims surrounded by blankets of material fairly obviously thrown out from them, to be explosive volcanoes. These people tacitly assumed that the Moon was at least as volcanically active as the Earth, perhaps more so given that there were so many craters. Others argued that the craters must be the result of the impacts of meteoroids, comets and asteroids, the pieces of interplanetary debris left over from the formation of the planets. By the mid-1970s astronauts had

walked on the surface of the Moon and brought back many rock samples, and satellites orbiting the Moon had surveyed most of its surface. It became clear from putting all of the information together that the craters on the Moon were virtually all due to impacts, most of which took place in the first 500 million years after the Moon formed. Only a handfull of crater-like features were produced by volcanic processes. Furthermore, the very fact that the Moon's surface is saturated with well-preserved impact craters and basins from the time just after its formation is proof of the fact, confirmed by all other lines of evidence, that the Moon never developed even a trace of the plate tectonic processes that play such an important part in removing heat from the Earth.

However, this does not mean that the Moon has not been volcanically active. In fact, all of the dark areas on the Moon's surface that can easily be seen from the Earth are regions where basaltic lavas have been erupted onto the floors of large impact basins. These lava-filled basins are called **maria** (singular: **mare**), latin for "seas", because the earliest telescopic observers thought that they were expanses of water. The basins were formed by asteroid impacts during the first few hundred million years of lunar history and they were flooded by large lava flows (Fig. 13.2), some more than 200 km long, about 500 million years later. The delay was because it took this long for heat liberated by the decay of radioactive elements to warm up the interior of the Moon to the point where extensive melting took place in the mantle. Large eruptions took place preferentially inside the impact basins partly because the dikes carrying magma from the mantle melt sources had a shorter distance to travel to reach the floors of the basins than elsewhere. Other factors may have played a role in concentrating large eruptions into basins. These include the fact that to compensate for the removal of part of the crust when a large basin forms, the mantle beneath the basin rises to some extent, which changes the temperature profile and helps to encourage subsequent partial melting.

Not all eruptions took place inside impact basins: a few occur on the surface of the old cratered anorthosite crust. Also, there are many linear valleys, usually called **linear rilles**, on the Moon

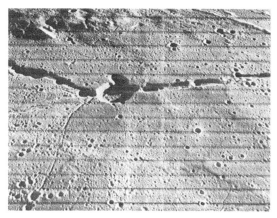

Fig. 13.3 The central part of the lunar linear rille Hyginus. The numerous collapse craters aligned and elongated along its length imply a strong volcanic association. (Part of Lunar Orbiter III frame 73M; NASA image.)

Fig. 13.2 Some of the lava flow units that flooded the interior of the Mare Imbrium impact basin on the Moon. The image is ~32 km wide. (Part of Apollo 17 Hasselblad frame #AS17-155-23714. NASA image.)

which are examples of **graben**, depressions where the crust has subsided between two parallel, nearly linear faults. These features form where tensional forces stretch the crust until the rocks break. In some cases there are minor volcanic features – domes, cones, and small flows – associated with these rilles (Fig. 13.3), and where these occur the rilles tend to be deeper, implying a greater horizontal extension of the crust than usual. This is easy to understand as the result of the need to make space for a dike nearing the surface. The implication is that much of the crust of the Moon was invaded by dikes that stalled not far below the surface as intrusions because they could not quite reach the surface to produce major eruptions. They were, however, able to force a small amount of magma to the surface to form the minor features that we see.

The duration of the Moon's volcanically active phase, deduced by using the proportions of radioactive elements in rock samples returned by the Apollo missions to measure the time since the lavas froze on the surface, was only about 1000 million years, less than one-quarter of lunar history. The reason it was so short is the small size of the Moon compared with the Earth. A planetary body

produces heat in proportion to the number of radioactive atoms it contains, which is proportional to its mass and hence to its volume, in turn proportional to its radius cubed. When that heat reaches the surface (whether it gets there volcanically or by conduction), the rate at which it is lost by radiation into the surrounding space is proportional to the surface area which is in turn proportional to the radius squared. Thus the ratio (heat produced/heat lost) is proportional to (radius cubed/radius squared), i.e., is proportional to the radius. So large planets have more trouble getting rid of their heat than small planets. Conversely, small planets lose heat efficiently and, as a result, are not volcanically active for very long. This idea acts as a guide to what we should expect on other bodies.

Although the most common volcanic features on the Moon are the long lava flows flooding the impact basins, there are other features interpreted to be evidence for large-volume eruptions: the **sinuous rilles** (Fig. 13.4). The inference is that in some places the eruption rate of lava was so great that the flows were turbulent rather than laminar, and the constant stirring of the lava increased the efficiency with which heat was transferred to the underlying ground. Some of these eruptions continued for long enough – several weeks – that the ground, often consisting of older lava flows, was heated to its solidus and began to melt. The molten

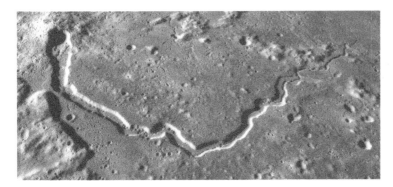

Fig. 13.4 The lunar sinuous rille called *Vallis Schroeteris* (Schroeters Valley), interpreted to be a channel formed when a hot turbulent lava flow melted the ground beneath it, meanders for ~180 km across the lunar surface with an average width of ~3 km and an average depth of ~1000 m. (Apollo 15 metric frame #AS15-M-2611; NASA image.)

rock was incorporated into the flow and so the flow began to melt a channel into the surface, eroding it at a rate of a few centimeters per hour to begin to form a sinuous channel. In many cases the eruption went on for several months, so that typically a 10 m thick flow was eventually traveling down the floor of a 100 m deep channel. About 30 such channels can be traced for many tens of kilometers across the lunar surface. The lengths of these channels, just like the lengths of more normal lava flows, can be used to deduce the volume eruption rate, and imply rates of up to 10,000 m^3 s^{-1}, similar to the rates implied by the longest lava flows.

Not all eruptions on the Moon involved large magma volumes or took place at high eruption rates. In an area called the Marius Hills there are more than 250 small (up to ~20 km in diameter), low (50–500 m high) features (Fig. 13.5), commonly called domes but actually more reminiscent of shield volcanoes on Earth. One possible explanation for these features is that they represent eruptions from a reservoir of magma that accumulated at the base of the crust. This magma cooled and crystallized to some extent before erupting with a higher viscosity and hence at a lower effusion rate than magma rising directly from the mantle to the surface. Elsewhere on the Moon, near the craters Gruithuisen and Mairan, there are six small (~10 km diameter and 1 km high) domes that appear to be the result of the extrusion of small amounts of an even more evolved type of magma. Although of great interest because the magma is so different from that elsewhere, these eruptions represent a vanishingly small fraction of the Moon's total volcanic activity.

Fig. 13.5 The area of the Moon known as the Marius Hills, a region in *Oceanus Procellarum*. Each "bump" on the surface is a small shield volcano, typically 5 to 10 km wide and 100 to 200 m high. (Part of Lunar Orbiter IV image 157H2; NASA image.)

Deposits from explosive eruptions on the Moon are very much rarer than on Earth. This is mainly because the Moon is completely depleted in volatile compounds such as water and carbon dioxide which are common in the Earth's mantle. However, there was a source of gas available to magmas erupting on the Moon but it had an unusual origin. The lunar mantle contains small amounts of free carbon, and carbon atoms can react with the oxides of certain metals to form carbon monoxide. The reaction only works at low pressures, so it would only have occurred in the magma near the tip of a new dike

Fig. 13.6 The ~120 km diameter lunar crater *Alphonsus*. The small, dark, elongate craters on the floor surrounded by dark haloes are interpreted to be sites of transient Vulcanian explosions. (Apollo 16 metric frame #AS16-M-2478; NASA image.)

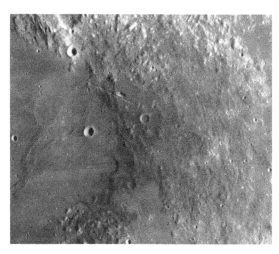

Fig. 13.7 The region around the lunar sinuous rille *Rima Bode*. The low reflectivity of the surface in the ~150 km diameter region occupying the middle one-third of the image is due to the mixing into the fragmental surface regolith of a pyroclastic deposit of volcanic glass beads. (Part of Lunar Orbiter IV image 109H2; NASA image.)

that was propagating toward the surface or within about 3 km of the surface in a dike that had already opened and was erupting magma. The largest amounts of gas that could be formed in an ongoing eruption were a few hundred parts per million, i.e., at least ten times less than in typical basaltic eruptions on Earth. Even so, the great expansion of this gas into the almost perfect vacuum above the lunar surface more than compensated for its small amount and threw out pyroclasts to form cinder cones and ash blankets at least as large as those on Earth, in some cases up to 3 km in diameter. These features are called **dark halo deposits** (Fig. 13.6), because the reflectivity of the dark basaltic pyroclasts is less than that of the rocks on which they are deposited. In many cases these pyroclastic features must have been buried by the very large volumes of lava being erupted, but in short-lived eruptions they were sometimes preserved. The much larger amounts of gas generated in the low-pressure regions behind the tips of new dikes were able to throw pyroclasts out to even greater distances, up to several tens of km, to form *dark mantle deposits* (Fig. 13.7). Pyroclasts from a few of these deposits were collected by the Apollo astronauts, and consist of submillimeter-sized glass beads.

The small size is due to the fact that almost every gas bubble that formed in a magma on the Moon eventually expanded and burst when exposed to the vacuum at the surface, thus tearing the magma

up into an enormous number of tiny droplets. If these droplets were thrown to great distances from the vent they were shaped into spheres by surface tension and then rapidly chilled to become glassy solids before landing. At the other extreme, if they were not thrown far, then the large numbers and small sizes meant that they formed fire-fountains in which droplets in the outer parts of the fountains shielded those in the inner parts and prevented them from radiating away their heat into space. Most of the droplets thus landed without having cooled at all, and formed hot ponds feeding lava flows instead of forming the spatter or cinder cones so common on Earth. The Moon has given us a great deal of insight into this pattern of behavior in explosive eruptions on planets with little or no atmosphere, and we shall come back to it when the eruptions on Mars and Io are considered.

13.5 Mars

A succession of spacecraft missions to Mars has shown that about 60% of the surface area of Mars, mainly in the southern hemisphere, preserves a very ancient highland terrain covered with impact craters. The rest of the surface, mainly the northern hemisphere, is lower in average elevation and consists of extensive plains made mainly of lava flows and finer grained windblown materials. The current martian atmosphere is thin, with a surface pressure about 200 times smaller than that of Earth. This, combined with the fact that Mars is further from the Sun than Earth, causes the mean surface temperature to be about −60°C; only in a few places for a few hours on summer afternoons does the temperature rise above the freezing point of water. The low atmospheric pressure means that water boils at between 1 and 2°C, and so the surface of Mars is currently both very cold and very dry.

All the available evidence implies that Mars still has plenty of water, present as ice in pore space and fractures in the outer few kilometers of the crust and as water trapped in pore space and fractures beneath the icy layer. In the lowlands surrounding the north polar cap there are deposits, currently being eroded by the wind, that may be the remains of the mud that collected on the floor of an ancient ocean that surrounded the north pole. The presence of such an ocean would imply that the atmospheric pressure was higher in the early history of Mars. The ghosts of old impact craters show through these sediments in places, and add to the evidence from the craters in the southern highlands that there has been little or no large-scale disturbance of the original crust of Mars, in other words that plate tectonics has never occurred on Mars.

The bulk of the volcanically erupted material on Mars is concentrated into two major provinces, Tharsis and Elysium, which contain a series of giant shield volcanoes (Fig. 13.8). The four largest of these, Olympus Mons, Ascraeus Mons, Pavonis Mons, and Arsia Mons, are in Tharsis and are typically 20 km high and 600 km wide. The three volcanoes forming the Elysium group, Hecates Tholus, Elysium Mons, and Albor Tholus, are much smaller, with the largest, Elysium Mons, being about half as tall and wide as the Tharsis shields. To the north of Tharsis lies the volcano Alba Patera, much lower (about 4 km tall) but wider (about 1000 km) than the four Tharsis shields, and scattered around Tharsis is a group of volcanoes (Uranius Paters, Ceraunius Tholus, Biblis Patera, Tharsis Tholus, and Jovis Tholus) that are apparently smaller than their

Fig. 13.8 The summit region of *Arsia Mons*, one of the giant shield volcanoes in the Tharsis region of Mars. The summit caldera dominating the image is ~110 km wide, and the entire volcano is more than 500 km in diameter and ~19 km high. The complex depressions on the SW and NE sides mark rift zones that extend in these directions. (Mars Odyssey THEMIS image courtesy of NASA/JPL/Arizona State University.)

giant neighbors but are partly buried by their lava flows so that it is hard to judge true sizes. There are also a few much older and heavily eroded volcanoes clustered around the Hellas impact basin in the cratered southern highlands.

The large sizes of the martian volcanoes and their grouping into two main provinces raise issues about the state of the martian mantle. If plate tectonics never developed on Mars, the crust has always been stationary relative to the mantle. This means that the upper boundary control on mantle convection is not the same on Mars as on Earth and, coupled with the smaller size of Mars, this appears to have significant consequences. Computer simulations of patterns of mantle convection predict that the Earth should have 20 to 30 major areas of upwelling in the mantle, and this is consistent with the number of volcanic hot spots that we see. These same computer models predict that Mars should have only a very small number of such regions of upwelling, perhaps only two or three, and that they should be more extensive than those on Earth. It is very tempting to identify Mars' two major volcanic provinces, Tharsis and Elysium, with these mantle hot spots. If this is the case then it would be expected that they might have been active for all of Mars' history, and this is supported by the finding that the range of ages estimated by counting impact craters for the various parts of Tharsis does span a very large fraction of martian geological time.

The martian shield volcanoes have morphologies very similar, apart from their size, to those of basaltic shield volcanoes on Earth. Their surfaces appear to be dominated by lava flows and they all have one or more collapse calderas at their summits (Fig. 13.8). All of the remotely sensed spectroscopic evidence from orbiting spacecraft suggests that the compositions of the volcanoes are basaltic to andesitic. Although no rock samples have been collected from Mars we are certain that some meteorites come from Mars, because the gases trapped in them are identical to the martian atmosphere sampled by the two Viking spacecraft that landed on the surface. These meteorites are volcanic rocks with essentially basaltic chemistry. As with the overall sizes of the volcanoes, the scales of individual features are large: whereas lava flows on the Hawaiian volcanoes tend to be 5 to 20 km long, those on Mars range from 30 to 300 km in length. Terrestrial basaltic calderas are rarely more than 3 km in diameter and 200 m deep, but martian examples are typically 20 to 40 km wide and up to 2000 m deep.

Much of this difference can be understood in terms of the differing environmental conditions, especially the atmospheric pressure and the gravity. For example, there are few pyroclastic deposits, and one possible reason for this is that we expect them to have a much greater dispersal than on

Earth. At the top of a martian volcano 20 km tall, the atmospheric pressure is only ~13% of its value at the foot of the volcano, where it is already 200 times less than on Earth. This 1500-fold difference has a profound effect on the eruption of any magma containing enough volatiles to cause it to fragment as it nears the surface (and martian basalts are expected to have similar amounts of gases to basalts on the Earth). The greater expansion of the released gas leads to greater fragmentation of the magma, making smaller pyroclast sizes, and much more acceleration of the erupting materials so that they reach much greater speeds. Finally the lower acceleration due to gravity means that pyroclasts thrown out to form cinder or spatter deposits will travel to even greater ranges. Table 13.2 compares the eruption of a basalt containing 0.25 wt% water on Earth at sea level and on Mars at the top of a Tharsis shield volcano. Eruption speeds on Mars are twice as large as on Earth and pyroclast ranges are more than 11 times greater. This means that the eruption of similar volumes of material on the two planets would lead to a cinder cone more than 130 times less high on Mars than on Earth, and thus very much harder to identify in a spacecraft image.

Similar issues relate to dispersal of clasts from martian eruption plumes. Computer models of the rise of Plinian eruption clouds predict that on Mars the clouds produced by a given eruption rate will rise about five times higher than on Earth, and so the martian winds (which are typically twice as strong as ours) should disperse materials over a much greater area. The only place where a fall

Table 13.2 Comparison of conditions in eruptions of basalt containing 0.25 wt% water on Mars and Earth.

	Earth	Mars	Units
Depth at which gas starts to exsolve	133	348	m
Depth at which magma fragments	34	90	m
Final amount of gas exsolved	0.2285	0.2499	wt%
Eruption speed of gas	66.4	139.6	m s^{-1}
Maximum range of ejected pyroclasts	450	5210	m

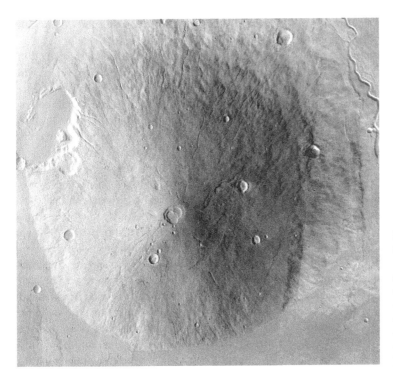

Fig. 13.9 The ~180 km diameter martian shield volcano *Hecates Tholus*. The area to the west of the summit caldera complex has fewer small impact craters per unit area than other parts of the volcano and may represent a recent pyroclastic fall deposit. (Mosaic of infrared Mars Odyssey THEMIS images courtesy of NASA/JPL/Arizona State University.)

deposit from a Plinian eruption has been identified with much confidence is on the western flank of Hecates Tholus (Fig. 13.9). The inference is that the eruption cloud was about 70 km high and the eruption rate was about 10^7 kg s^{-1}. However, there is a large area to the west of Tharsis where the surface is known to be a very poor reflector of radar signals, and some scientists think that this whole area may be blanketed by tens of meters of very fine pyroclastic dust. Indeed, it is just possible that the dust which is present everywhere on Mars, being sculpted into dunes by the wind, is not the result of the slow chemical weathering of solid rocks but is just an endlessly reworked planet-wide accumulation of primary volcanic pyroclastic material.

There is one caveat to this speculation. It may not be safe to extend models of eruption clouds on Earth to the current martian environment. The low surface pressure means that beyond a height of somewhere between 10 and 20 km some of the simplifying assumptions used in these models may not be valid. This suggests that very high eruption clouds may never have formed on Mars in very high discharge rate eruptions. Instead, something closer to a hybrid may have existed between the fountaining over the vent that on Earth feeds pyroclastic density currents, and the nearly ballistic dispersal of pyroclasts that dominated eruptions on the Moon and is common now on Io. No attempts have yet been made to develop theoretical models of such eruptions.

The difficulty of detecting both widespread pyroclastic fall deposits and the martian equivalents of near-vent cinder, spatter, or scoria deposits has led to the suggestion that, even if pyroclastic materials are not distributed planet-wide, at least the shield volcanoes may contain great quantities of pyroclastics interspersed with the lava flows that are much easier to identify. This would be hard to verify without seismic data or drill cores, but it would not be surprising given that the low atmospheric pressure would be expected to enhance explosive activity.

The average acceleration due to gravity on Mars is only 38% of that on Earth. As mentioned earlier, this means that at a given distance below the surface the pressure due to the weight of the overlying rocks is nearly three times less. However, the strengths and other elastic properties of martian rocks are essentially the same as those on Earth, and this leads to some interesting consequences. Some of these can be seen in Table 13.2, where the depth at which magmatic gases start to exsolve and the depth at which magma undergoes fragmentation are both seen to be about three times greater on Mars than Earth. Other consequences are that the depths at which magmas reach neutral buoyancy are also about three times greater on Mars and so are the vertical extents of magma reservoirs. It is hard to predict how the gravity should affect the horizontal extent of a magma reservoir; this may have more to do with the rate of supply of magma from the mantle in the early stages of reservoir formation. However, we do have observations to help: the widths of the calderas at the summits of martian volcanoes are typically ten times as large as those on Earth, meaning that the reservoir widths are also about 10 times greater, thereby making the typical reservoir volumes 300 times greater. The widths of dikes are likely to be about twice as great on Mars but the excess pressures in magma chambers are likely to be similar to those on Earth, and this leads to typically greater magma flow speeds in dikes (by a factor of at least two), and hence the potential for greater magma effusion rates (by a factor of up to about ten). This in turn implies greater lengths of lava flows (also by a factor of about ten, which is consistent with what is observed) and, coupled with the much larger volumes of magma stored in reservoirs, gives the potential for extremely large compound lava flow fields, which are indeed what are seen to dominate most of the shield volcanoes. A second indicator of long-lived high-effusion-rate eruptions exists in the form of sinuous rilles of the kind first identified on the Moon. These are not found everywhere, but are particularly common on the volcano Elysium Mons.

A final aspect of the sizes and ages of martian volcanoes deserves mention. The total volume of any one volcano divided by the length of time for which it was active gives the average rate at which magma has been supplied to it from the mantle. Typical figures for the Tharsis shields are 10^6 km^3 (from topographic data) and about 2 billion years (from impact crater counts, and subject to an error that could be as large as a factor of two). This implies an average supply rate of 0.015 m^3 s^{-1}, which is about 400 times less than the current rate at which magma is supplied to the Hawai'I hot spot on Earth. This low supply rate could not possibly maintain a magma reservoir of the size inferred above from the caldera sizes against heat loss by conduction. Even worse, this supply rate could not initiate the accumulation of such a reservoir – each batch of new magma would freeze long before it was joined by a subsequent batch. For example, if the first magma batch formed a sill 30 km in diameter (the full width of the eventual reservoir), the elastic properties of the host rocks would require the sill to be about 30 m thick. This represents a volume of about 20 km^3. Thus at an average supply rate of 0.03 m^3 s^{-1} the next magma batch would be expected to arrive 23,500 years later. But the time taken to cool a sill 30 m thick to the point where it is all solid is only about 25 years! Worse yet, the martian shield volcanoes typically have several summit calderas, and the way these overlap and intersect one another implies that each new one formed from the collapse of the roof of a magma reservoir that was created in a new location after an earlier reservoir had ceased to be supplied with magma and had frozen. Thus the creation of new reservoirs in a given volcano has to occur not once but many times. Also the time taken for an old reservoir of the size seen on Mars to cool to the point where the density and stress distributions force any new reservoir to form at a different location is two to three million years.

The only simple way to reconcile these observations and calculations is to assume that the magma supply to any one volcano from the underlying mantle hot spot is episodic rather than continuous. To make all of the steps in the cycle work correctly we have to assume that the volcanoes each have active periods lasting about 1 million years alternating with dormant periods of about 100 million years. There is an intriguing consequence. The most recent crater counts made on the highest resolution images available from current spacecraft in orbit around Mars show that some lava flows may be at most a few million to a few tens of million years old. In the past this has been taken to mean that the martian volcanoes are dead, and that humans have started exploring Mars too late to see any volcanic activity. But given the time scale of the

Fig. 13.10 Part of the *Memnonia Fossa* graben on Mars. The linear hills aligned roughly east–west on the floor of the graben may be outcrops of the top of the dike that caused the subsidence of the graben floor. The gap in the north wall of the graben is ~5.5 km wide; it was eroded by the overflow to the north of water from deep underground that filled the valley. (Part of THEMIS visible image #V04762003, courtesy NASA/JPL/ASU.)

cyclicity that is emerging, plus the fact that being in an active phase does not have to mean being active all the time, it is just as likely that all of the volcanoes are currently in dormant phases or between eruptions. Humans may yet see eruptions on Mars, although the long time scales imply that an eruption during any one person's lifetime is not very likely.

There are two consequences of the idea that the magma supply from martian mantle plumes is episodic. First, swarms of graben can be seen radiating from many of the Tharsis volcanoes. Some scientists argue that these are mainly created by tension due to the way the developing volcanic edifices stress the underlying lithosphere. However, many of the graben have collapse craters or small volcanic features associated with them, and some show low mounds aligned along their floors (Fig. 13.10). It is tempting to assume that the mounds are the tops of dikes which formed the graben, the dikes being fed from the central magma reservoir under the volcano at the center of the graben swarm. Graben creation would then be most logically associated with the episodes of upsurge in magma production in the mantle which, as seen above, may occur at 100 million year intervals. Many of the enormous valleys formed by water floods that are seen in various places on Mars have sources located either within graben of this kind or within nearby complex areas of collapsed terrain that appear to be underlain by sills fed by the dikes. Indeed, in some of these places deposits occur that appear to be the products of hydrovolcanic explosions. Thus the volcanic episodes may be the direct cause of the water-release events. However, they do this not by melting cryosphere ice but by fracturing the cryosphere and releasing some of the vast supply of water trapped in aquifers beneath it. And if future volcanic eruptions are possible, so are large catastrophic water flood events. The sites of these volcanically triggered water floods are the most likely places to look for evidence of living organisms that might be surviving in the underground water reservoirs.

13.6 Venus

Of all the terrestrial planets, Venus is closest in size to the Earth, and should be at a similar stage in its thermal and volcanic history. We cannot see the surface directly in visible light from orbiting spacecraft because the complete water droplet cloud cover in the upper part of the dense carbon dioxide atmosphere blocks light transmission. However, three missions have taken radar scanners to Venus and produced radar images of its surface (Fig. 13.11) with high enough resolution (~100 m)

Fig. 13.11 Mosaic of RADAR scans of a 210 km by 140 km area of volcanic plains on Venus from NASA's Magellan spacecraft. The dark patches are relatively young lava flows superimposed on older, lighter flows. Darker areas are rougher and lighter areas area smoother in these images. (NASA Magellan RADAR image.)

to show us that gently undulating volcanic plains occupy about 80% of the surface area while mountainous regions occupy the rest. Most of the plains consist of various kinds of lava flows, more than 50 of which are classified as flood lavas with

flow lengths averaging 350 km, the longest having double this length. Scattered around the plains are more than 20,000 small dome-like volcanoes up to 15 km in diameter; these are grouped into more than 500 clusters with the clusters averaging 150 km in diameter. Additionally about 270 features are classified as intermediate-sized volcanoes averaging 25 km in diameter. Finally, many of the mountains are large shield volcanoes, of which there are more than 150, typically 400 km in diameter and 2 to 4 km high (Fig. 13.12). Some of the flows down the flanks of these volcanoes can be several hundred kilometers long. More than 80 of the large shield volcanoes have summit calderas, averaging about 60 km in diameter. These calderas are much wider than those on Earth – in fact bigger than most of those on Mars. However, the acceleration due to gravity is only a little less on Venus than on Earth, so this underlines the suggestion made earlier that magma reservoir width is not controlled by gravity.

The mountainous areas on Venus that are not volcanoes generally form linear belts that seem to

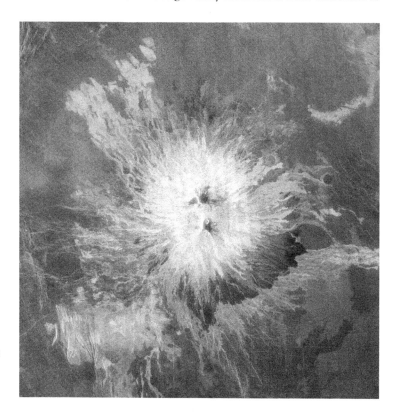

Fig. 13.12 The shield volcano Sapas Mons on Venus stands 1.5 km above the surroundings plains and lava flows spread out down its flanks in all directions for more than 150 km. (NASA Magellan RADAR image.)

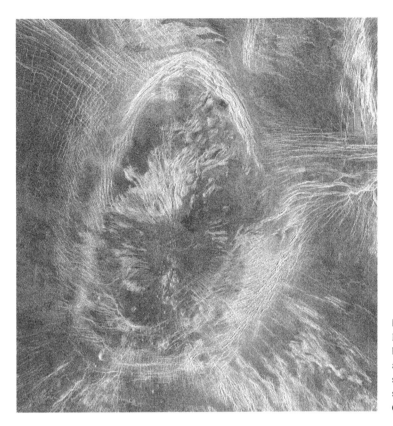

Fig. 13.13 *Bahet Corona*, in the Fortuna region of Venus, is about 230 by 150 km in size. Numerous tectonic and volcanic features are visible, suggesting that coronae and related structures form above mantle plumes. (NASA Magellan RADAR image.)

be produced by compression, although in many places linear sets of fractures are seen implying that tensional forces have also been at work. Embedded in these upland areas are terrains described as **tesserae**, from the latin for tiles. These are localized regions of very intense deformation and fracturing, and may represent the oldest rocks visible on the surface of the planet.

Despite the abundance of volcanic features, nowhere on Venus are there clear analogs of the Earth's spreading ridges and subduction zones. The best interpretation is that the surface of Venus has been fractured into a series of plates, but that the plates simply jostle against one another rather than being systematically enlarged from spreading ridges or destroyed by subduction. This process therefore does very little to help the interior of Venus to lose heat. Instead, heat must be lost by conduction through the lithosphere and by eruptions of magma at volcanic centers, presumably loc-

ated above mantle hot spots. In this regard there are more than 450 features on Venus that have no exact equivalent on Earth (Fig. 13.13). These are the **coronae** (latin for crowns), **arachnoids** (because of their alleged resemblance to spiders in their webs!) and **novae** (the analogy is to an exploding star). These features range from less than 100 to as much as 2000 km in size and may represent different stages in the evolution of a single kind of structure located over a mantle plume that penetrates to relatively shallow levels. Alternately they may represent different ways in which the crust responds to the deformation caused by such a plume. All of these features are circular to oval in shape, and have a central plateau or dome surrounded by a depressed moat that has annular (and quite often also radial) fractures associated with it. Lava flows are commonly found in the moat or associated with the fractures, and where the fractures radiate out for many hundreds of

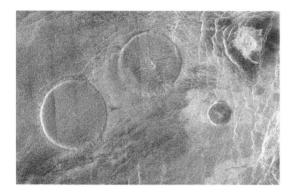

Fig. 13.14 So-called "pancake domes" in the Eistla region of Venus, probably formed by the eruption of viscous lava. These large examples are up to 65 km in diameter and ~600 m high. (NASA Magellan RADAR image.)

kilometers it is suspected that they represent the traces of swarms of giant dikes.

The dense atmosphere that makes visible observations difficult effectively traps heat from the Sun so that most of the surface is hotter than 400°C, hot enough to melt lead. These high temperatures, coupled with the high atmospheric pressure (~9 MPa in the lowlands and ~4 MPa on mountain tops) and corrosive nature of the atmosphere (traces of HCl and HF are mixed with the carbon dioxide) are very damaging to any probes landing on the surface of Venus. Nevertheless the Soviet Union landed a total of nine spacecraft on the surface plains, six of which survived long enough to make various kinds of compositional measurements that confirmed the presence of a range of different types of basaltic rocks. This is very much what would be expected if most of the volcanism consists of the eruption of mantle material melted in rising plumes.

There is evidence that other types of magma have erupted on Venus. There are about 80 near-circular, "pancake"-shaped domes (Fig. 13.14), with diameters in the range 20–30 km and thicknesses up to ~1 km, that appear to consist of lava with the viscosity of dacite or even rhyolite. Given that the high atmospheric pressure reduces volatile exsolution from Venus magmas, it is possible that these lavas would have erupted explosively on Earth to form ignimbrite sheets. At the other extreme, on the plains of Venus there are a number of long (hundreds of kilometers) winding channels called **canali**. These are typically only a few kilometers wide but the longest extends for more than 6000 km! Of the many suggestions about their origin, the most likely seems to be the long-duration eruption of a large volume of very low-viscosity lava.

How can the predominance of volcanic features be reconciled with the absence of plate tectonics on Venus? The thick atmosphere shields the surface of the planet from all but the largest impacting meteorites. But enough have penetrated to the surface and formed craters for us to estimate that the average ages of the plains are in the range 600 to 800 Ma. In fact, some interpretations of the crater distributions imply that "all" of the volcanic surfaces have about this age. Thus it has been suggested that, with no plate tectonics, the relatively inactive Venus lithosphere slowly heats up with time until a massive episode of flood volcanism almost completely resurfaces the planet in a geologically short time interval. This exhausts the magma supply and re-sets the system to start another several hundred million year cycle of heating. It should be stressed, however, that this is not the only interpretation of the impact crater distribution, and some volcanic areas may be much younger than this model implies.

13.7 Mercury

Our knowledge of the surface of Mercury is very restricted because so far only one spacecraft has visited the planet. Mariner 10 made a series of three fly-bys in 1974 and imaged a total of just over half the surface at resolutions between about 100 and 4000 m per pixel. This was good enough to show that the surface is very old and dominated by impact craters and basins, just like the Moon and the highlands of Mars. However, there seems to be a predominance of relatively flat ground between many of the craters, giving rise to the classification of much of the surface as "intercrater plains" (Fig. 13.15). The inference is that this material consists of lava flows, but this is by no means proven.

The highest resolution images show lobate boundaries in some areas that may possibly be the

Fig. 13.15 Possible smooth volcanic plains on the surface of Mercury. The undulating ridges are probably thrust faults formed after the flows were erupted. The largest impact crater in the image is ~100 km in diameter. (Mariner 9 frame FDS 167, courtesy NASA/JPL/Northwestern University.)

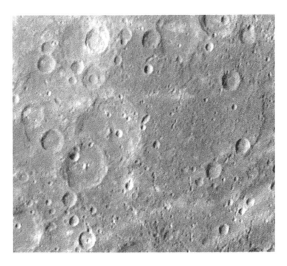

Fig. 13.16 The arcuate scarp to the far right in this 400 km wide image of plains on Mercury may be a lobate lava flow front. (Mariner 9 frame FDS 166738, courtesy NASA/JPL/Northwestern University.)

edges and fronts of lava flow units (Fig. 13.16), but elsewhere similar features seen at lower resolution have been traced for many hundreds of kilometers and are interpreted to be the traces of thrust faults produced by a general shrinking of the crust when the large metal core of Mercury froze at some point in its geological history.

We know from the mean density of Mercury that it has an iron core that is unusually large for the size of the planet. This has led to the suggestion that Mercury, like the Earth, was involved in a giant collision with another planet-sized body in its very early history. The event presumably occurred after Mercury had at least partly differentiated into a core, mantle and crust, and the impact stripped off much of the crust and mantle but left the core undisturbed. Mercury did not retain part of the debris, as the Earth did to form its Moon, but lost it all.

If this suggestion is correct, it has some important implications for what can be seen on the surface. If much of the evolution of Mercury into a mantle and crust had finished when the impact took place, then at least a large part of what can be seen on the surface should in fact be olivine-rich rocks that normally would be associated with the mantle. Alternately, if significant further evolution of the mantle occurred after the giant impact, perhaps what we should see is a relatively thin basaltic crust. Our only evidence comes from simple spectroscopic data from the Mariner 10 spacecraft, augmented by spectroscopic observations from Earth-based telescopes that are hindered by Mercury's small size, great distance from Earth, and closeness to the Sun. The main conclusions from these data are that Mercury's surface rocks contain smaller amounts of iron and titanium than those of the Moon, and are more akin to the lunar highland rocks than to the basalts of the maria. Clearly our understanding of the possible volcanic history of Mercury is very primitive. Two spacecraft, Messenger and BepiColombo, are planned to arrive at Mercury in the period 2010 to 2020, and we shall have to wait for results from these probes before much progress can be made.

13.8 Io

The intense tidal flexing of Io by Jupiter produces internal heating at a prodigious rate. As a result Io is very vigorously volcanically active. The activity was detected by the two Voyager spacecraft that flew by Io in 1979 and monitored in depth between

Fig. 13.17 This comparison image shows changes in lava flow field being formed from the volcano Prometheus on Io. The right hand frame is the ratio of two images taken by the Galileo spacecraft on October 11, 1999 and February 22, 2000. Dark areas in the ratio image indicate fresh lava eruptions during the ~14 week interval. Each Galileo spacecraft image frame is ~65 km wide. (Image courtesy of NASA/JPL/University of Arizona.)

1996 and 2002 by the Galileo probe, which obtained visible and infrared images and took temperature data. More than 300 vents have been identified on the surface, and at least ten of these are likely to be active at any one time. Some appear to continue to erupt for at least several years whereas other eruptions are over in at most a few days. Some of the long-lived eruptions feed complex lava flow sheets (Fig. 13.17), the great lengths, small thicknesses, and high eruption temperatures of which suggest that they are unusually hot basalts. Other long-lived eruptions are explosive and give rise to what are usually called eruption plumes (Fig. 13.18), although more accurately they are giant lava fountains since there is no significant atmosphere on Io to support a convecting eruption cloud.

A major feature of Io is the brightly colored surface. The colors (red, orange, yellow, white, even a tinge of green) are due to deposits of sulfur and sulfur compounds, especially sulfur dioxide, together with small amounts of metal salts. What are not seen are any clear indications of water or carbon dioxide, the two commonest volcanic volatiles. Io does not have any significant atmosphere other than the gases emitted from the volcanoes, and so is prone to losing those gases into space. Indeed, an annular ring of atoms of sulfur,

Fig. 13.18 Near the north pole of Io a 290 km high eruption plume rises from one of the vents of the Tvashtar volcanic system. A smaller plume from the volcano Prometheus is visible near the equator on the left side of the image, and the bright spot on the night-side of Io not far from the south pole is the top of a plume from the volcano Masubi, just catching the sunlight. This image was taken by the Long Range Reconnaissance Imager on the New Horizons spacecraft as it traveled through the Jupiter satellite system on its way to Pluto. (NASA image.)

oxygen, sodium, and other volatile elements accompanies Io in its orbit around Jupiter. The ease with which gases are lost from an atmosphere depends on their molecular weight, and the values for water, carbon dioxide, diatomic sulfur (the commonest form), and sulfur dioxide are close to 18, 44, 64, and 64, respectively. The inference is that over the course of geological time Io has erupted and lost essentially all of the water and carbon dioxide from its mantle. It may well have erupted most of the sulfur-related compounds onto the surface, but they have not yet been completely lost but instead are recycled through volcanic activity. The recycling is not in the form of subduction back into the mantle, for there are no signs of plate tectonics on Io, but instead consists of sulfur and sulfur dioxide collecting on the surface as solids, being buried under layers of pyroclasts and lavas, being heated by the geothermal gradient, and melting to form "aquifers" in the crust. The dikes feeding new eruptions sometimes cut through these aquifers and the volatile-poor magmas in the dikes absorb the aquifer liquids. The liquids then boil to vapor and drive the violently explosive eruptions just as if they were expanding magmatic volatiles. The heights of the largest plumes, commonly up to at least 300 km, and their widths, up to 1200 km, imply eruption speeds of more than 1000 m s^{-1}, and the analysis in Chapter 6 shows that this could be achieved only if the magma contained about 30 wt% of volatiles. This is vastly more than could be dissolved in any possible juvenile magma from the mantle.

Some plumes seem to be caused not by dikes absorbing liquid volatiles but by lava flows advancing over a surface rich in solid volatiles. A finite time is taken for enough heat to be conducted downward to melt and evaporate the solids, and so jets of gas punch holes through many places in the lava flow after its front has gone by and combine to form a plume. This is somewhat akin to the formation of pseudocraters around rootless vents in terrestrial lava flows crossing marshy ground, and also can be thought of as a kind of hydromagmatic activity, although the volatile involved is not water.

There are many calderas on Io (Fig. 13.19), typically with diameters of several tens of kilometers, implying that beneath them are magma reservoirs of similar size. Attempts have been made to infer

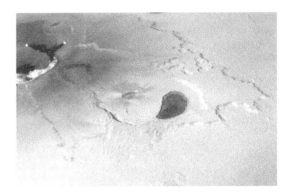

Fig. 13.19 Near the left edge of this 250 km wide image a mildly explosive eruption is taking place in one of the chain of calderas forming the Tvashtar Catena on Io. The image is a composite of data obtained through near-infrared, clear, and violet filters on the Galileo spacecraft. (Image courtesy of NASA/JPL/University of Arizona.)

whether these magma reservoirs have formed at neutral buoyancy levels in the crust, but this is hindered by uncertainties about the volatile content, if any, of the ascending magmas and also the density structure of the crust. The low value of the acceleration due to gravity on Io would lead us to expect that levels of neutral buoyancy might be at depths of order 15 km and that magma reservoirs might have large vertical extents.

Many calderas are clearly the sites of eruptions from fissures, both explosive and effusive (Fig. 13.19). Attempts have been made to estimate eruption rates by analyzing the sizes and dynamics of plumes, by measuring the heat release rate from eruption sites, and by measuring the increases in areas of lava flows with time. The latter is not easy because the Galileo spacecraft only made close approaches to Io every few months. Also, the resolution of the images was not good enough to allow flow thicknesses to be estimated accurately. Nevertheless, there is some convergence on eruption rates up to 10^5 m^3 s^{-1} with occasional bursts up to 10^6 m^3 s^{-1} and erupted volumes up to 100 km^3. The high eruption rates are consistent with the high temperatures and low viscosities of the magmas and, if the lava flows are being fed from magma reservoirs behaving elastically, the arguments in Chapter 4 imply that the reservoirs should have volumes of up to 30,000 km^3. This is possible with ~50 km diameter magma reservoirs if they have vertical extents of ~15 km and, although impressive, this would be consistent with the effects of the low gravity on reservoir geometry.

13.9 Europa

Europa was extensively imaged by the Galileo spacecraft at resolutions as good as 200 m per pixel. The high reflectivity of the surface of Europa is consistent with the spectroscopic evidence that the surface consists almost entirely of water ice (Fig. 13.20). The ice appears to be "contaminated" to some extent with dissolved salts such as magne-sium sulfate, and with small amounts of rock and metal dust from the impact of meteorites, but it is clear that it is not a silicate surface. Nor is it a rocky body covered with a thin layer of ice, because there are no large mountains. There is even a near-total lack of large and small impact craters. The absence of large craters might not be surprising, because ice is a weak material that flows under stress (as in glaciers on Earth), and over hundreds of millions of years the rims of large craters and basins would sink, and the floors rise, until almost nothing was visible. However, the lack of small craters, which form much more frequently and deform propor-tionately less, implies that the surface is geolog-ically young.

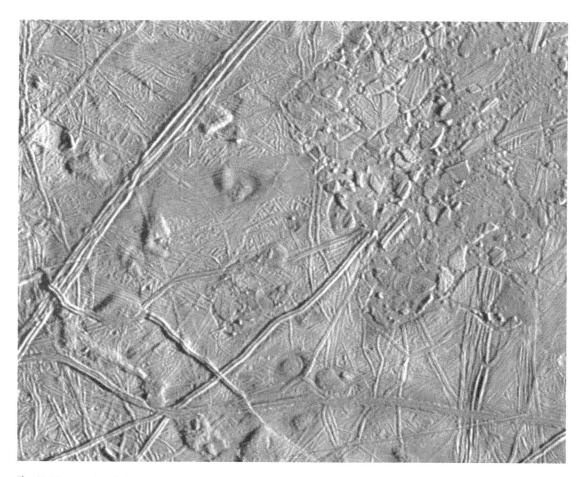

Fig. 13.20 A 100 by 140 km area of Europa showing complex ridges and fracture patterns cutting the ice layer that forms the surface. The texture in some places seems to imply that the ice has melted and re-frozen. (Image from the Galileo spacecraft's Solid State Imaging system courtesy of NASA.)

The mass and diameter of Europa are known very accurately, thus giving us its mean density, and we also have some idea of the distribution of density within it from the tracking of the Galileo spacecraft during close approaches. The most likely internal structure is a differentiated silicate body with a water ocean about 100 km deep, the outer 5–10 km of which are frozen to form the observed ice surface. The young age of that surface implies that water frequently escapes onto the surface from beneath the ice to cover up impact craters, and indeed there are enormous numbers of long thin fractures from which water appears to have flowed a short distance before freezing. The rheological changes in these water flows as ice crystals form in them should theoretically have some similarities to the development of a yield strength in a cooling silicate lava flow, but this has not yet been studied in detail. There are also patches of crust on Europa where raised blocks of ice with old fracture patterns lie near one another and can be fitted together like a jigsaw. These seem to be places where the crust melted nearly all the way to the surface, and the unmelted top layer fractured into icebergs that drifted apart in water that welled-up between them before freezing again.

All this evidence points to vigorous thermal convection in the ocean under the ice, and possibly even slow convection within the ice layer itself. The same calculations of tidal flexing that confirm the massive heat source inside Io also imply a smaller but still significant source inside Europa, and it is not out of the question that volcanic activity takes place in the silicate body beneath the ocean even today. As seen earlier, the high pressure would limit the activity to the eruption of lava with no exsolved gas bubbles. This would mean that there was little opportunity for a complex density profile to evolve in the silicate crust, and so no neutral buoyancy level at which magma would preferentially accumulate to form magma reservoirs. Thus any magma would have to travel directly to the surface from its source zone in the mantle, or would stall as an intrusion at some depth dictated by the amount of cooling that it had experienced as it rose; this is very similar to what happened on our Moon. We might speculate that if water is the key to the Earth having plate tectonics, the process might have developed on Europa. Unfortunately the ocean is so deep that there is little chance that surface features on the ice would directly reflect what is happening 100 km below.

13.10 Differentiated asteroids

Of the few asteroids yet visited by spacecraft, none has been a body that we think evolved to the point of having volcanic eruptions. Most of our information about asteroids is obtained from the meteorites broken off them, and usually it is not known which asteroid has produced a given meteorite. In a few cases, however, there are good enough spectroscopic observations from Earth telescopes (and the Hubble space telescope) to be sure that certain groups of meteorites come from a particular body. By far the strongest case can be made for the Howardite, Eucrite, and Diogenite groups of meteorites coming from one of the largest asteroids, the 520 km diameter 4 Vesta (Fig. 13.21). These meteorites include all of the kinds of rocks one would expect to get from a dif-

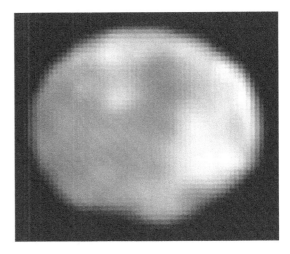

Fig. 13.21 The best available image of the asteroid 4 Vesta, taken by the Hubble Space Telescope. Note the asymmetric shape and the large depression, evidence of a large impact cratering event. The image has been digitally restored to a resolution of ~10 km per picture element. (Image credit: Ben Zellner (Georgia Southern University), Peter Thomas (Cornell University) and NASA.)

ferentiated asteroid: pieces of vesicular surface lava flow and slowly cooled magma that intruded into the crust; pieces of olivine-rich rock representing the mantle after basaltic melts have been removed from it; and fragments of the surface **regolith**, the outermost layer of rocks broken up and mixed by large numbers of small impact craters. The Hubble image of 4 Vesta shows an irregular surface with one giant crater punching through the crust into the mantle (Fig. 13.21). This is exactly the configuration to be expected after analyzing the meteorites.

There are other groups of meteorites where all the members of the group have the same trace-element chemistry and isotope ratios, and have clearly all come from the same differentiated body. The Aubrite meteorites are the residual mantle of an asteroid that was rich in volatiles, but there are no meteorites found that would represent the crust of this body. We infer that explosive eruptions occurred when basalts from the mantle of this body reached the surface, and that these were so vigorous that all of the pyroclastic droplets were erupted with speeds greater than the escape velocity from the asteroid. Thus no significant basaltic crust ever formed on this body. The Ureilites are another group of rocks from an asteroid mantle. The detailed chemistry of these meteorites is throwing light on the mechanisms whereby basaltic partial melts separate from their source rocks, either quickly by forming dikes or slowly by percolating along mineral grain boundaries. In some meteorites, e.g., the Acapulcoite, Lodranite, and Angrite groups, veins are present that represent cracks containing melt and sometimes gas bubbles; these seem to be small dikes caught in the act of migrating to the surface to feed eruptions. Needless to say volcanic rocks from asteroid mantles are invaluable in trying to understand process in planetary interiors: it is only very rarely that eruptions on Earth bring up mantle rocks from depths corresponding to those of many meteorite samples from asteroids.

13.11 Summary

• All of the other planetary bodies with compositions roughly similar to that of the Earth that we might expect to have volcanic activity now, or to have been volcanically active in the past, appear to do so, although the evidence is still poor for Mercury due to lack of data.

• None of these other volcanically active planets appear to have, or to have had in the past, a plate tectonic system like that of the Earth, and we still have no complete understanding of why this is so. The inference is that the volcanism is driven by mantle plumes forming hot spots, and the presence of giant shield volcanoes on Mars and Venus is consistent with this.

• The commonest volcanic rock type erupted on all of the bodies for which there are good data is some variety of basalt, as we would expect for primary melts from planetary mantles.

• Theory suggests that the lower the acceleration due to gravity on a body the wider the dikes that can form in its lithosphere, the larger the magma reservoirs that can form, and the larger the length and volume of typical lava flows. This prediction seems to be borne out by the great sizes of lava flows on Mars, Venus, the Moon, and Io.

• The lack of atmospheres on the Moon and Io, and the low atmospheric pressure on Mars, mean that vigorous explosive eruptions are more likely on these bodies. Clear evidence for this is seen on the Moon and Io but is less clear for Mars, where complex surface processes due to the presence of the thin atmosphere – wind and weathering – may help to hide the resulting features.

• Some of the most interesting evidence about what happens deep within the Earth, which we cannot reach directly, comes from examining the fragments of the interiors of differentiated asteroids that have been broken up by impacts and delivered to Earth as meteorites.

13.12 Further reading

Cattermole, P. (1996) *Planetary Volcanism*. Wiley, Chichester.

Davies, A. (2007) *Volcanism on Io*. Cambridge University Press, 376 pp. ISBN: 9780521850032.

Lopes, R.M.C. & Gregg, T.K.P. (Eds) (2004) *Volcanic Worlds: Exploring the Solar System Volcanoes*. Springer-Praxis, Chichester.

Frankel, C. (2005) *Worlds on Fire*. Cambridge University Press, 366 pp. ISBN: 9780521803939.

Wilson, L. (1999) Planetary volcanism. In *Encyclopedia of the Solar System* (Eds P. Weissman, L. McFadden & T. Johnson), pp. 877–97. Academic Press, San Diego, CA.

Zimbelman, J.R. & Gregg, T.K.P. (Eds) (2000) *Environmental Effects of Volcanic Eruptions – from Deep Oceans to Deep Space*. Kluwer Academic Press, Dordrecht.

13.13 Questions to think about

1 What effects do differences in atmospheric pressure between the various volcanically active bodies in the Solar System have on eruption styles?

2 What effects do differences in the acceleration due to gravity between the various volcanically active bodies have on eruption styles?

3 Which bodies in the Solar System are known to be volcanically active at the present time?

Answers to questions

CHAPTER 1

1 All magmas coming from the mantle appear to contain some dissolved volatiles that can be released as gases at low pressure. Explosive activity happens when a sufficiently large number of gas bubbles are formed. Thus truly effusive eruptions require that not too much gas is released in the erupting magma. On the deep ocean floors the water pressure is so high that gases stay dissolved in virtually all magmas, suppressing explosive activity. If magma is stored at shallow depth it gets the chance for gas bubbles to escape into the surrounding rocks before the magma erupts. If magma comes directly to the Earth's surface the key issue is just how much gas it contains.

2 Common types of activity on human time scales are Strombolian, Hawaiian, Vulcanian, subPlinian, Plinian, hydromagmatic, deep marine, and subglacial eruptions. Very much rarer are flood basalt, ultra-Plinian, ignimbrite-forming, and diatreme-forming eruptions. This is very fortunate, because the less frequent eruptions are those that are more violent, and likely to spread larger volumes of volcanic materials over larger areas more quickly.

CHAPTER 2

1 Magmas formed at mid-ocean ridges and hot spots are generally some or other variety of basalt. They are produced by partial melting of mantle rocks as a result mainly of decompression of rocks in the rising parts of mantle convection systems. Subduction zone magmas are generally more silica-rich than basalts and have a wide range of compositions.

They are produced when ocean floor rocks are heated by contact with the hotter rocks into which they are being pulled down and helped to melt by the presence of water in ocean floor sediments being carried along with them.

2 First, melts are almost always less dense than their pre-melting parent rocks and also the unmelted solid residues remaining after melting occurs. Thus buoyancy – the effect of gravity on the difference in density – acts to make them rise. Second, the parent rocks are commonly subjected to nonuniform stresses by processes such as mantle convection and plate movements, and these tend to drive the melts upward as compaction occurs.

CHAPTER 3

1 The rheology of rock typically depends on temperature, pressure, and the strain rate – the rate of deformation – imposed on the rock. It is difficult, but not impossible, to subject rocks in the laboratory to temperatures like those in the mantle. It is significantly more difficult to reproduce the mantle pressures, and it is impossible to carry out experiments for the lengths of time needed to reproduce the mantle strain rates.

2 At great depths, where the mantle is partly molten in the rising parts of convection cells, both the host rock and the melt forming within it will be rising. As melt percolates between mineral grains and becomes concentrated in the upper part of a convection cell, the region of melt concentration behaves as a diapir, i.e., a body of relatively low-viscosity fluid rising through a layer of relatively high-viscosity fluid. When the high-viscosity fluid can no longer deform fast enough, it starts to behave as a brittle material, and a fracture forms in it. The low-viscosity fluid flows into the fracture to form a dike.

3 Diapir movement is controlled by the slow deformation of the high-viscosity mantle rocks surrounding the diapir. Dike tips propagate as brittle fractures and so can potentially travel at very high speeds; the actual speed is controlled by the ability of the low-viscosity magma within the dike to flow. Mantle rock viscosities are ~10^{20} times larger than molten basalt viscosities, hence the enormous difference in speed.

4 If the magma is rising mainly as a result of buoyancy, it can become trapped when it is no longer less dense than the rocks surrounding it. The base of the crust, where there is generally a significant decrease in rock density from mantle to crust, is a likely density trap for many magmas in continental areas and for high-density magmas in oceanic environments. If magma is rising in dikes the situation is more complex because, not only must the magma have a net positive buoyancy integrated over the entire vertical length of the dike, but also the stress intensity at the propagating dike tip must be large enough to overcome the effective fracture toughness of the host rocks.

CHAPTER 4

1 First, a caldera may be present, implying that at some stage the surface rocks collapsed or subsided into space left by an eruption from the shallow magma reservoir. Second, the composition of the erupted magma may suggest that it has been cooled by a significant amount since it left the mantle. Third, tilt-meters around the summit of the volcano may record patterns of deformation in the form of inward or outward tilting that imply deflation or inflation of a body of fluid beneath the surface. Fourth, seismometers on the volcano recording very distant earthquakes may detect the surface shadow of a zone beneath the volcano through which S waves cannot travel, implying the presence of fluid.

2 We very strongly suspect that many kinds of volcano have some sort of magma storage system at fairly shallow depth beneath them, and that basaltic

volcanoes in particular grow laterally by sending out dikes radially into rift zones. We would therefore look for the now-frozen remains of these features: laccoliths or very large sills as evidence of the main intrusions, and swarms of dikes in the vicinity.

3 Basaltic volcanoes at intraplate hot-spots are fed more or less continuously on a time scale of years by magma from the mantle, and so any given batch of magma is not likely to stay in the magma reservoir beneath the summit for long enough to undergo a great deal of chemical evolution. In contrast, basaltic volcanoes at mid-ocean ridges receive major inputs of magma much less frequently, perhaps at intervals of several decades, and so have time to develop more evolved melts in their storage zones as a result of cooling and fractional crystallization of stagnant magma.

4 In the order given in Table 4.1 (Pinatubo, Fernandina, Katmai, Krakatau, Rabaul, Taupo, Santorini, Toba, Yellowstone and La Garita), the caldera areas are about 4.9, 24, 8, 50, 118, 960, 55, 1890, 2200, and 2100 km^2, the vertical depths of erupted magma are 0.92, 0.004, 1.52, 0.2, 0.093, 0.036, 0.45, 0.8, 0.68, and 2.42 km, and the percentages of the total magma content are 18, 0.08, 30, 4, 1.9, 0.7, 9, 16, 14, and 48. Only Fernandina is marginal; all of the others are well above 0.1% and would have been expected to show some amount of caldera collapse, as was observed.

CHAPTER 5

1 Carbon dioxide is much less soluble than water, and its solubility decreases faster as the pressure decreases, so magmas rising from the mantle start to exsolve mainly carbon dioxide long before they exsolve much water.

2 The main factor is the presence or absence of crystals to act as nuclei on which vapor molecules can congregate to avoid the limitations of surface tension.

3 The three processes are: (i) diffusion of volatile molecules through the liquid to reach gas bubbles; (ii) expansion of bubbles as the pressure decreases in response to Boyle's law; (iii) coalescence, i.e., joining together of bubbles that come into contact.

4 The lower viscosity of basaltic magma allows bubbles to rise more quickly through the liquid

than in more evolved magmas so the large bubbles can overtake smaller ones more easily.

5 The main difference is the extent to which the gas bubbles derived from the volatiles in a given volume of magma stay uniformly distributed within that volume as it rises. Intermittent activity occurs at the surface vent when large numbers of small bubbles coalesce into giant bubbles or slugs filling almost all of the width of the dike or conduit.

CHAPTER 6

1 First, the continuity eqn 6.1 shows that any increase in gas volume, and hence decrease in the bulk density, of the magma in a dike of constant shape can be compensated for only by an increase in speed. Second, the energy eqn 6.4 shows that any decompression of the magma, especially of the gas, provides energy that contributes to increasing the speed.

2 The main effect is to reduce friction with the dike walls. Before fragmentation the fluid in contact with the wall is mainly liquid magma; after fragmentation the fluid in contact with the wall is mainly gas which has a vastly smaller viscosity.

3 First, because gas expansion is the main driving force, the larger the proportion of gas in a magma the higher its eruption speed. Second, a high gas content causes bubble nucleation to occur earlier, and hence deeper in the dike, ensuring that the gas experiences a larger pressure change which releases more energy. Third, earlier bubble nucleation means earlier fragmentation of the magma and hence more of the magma rise takes place under low wall-friction conditions.

4 Many explosively erupting magmas accelerate in the dike to the point where the rise speed is equal to the speed of sound in the gas–pyroclast mixture. Unless there is some part of the dike where the walls slope outward toward the surface to make a de Lavalle nozzle the mixture cannot go any faster and so it cannot decompress down to atmospheric pressure until it gets out of the vent.

5 At first the magmatic material shares its momentum with the air that it entrains and so slows down. But heat from the magma warms the air and produces buoyancy that drives the mixture upward and increases its speed. Eventually all the thermal energy is traded for potential energy and the mixture slows down as it approaches its maximum height.

6 The main control is the rate at which heat is added to the atmosphere: the plume height is proportional to the fourth root of the heat release rate. Heat release rate is proportional to mass eruption rate and magma temperature. There is also an influence of atmospheric temperature profile and humidity (hence latitude and season).

7 Large pyroclasts have a large terminal fall velocity through the gases in an eruption cloud and so can never be carried to great heights. Small clasts can easily be carried up to great heights, but also some of them get swept to the edge of the eruption cloud, where the rise speed is small at all heights, and fall out from there.

8 Two factors encourage this evolution: erosion of the conduit and vent, resulting in a larger mass eruption rate, causes less air entrainment to occur, providing less buoyancy to the eruption column. Also late-erupted magma may be poorer in volatiles than magma erupted at the start of the eruption, and thus will have a smaller eruption speed, also reducing air entrainment.

CHAPTER 7

1 The first way involves only the gases released from the magma and occurs when the gas becomes nonuniformly distributed in the magma as a result of small gas bubbles joining together to form large bubbles which burst intermittently at the surface of the magma column. The second way involves the magma coming into contact with and boiling external fluids – essentially always water on the Earth. The water may infiltrate a column of magma that has stalled in the conduit or the magma may flow over wet ground or into shallow water.

2 The speed at which pyroclasts are ejected depends on the pressure reached in the trapped gas that is driving the explosion before the "lid" trapping it breaks and also on the relative amounts of gas and solid material. The greater the pressure, and the greater the proportion of gas, the greater the speed.

3 The distances traveled by pyroclasts thrown out from an explosion at a given speed depend on the sizes of the drag forces acting on them. Water has both a larger density and a larger viscosity than air, and so ejected clasts will travel to much smaller distances under water. Also, if the explosion happens under deep water, the weight of the water will reduce the pressure differences between the compressed trapped gases and the surroundings, thus reducing the violence of the explosion.

4 The steady case just needs us to insert $M_f = 2 \times 10^5$ kg s^{-1} into eqn 6.7 to find $H = 4.99$ km. For the intermittent case we see that if the average rate is 2×10^5 kg s^{-1} and explosions occur every 10 seconds, the mass per explosion must be 2×10^6 kg. Inserting this for M_e in eqn 7.5 we find $H = 1.58$ km, a much lower plume.

5 A "real" vent is the site of volcanic activity and is directly physically connected to the dike feeding magma to the surface. A rootless vent is the site of activity (e.g., explosive activity when lava flows over wet ground) but is not directly connected to any magma pathway beneath the surface.

6 The key issue is that the hot magma and cold water interact in the right proportions to maximize the conversion of the thermal energy of the magma to the kinetic energy of the eruption products.

CHAPTER 8

1 If you measured the product of the size and density of the largest clast and plotted this on Fig. 8.6, you could infer a mass flux and hence an eruption cloud height. However, you would not know if the exposure site was downwind from the vent or in some other direction. So you might have plotted your data point at too great a distance from the vent, and this would have caused you to overestimate the eruption rate. So what you get from your single exposure is a "maximum" estimate of the eruption rate and the corresponding eruption cloud height.

2 If there is no wind, all the clasts released at a given height in the eruption cloud (meaning all clast sizes up to the largest that can be supported at that height) and hence at a given distance from the vent drift vertically down to the ground to accumulate in the same place. If a wind is blowing, smaller particles are moved further downwind than larger clasts because they fall at a lower speed. A given location on the ground will now accumulate some relatively large clasts released high in the cloud and not blown very far, and some smaller

clasts that were released from lower in the eruption cloud but blew further. These small clasts must now cover a larger area than if they had not been blown so far (or moved at all) and so there will be fewer of them per unit area on the ground. Thus, in the downwind direction at least, the deposit will be relatively rich in larger clasts near the vent and relatively rich in small clasts far from the vent.

3 The kinetic energy per unit mass represented by the eruption speed is $0.5 \times 100^2 = 5000$ J kg^{-1}. The added potential energy per unit mass that the flowing material gains due to moving vertically by 1000 m under an acceleration due to gravity of 10 m s^{-2} is $10 \times 1000 = 10^4$ J kg^{-1}. Adding the two energies gives 15×10^3 J kg^{-1}. Converting to the corresponding velocity U, i.e., $0.5\ U^2 = 15 \times 10^3$, gives $U = 173$ m s^{-1}. This is a maximum possible speed because it neglects any energy lost due to friction with the ground while traveling down the slope.

4 If the pyroclastic density current travels out over the water its bulk density is presumably less than the density of the water. The hot particles at the base make contact with the water and boil some of it to steam that enters the body of the flow. The clasts at the base of the current that make the best contact with the water get cooled, and some liquid water may get sucked into the vesicle spaces within them as the gas in the vesicles chills and contracts, so these clasts may get waterlogged and may sink into the water. Meanwhile, the throughput of added hot steam makes the clast concentration in the body of the flow smaller, so large clasts are less well supported by grain–grain contacts and begin to migrate to the bottom of the flow, preferentially coming into contact with the water. Also as the steam escapes it will carry some of the smallest clasts away with it into the overlying phoenix cloud. Thus the body of the flow may lose large clasts at the base and small clasts at the top and get thinner, eventually becoming so thin that all that is left is a layer of cool pumice clasts floating on the water surface and a phoenix cloud dispersing on the wind.

CHAPTER 9

1 The rootless flow must form by the accumulation of clots of magma that have been transported through a lava fountain. In doing so the clots must have lost some heat by radiation, so the resulting flow will be cooler and more viscous, will flow more slowly, and will be thicker. However, if it is fed at the same volume flux it will travel a similar distance.

2 Although lava flowing in a tube is heavily insulated against heat loss, it still loses some heat. So the tube-fed flow units will probably not advance to the same distance that they would if they were fed directly from the vent. Thus the flow field involving the tube-fed flows will probably have a slightly smaller area (although of course some of its flow units will be thicker to compensate).

3 Bubbles, like crystals, give a lava an inherent yield strength. They also make it less dense. Equations 9.3 and 9.4 show that both factors mean that the vesicular lava will form much deeper and wider levées. If the lava fills the central channel to the top of the levées, eqns 9.6 and 9.7 show that it will flow faster (the greater depth squared will more than outweigh the lower density in eqn 9.6 and the density does not matter in eqn 9.7). If the lava is deeper and is flowing faster, then eqn 9.8 shows that for the same mass flux the width of the channel will be smaller in the case of the vesicular lava.

CHAPTER 10

1 The high pressure of the overlying water reduces gas exsolution, and the presence of exsolved gas is the single most important factor driving explosive activity.

2 Although there are many factors affecting the details, the fact that evolved magmas in general contain more dissolved volatiles than basaltic magmas is the basic reason.

3 The main issue is whether gas can become concentrated into parts of the magma from which it has exsolved. An example is when coalescence of gas bubbles takes place in a low-viscosity magma rising slowly enough for large bubbles to overtake smaller ones and grow very large by sweeping up many small bubbles.

4 More than one scheme has evolved because sometimes it is the rate at which magma is erupted that matters most and in other cases it is the volume of material.

5 First, in general, the larger the volume of a magma reservoir, the larger the maximum volume of magma that can be erupted from it; but a major caveat is that if the rocks above the reservoir fail in an inelastic way, so that caldera collapse can occur, then a much larger fraction of the reservoir contents can be erupted than when the rocks behave elastically. Second, the larger the reservoir, the longer it is likely to take after an earlier eruption before the conditions needed to trigger a later eruption are reached. Thus there tends to be a direct relationship between repose time and erupted volume.

CHAPTER 11

1 It would be safer to live 2 km away from a basaltic volcano than a similar distance away from a rhyolitic volcano. This distance is greater than the range of most fire fountain deposits and is far enough away that you should have an adequate warning of any approaching lava flow. However, 2 km is well within the range of even a small pyroclastic density current, and is close enough that a thick and quite coarse air fall deposit could accumulate from even a modest eruption column.

2 These types of eruption produce large volumes of relatively fine-grained materials (whether as fall or flow deposits) which can be mobilized by rain or snow to cause mudflows from months to years after an eruption.

3 The main problem is that to monitor the whole planet the satellites have to be in near-polar orbits, but the ~90-minute orbital period means that a given satellite cannot overfly the same point on the Earth's surface until about 16 days after its earlier visit. Near-continuous monitoring of one location would require very large numbers of satellites.

CHAPTER 12

1 The high eruption cloud carries small particles to great heights, and these take a long time to fall to the ground, so their opportunity to have physical and chemical effects is maximized. Co-ignimbrite clouds from pyroclastic flows rarely reach such great heights as Plinian eruption clouds.

2 High eruption clouds can deposit water vapor, sulfur dioxide and small silicate particles at great heights. First, the silicate particles act as nuclei on which the vapor condenses and the sulfur dioxide then dissolves and oxidizes to form aerosol droplets of sulfuric acid that reflect sunlight and cause cooling. Second, the water vapor causes changes in the equilibrium of chemical reactions and can influence ozone stability.

3 Although the eruption plumes from small-scale basaltic eruptions do not rise very high, the plumes from high eruption-rate flood-basalt eruptions probably do penetrate the stratosphere, and these eruptions release more sulfur dioxide than typical evolved magmas and also continue for much longer.

CHAPTER 13

1 The lower the atmospheric pressure, the greater the amount of volatiles released from a magma reaching the surface (volatile solubility is pressure-dependent). Also, the lower the atmospheric pressure, the greater the amount of expansion of whatever volatiles are released from the magma. This leads to a higher eruption speed in the material leaving the vent, more efficient entrainment of atmospheric gas, and generally a greater dispersal of pyroclastic products.

2 There are two main effects. The first is that the lower the acceleration due to gravity the smaller will be the pressure or stress at a given depth due to the weight of the overlying rocks. This leads to wider dikes, thicker sills and more vertically extensive magma chambers. The second effect concerns the distribution of pyroclasts. Clasts of a given size and density incorporated into an eruption cloud will have smaller weights, and so can be carried up to greater heights and blown further in a given wind. Similarly, clasts ejected more nearly ballistically from a vent can travel to greater ranges.

3 We have only actually observed volcanic eruptions on the Earth and on Jupiter's satellite Io. Some of the lava flows on Venus and Mars are very fresh in appearance and we think that they were erupted geologically recently. This idea is supported by theoretical arguments as well as observations, and so we expect to see further eruptions on these planets some time in the future. In contrast, both theory and observation suggest that our Moon and the planet Mercury are too small to retain significant heat sources and are volcanically dead.

Index

Page numbers in *italics* refer to figures and those in **bold** refer to tables.

Printed and bound by CPI Group (UK) Ltd, Croydon, CR0 4YY

27/10/2024